Math Unlimited

Essays in Mathematics

MATH UNLIMITED

Essays in Mathematics

Editors:

R. Sujatha
Department of Mathematics,
Tata Institute of Fundamental Research, Mumbai, India

H.N. Ramaswamy
University of Mysore, India

C.S. Yogananda
Sri Jayachamarajendra College of Engineering,
Mysore, India

CRC Press
Taylor & Francis Group
an informa business
www.crcpress.com

6000 Broken Sound Parkway, NW
Suite 300, Boca Raton, FL 33487
270 Madison Avenue
New York, NY 10016
2 Park Square, Milton Park
Abingdon, Oxon OX14 4RN, UK

Science Publishers
Jersey, British Isles
Enfield, New Hampshire

Published by Science Publishers, an imprint of Edenbridge Ltd.

- St. Helier, Jersey, British Channel Islands
- P.O. Box 699, Enfield, NH 03748, USA

E-mail: *info@scipub.net*   Website: *www.scipub.net*

Marketed and distributed by:

| 6000 Broken Sound Parkway, NW Suite 300, Boca Raton, FL 33487 |
| 270 Madison Avenue New York, NY 10016 |
| 2 Park Square, Milton Park Abingdon, Oxon OX14 4RN, UK |

Copyright reserved © 2012

ISBN: 978-1-57808-704-4

Library of Congress Cataloging-in-Publication Data

Math unlimited : essays in mathematics / [edited by] R. Sujatha, H.N. Ramaswamy, C. S. Yogananda.
　　p. ; cm.
　Summary: "This collection of essays spans pure and applied mathematics. Readers interested in mathematical research and historical aspects of mathematics will appreciate the enlightening content of these essays. Highlighting the pervasive nature of mathematics today in different areas, the book also covers the spread of mathematical ideas and techniques in areas ranging from computer science to physics to biology"-- Provided by publisher.
　　Includes bibliographical references and index.
　ISBN 978-1-57808-704-4 (pbk.)
1. Mathematics. I. Sujatha, R. II. Ramaswamy, H. N. III. Yogananda, C. S., 1960-
　QA7.M44425 2011
　510--dc23

　　　　　　　　　　　　　　　　　　　　　　　　　　　　2011023966

Preface

Mathematics, as a highly structured edifice, is a shining example of a collective human endeavour that transcends boundaries of time and space. Since antiquity, civilizations around the world have grappled with fundamental mathematical questions and concepts. Today, the subject has made inroads into numerous other areas and has proven to be a surprisingly powerful and effective tool in other sciences. Simultaneously, Mathematics, both in its pure abstract form and in applications, continues to expand, with advances at the frontiers of research. Some of these advances are in turn spurred by queries originating in other sciences!

The goal of this volume is to present a glimpse of these aspects in the form of essays. It is intended largely for students at the Masters level or graduate level and others with an interest in, and some exposure to, advanced Mathematics. It aims to capture both the pervasive nature of the subject and the larger picture in selected areas within the subject. Thus the contributions range from essays within pure Mathematics and applications—Opinion Polls and Finance, the role of mathematical ideas in Computer Science, Physics and Biology, as well as historical aspects of Mathematics, which bear eloquent testimony to the lasting nature of key mathematical ideas. As the reader may notice at first glance, certain topics are conspicuous by their absence, for example, Chemistry and Economics. Perhaps, they will form the material for another volume.

The essays are divided into specific broader areas and we briefly describe them below. The first and second collection of essays deal with topics in Pure and Applied Mathematics. They also serve to illustrate the timelessness of mathematical ideas and their constant evolution. For instance, matrices, as we know them today, were formally developed in the middle of the nineteenth century by western mathematicians. However, their implicit use in solving systems of linear equations goes back to the third century B.C. in China! Matrices are important mathematical tools today, with determinants and eigenvalues having a range of applications in diverse areas. One of the essays provides a brief glimpse of the role that eigenvalues play in principal component analysis. The essay on Group Theory aims to present the larger picture beyond the study of the basic theory that students are exposed to in their undergraduate years. Group theory is one example of an area of mathematics that witnessed rapid developments in the last century, once the axiomatics of the subject were formally put in place. Nevertheless, there are numerous open problems in this area.

The celebrated mathematician Carl Friedrich Gauss called Number Theory the 'Queen of Mathematics'. From the study of basic prime numbers and its generalizations, this area has grown tremendously with links to other modern subjects of research within Mathematics. The essays on Number theory provide a veritable feast to the reader in explaining some deep concepts within this beautiful subject. They simultaneously explain how many problems that can be simply stated without requiring any mathematical sophistication are among the oldest problems of humankind, and deepest among unsolved problems in Mathematics. The attempts towards resolution of such problems have in turn led to powerful new techniques and theories arising within Mathematics. The essays also highlight the connections that Number Theory has with other areas of Mathematics like Geometry and Complex topology. The hallmarks of Mathematics are precision and rigour. This is illustated beautifully, for example, in the essay that describes how an intuitive concept like 'Curvature' has a precise mathematical meaning. All the complexity associated with the notion of curvature, can in fact be beautifully encapsulated in mathematical ideas, which owe their origins to the work of Euler, Riemann and Gauss. Today it spreads its dominions over Riemannian Geometry, which has deep links with the developments in Physics in the last century, especially Einstein's Theory of Relativity.

Applied Mathematics by itself would warrant a separate volume. Galileo Galilei averred that Mathematics is the key and door to all the sciences. In recent centuries, this is borne out by the widespread use of differential equations in trying to understand, simulate and model a variety of phenomena that occurs in nature. Statistics and Probability have now become integral parts of the quest towards understanding behaviour, formulating economic theories, prediction models, etc. The celebrated Central Limit Theorem in classical Probability theory asserts that that averages taken over large samples of well-behaved independent, identically distributed random variables converge on expectation values. The theory of large deviations deals with rates at which probabilities of certain events decay, as a natural parameter in the problem varies. S.R.S. Varadhan, an Abel Laureate, and considered the founder of the theory of Large Deviations discusses this concept in his essay. Some of the mathematical concepts that underpin ideas in the new area of Financial Mathematics is dealt with in one of the essays, while another explains those behind modern day exit polls and predictions of results in elections.

Today's age is a Digital Age. The development of Internet Technology has been incumbent on Theoretical Computer Science, which in turn borrows from older abstract formal concepts which are intrinsically mathematical in nature. The study of primes, carried out over centuries, by mathematicians across the globe and in different civilizations, today has profound implications for internet security. With the enormous amount of real time data now being generated by the ubiquitous use of computers, new mathematical theories need to be developed to meaningfully glean information from this voluminous data. The collection of essays on Mathematics in Computer Science discusses some aspects of the role of mathematical ideas in Automata theory, Quantum Computing and Complexity. Another essay outlines some specific contributions of the renowned theoretical Computer Scientist and Nevanlinna Prize winner, Robert Endre Tarjan.

The famous physicist and Nobel Laureate Eugene Wigner was deeply perceptive when he remarked on the 'unreasonable effectiveness of mathematics in the natural sciences'. Physics and Mathematics have had a symbiotic relationship that spans over five centuries. The last half of the twentieth century witnessed the remarkable explosion in the use of mathematics in Physics, through the Theory of Relativity, Quantum Physics and String theory. Mathematical techniques and results proved in the abstract earlier on by mathematicians proved to be exactly what the Physicist needed for these new theories! In recent years, this phenomenon has curiously and interestingly been reversed, with cues and questions coming from Physics, and which the mathematicians are trying to place in a generalised abstract framework that will surely help in the advance of both Physics and Mathematics. The essays on this subject provide a perspective on this. It is interesting to speculate on how great minds like Hermann Weyl and Harishchandra would have viewed this 'reverse osmosis' of insight from Physics driving research in Mathematics!

Applications of Mathematics in Biology is perhaps the youngest among the subjects considered in this volume. In a rather relatively short span of twenty odd years, this area has flowered and is now growing at a breathtaking pace. To be sure, advances in technology have been an important contributing factor, as has the explosive growth of computing power. Mathematics has been the key element in harnessing these advances to provide a framework for spurring further research that delves into areas of biology that straddle the workings of the Brain, unravelling the genetic code, evolutionary biology, cognition, etc. The essays on some of these themes present a sketch of these nascent and fascinating areas. This century is surely poised to witness some remarkable advances on these new and unexplored frontier areas of interdisciplinary research.

The final collection brings the reader back firmly into the realm of the basic principle of this volume, namely the intrinsic value of mathematical ideas. These essays serve as reminders that the growth and evolution of mathematical ideas stretch back to antiquity, displaying a remarkable continuity in thought across Time and Civilizations, and a careful reading of the essays should make this amply clear to the reader. We have strived to give a comprehensive list of references to the essays so that the interested reader can delve deeper in the areas that (s)he chooses to. A diligent reader should also perceive the beauty of ideas and an inner unity within the subject. The French painter Pierre August Renoir firmly believed that Art is Eternal. We hope that these essays succeed in impressing upon the reader the eternity of mathematical thought as well!

Contents

Part I
Mathematics for its Own Sake

CHAPTER 1

Group Theory—
What's Beyond

B. Sury

Stat-Math Unit, Indian Statistical Institute,
8th Mile Mysore Road, Bangalore 560059, India.
e-mail: sury@isibang.ac.in

Birth and infancy of group theory

Major progress in group theory occurred in the nineteenth century but the evolution of group theory began already in the latter part of 18th century. Some characteristic features of 19th century mathematics which had crucial impact on this evolution are concern for rigor and abstraction and the view that mathematics is a human activity without necessarily referring to physical situations. In 1770, Joseph Louis Lagrange (1736–1813) wrote his seminal memoir *Reflections on the solution of algebraic equations*. He considered 'abstract' questions like whether every equation has a root and, if so, how many were real/complex/positive/negative? The problem of algebraically solving 5th degree equations was a major preoccupation (right from the 17th century) to which Lagrange lent his major efforts in this paper. His beautiful idea (now going under the name of Lagrange's resolvent) is to 'reduce' a general equation to auxiliary (resolvent) equations which have one degree less. Later, the theory of finite abelian groups evolved from Carl Friedrich Gauss's famous "Disquisitiones Arithmeticae". Gauss (1777–1855) established many of the important properties though he did not use the terminology of group theory. In his work, finite abelian groups appeared in different forms like the additive group $\mathbf{Z}_n$ of integers modulo n, the multiplicative group $\mathbf{Z}_n^*$ of integers modulo n relatively prime to n, the group of equivalence classes of binary quadratic forms, and the group of n-th roots of unity. In 1872, Felix Klein delivered a famous lecture *A Comparative Review of Recent Researches in Geometry*. The aim of his (so-called) Erlangen Program was the classification of geometry as the study of invariants under various groups of transformations. So, the groups appear here "geometrically" as groups of rigid motions, of similarities, or as the hyperbolic group etc. During the analysis of the connections between the different geometries, the focus was on the study of properties of figures invariant under transformations. Soon after, the focus shifted to a study of the transformations themselves. Thus the study of the geometric relations between figures got converted to the study of the associated transformations. In 1874, Sophus Lie introduced his "theory of continuous transformation groups" what we basically call Lie groups today. Poincaré and Klein began their work on the so-called *automorphic functions* and the groups associated with them around 1876. Automorphic functions are generalizations of the circular, hyperbolic, elliptic, and other functions of elementary analysis. They are functions of a complex variable z, analytic in some domain, and invariant under the group of transformations $x' = \frac{ax+b}{cx+d}$ (a, b, c, d real or complex and $ad - bc \neq 0$), or under some subgroup of this group. We end this introduction with the ancient problem of finding all those positive integers (called congruent numbers) which are areas of right-angled triangles with rational sides. To this date, the general result describing all of them precisely is not proved. However, the problem can be re-stated in terms of certain groups called elliptic curves and this helps in getting a hold on the problem and in obtaining several partial results. Indeed, if n is a positive integer, then the rational numbers x, y satisfying $y^2 = x^3 - n^2 x$ form a group law which can be described geometrically. Then, n is a congruent number if and only if, this group is infinite!

1 Unreasonable effectiveness of group theory

In a first course on group theory, one does not discuss much beyond the Sylow theorems, structure of finitely generated abelian groups, finite solvable groups (in relation to Galois theory), and the simplicity of the alternating groups A_n for $n \geq 5$. For instance, nilpotent groups are hardly discussed and nor is there a mention of other non-abelian simple groups, let alone a proof of their simplicity or a mention of the classification of finite simple groups. Moreover, infinite groups are also hardly discussed although quite a bit could be done at that stage itself. For instance, free groups are barely discussed and profinite are rarely discussed. Even a second course in group theory usually does not give one an inkling as to its depth and effectiveness. Here, we take several brief de-tours to indicate this. Let us start with some familiar aspects first.

1.1 Classification of finite simple groups

The basic problem in finite group theory was to 'find' all possible finite groups of a given order. However, this is too difficult/wild a problem - even an estimate (in terms of n) of the number of possible non-isomorphic groups of order n is a rather recent result requiring deep analysis of several years. As simple groups are building blocks, the classification of all finite, simple groups is a more tractable problem. The biggest success story of recent years is an accomplishment of this task. In rough terms, the big classification theorem of finite simple groups (CFSG) asserts:

Every finite simple group is a cyclic group of prime order, an alternating group, a simple group of Lie type or, one of the 26 sporadic finite simple groups.

A self-contained proof at this point would take 10,000 pages and there are serious attempts to simplify several parts of it. For instance, the famous odd-order theorem which is an important part of the CFSG itself takes more than 2000 pages. The odd-order theorem is the beautiful assertion proved by Feit and Thompson: "Any group of odd order is solvable."

The CFSG is extremely useful in many situations but, often it is some consequence of the CFSG which is applied. For instance, a consequence of CFSG is that every nonabelian, finite simple group can be generated by two elements, one of which could be an arbitrary nontrivial element. Another such fact is that each element is a commutator. See Carter (1993) for further reading.

1.2 Platonic solids

The determination of finite subgroups of the group of 3-dimensional rotations plays an all-important role in proving that there are precisely five Platonic solids - the cube, the tetrahedron, the octahedron, the dodecahedron

and the icosahedron. Let us describe this more precisely. The symmetry group of a 3-dimensional figure is the set of all distance preserving maps, or isometries, of $\mathbf{R}^3$ which map the figure to itself, and with composition as the operation. To make this more concrete, we view each Platonic solid centered at the origin. An isometry which sends the solid to itself then must fix the origin. An isometry which preserves the origin is a linear transformation, and is represented by a matrix A satisfying $A^T A = I_3$. Thus, the symmetry group of a Platonic solid is isomorphic to a subgroup of the orthogonal group $O_3(\mathbf{R}) = \{A : A^T A = I\}$. Elements of $O_3(\mathbf{R})$ are either rotations or reflections across a plane, depending on whether the matrix has determinant 1 or -1. The set of rotations is then the subgroup $SO_3(\mathbf{R})$ of $O_3(\mathbf{R})$. Let G be the symmetry group of a Platonic solid, viewed as a subgroup of $O_3(\mathbf{R})$. If $R = G \cap SO_3(\mathbf{R})$, then R is the subgroup of rotations in G. We note that if $z : \mathbf{R}^3 \to \mathbf{R}^3$ is defined by $z(x) = -x$ for all $x \in \mathbf{R}^3$, then z is a reflection, z is a central element in $O_3(\mathbf{R})$, and $O_3(\mathbf{R}) = SO_3(\mathbf{R}) \times \langle z \rangle \cong SO_3(\mathbf{R}) \times \mathbf{Z}_2$. These facts are all easy to prove. Thus, $[O_3(\mathbf{R}) : SO_3(\mathbf{R})] = 2$. As a consequence, $[G : R] \leq 2$. The element z is a symmetry of all the Platonic solids except for the tetrahedron, and there are reflections which preserve the tetrahedron. Therefore, $[G : R] = 2$ in all cases, and $G \cong R \times \mathbf{Z}_2$ for the four largest solids. Thus, for them, it will be sufficient to determine the rotation subgroup R. The final outcome is given in the following table:

Solid	Rotation Group	Symmetry Group
tetrahedron	A_4	S_4
cube	S_4	$S_4 \times \mathbf{Z}_2$
octahedron	S_4	$S_4 \times \mathbf{Z}_2$
dodecahedron	A_5	$A_5 \times \mathbf{Z}_2$
icosahedron	A_5	$A_5 \times \mathbf{Z}_2$

It is no coincidence that the symmetry group of the octahedron and the symmetry group of the cube are isomorphic, as are the groups for the dodecahedron and icosahedron. There is a notion of *duality* of Platonic solids. If we take a Platonic solid, put a point in the center of each face, and connect all these points, we get another Platonic solid. The resulting solid is called the *dual* of the first. For instance, the dual of the octahedron is the cube, and the dual of the cube is the octahedron. By viewing the dual solid as being built from another in this way, any symmetry of the solid will yield a symmetry of its dual, and vice-versa. Thus, the symmetry groups of a Platonic solid and its dual are isomorphic. The groups of symmetries of regular polyhedra arise in nature as the symmetry groups of molecules. For instance, $H_3C - CCl_3$ has the symmetry group C_3, S_3 is the symmetry group of the ethane molecule C_2H_6 and S_4 is the symmetry group of uranium hexaflouride UF_6 etc.

1.3   Lattice groups and crystallographic groups

Given a lattice L in $\mathbf{R}^n$ (that is the group of all integer linear combinations of a vector space basis), its group of automorphisms is isomorphic to $GL(n, \mathbf{Z})$. Finite subgroups of such a group arise if we are looking at orthogonal transformations with respect to a positive-definite quadratic form (metric). The (finite) groups of orthogonal transformations which take a lattice in $\mathbf{R}^n$ to itself is called a *crystallographic class*. There are exactly 10 plane crystallographic classes (when $n = 2$) - the cyclic groups of orders $1, 2, 3, 4, 6$ and the dihedral groups of orders $2, 4, 6, 8, 12$. In 3-dimension, there are 32 crystallographic classes. Determination of these groups does not completely solve the problem of finding all the inequivalent symmetry groups of lattices. That is to say, two groups in the same crystallographic class may be inequivalent as symmetry groups (algebraically, two non-conjugate subgroups of $GL(n, \mathbf{Z})$ may be conjugate by an orthogonal matrix). The inequivalent symmetry groups are 13 in the plane case and 72 in the case $n = 3$. The corresponding polygons (resp. polyhedrons when $n = 3$) of which these groups are symmetry groups can be listed and a description of the actions can be given without much difficulty. In general dimension, one can use a general result of Jordan to show that there are only finitely many symmetry groups of lattices.

Related to the above-mentioned finite groups are certain infinite discrete groups called *crystallographic groups*. Atoms of a crystal are arranged discretely and symmetrically in 3-space. One naturally looks at groups of motions of 3-space which preserve the physical properties of the crystal (in other words, takes atoms to atoms and preserves all relations between the atoms). More generally, one can do this for n-space. Basically, a crystallographic group in n dimension is a discrete group of motions of $\mathbf{R}^n$ such that the quotient of this action is a compact space. For instance, the group $\mathbf{Z}^n$ is such a group. More generally, a classical theorem is:

Bieberbach's theorem. If G is a crystallographic group in dimension n, then the subgroup of all translations in G is a normal subgroup A of finite index in it.

This, along with a result of Jordan implies that the number of non-isomorphic crystallographic groups in a given dimension is finite (Jordan's theorem is that any finite group has only finitely many inequivalent 'integral' representations in any fixed dimension n). This solves the first part of Hilbert's 18th problem. In crystallography, it is important to know the groups in dimension 3. It turns out that there are 219 such groups (230 if we distinguish mirror images) and have a special naming system under the "*International tables for crystallography*". On the plane (that is, in dimension 2) there are 17 crystallographic groups known as *wallpaper groups*. They make very beautiful patterns and can be seen abundantly in Escher's paintings as well as architectural constructions (see Schattschneider (1978)). In 4-dimensions, there are 4783 crystallographic groups. There are also groups appearing in non-euclidean crystallography which involve non-euclidean geometry.

1.4 Reflection groups

The group of permutations of a basis of Euclidean space is a group of orthogonal transformations of the Euclidean space which is generated by reflections about hyperplanes. In general, groups of orthogonal transformations generated by reflections arise in diverse situations; indeed, even in the theory of crystallographic groups. One abstractly studies such groups under the name of Coxeter groups. A *Coxeter group* G is a group generated by elements $s_1, \ldots, s_n$ with relations $(s_i s_j)^{m_{ij}} = 1$ where $m_{ii} = 2$ and $m_{ij} = m_{ji} \geq 2$. If the so-called Gram matrix $a_{ij} = \cos(\pi/m_{ij})$ is positive-semidefinite, the corresponding Coxeter group is a crystallographic group and these have been classified. If the Gram matrix is positive-definite, this Coxeter group is finite; the irreducible ones are classified under the names $A_n, B_n, D_n, E_6, E_7, E_8, F_4, H_3, H_4, I_2(m)$. The subscripts denote the dimensions of the Euclidean space on which the group acts. This classification is the key to classifying the complex simple Lie algebras also. See Humphreys (1997) for further reference.

1.5 Galois groups versus polynomials

Lagrange's method of resolvents can be briefly described as follows: Let f be a polynomial, with roots $r_1, r_2, \ldots, r_n$. For any rational function F of the roots and the coefficients of f, look at all the different values which F takes when we go through the various permutations of the roots of f. Denoting these values by $s_1, s_2, \ldots, s_k$, he considers the (resolvent) polynomial

$$g(x) = (x - s_1)(x - s_2) \cdots (x - s_k).$$

The coefficients of g are symmetric functions of the roots of f; so, they are polynomials in the elementary symmetric functions in them. In other words, the coefficients of g are polynomials in the coefficients of f. Lagrange showed that the number k divides $n!$. *This is the origin of the so-called Lagrange's theorem in group theory.* For example, if f has roots r_1, r_2, r_3, r_4 (and so degree of f is 4), then the function $F = r_1 r_2 + r_3 r_4$, assumes 3 different values under the 4! permutations of the r_i's. So, the resolvent polynomial of a quartic (degree 4 polynomial) is a cubic polynomial. When Lagrange tried to apply this to quintics (the polynomials of degree 5), Lagrange found the resolvent polynomial to be of degree 6 (!) Even though he did not succeed in going further in the problem of quintics, his method was path-breaking as it was the first time that solutions of polynomial equations and permutations of roots were related to each other. The theory of Galois has its germ in Lagrange's method. Galois theory is now a basic subject taught in undergraduate programmes as it plays a role in all branches of mathematics. Here is a nice application of Galois theory to the study of polynomials over integers.

Let $f = a_0 + a_1 x + \cdots + a_{n-1} x^{n-1} + x^n$ *be a monic, integral polynomial of degree $n > 1$. Suppose that for every prime p, the polynomial f considered as a polynomial with coefficients in $\mathbf{Z}/p\mathbf{Z}$ has a root. Then, f must be reducible over* $\mathbf{Z}$.

Here is how it is proved. For a prime p, the decomposition type of f is the set of numbers $n_1 \leq n_2 \cdots \leq n_r$ where the polynomial f - when considered modulo p - is a product of irreducible polynomials of degrees $n_1, \ldots, n_r$. For an infinite set P of primes, one has often a well-defined notion of density. We briefly recall it although we will not need any details except the statement of the following theorem. The theory of the Riemann zeta function tells us that for any complex number s with Re $s > 1$, the infinite product $\prod_{p \; prime}(1 - 1/p^s)^{-1}$ converges to a finite number denoted by the sum $\sum_n 1/n^s$. Also,

$$\prod_{p \; prime} (1 - 1/p^s)^{-1} - \frac{1}{s-1}$$

has a finite limit as $s \to 1$ from the right. Then, one defines the density of a set P of primes to be the limit as $s \to 1^+$ of the ratio

$$\frac{\prod_{p \in P}(1 - 1/p^s)^{-1}}{\prod_{p \; prime}(1 - 1/p^s)^{-1}}$$

if this limit exists. For instance, finite sets P have density zero. The theorem of Dirichlet on infinitude of primes in arithmetic progressions has the analytic statement that if $P(c,d)$ is the set of primes in an arithmetic progression $dn + c$ with $(c,d) = 1$, then the density of $P(c,d)$ is $1/\phi(d)$. We will not need any properties of the density excepting the following statement which is a weaker form of the so-called Frobenius Density Theorem:

Let f be a monic, irreducible, integral polynomial. The set of primes p modulo which f has decomposition type $n_1, n_2, \ldots, n_r$ has density equal to $|N/Gal(f)|$ where $N = \{\sigma \in Gal(f) : \sigma$ has a cycle pattern $n_1, n_2, \ldots, n_r\}$.

Here $Gal(f)$ is considered as a permutation group on the roots of f. Recall an elementary fact about finite groups:

If H is a proper subgroup of a finite group G, then the union of all the conjugate subgroups of H cannot be the whole of G.

Let us now prove the assertion made in the beginning. Suppose f is an irreducible polynomial such that f has root modulo every prime. Then the Frobenius Density Theorem shows that every σ has a cycle pattern of the form $1, n_2, \ldots$ This means that every element of $Gal(f)$ fixes a root, say β. Since f is irreducible, the group $Gal(f)$ acts transitively on the roots of f. Thus, this group would be the union of the conjugates of its subgroup H consisting of those elements which fix the root β. But the above fact on finite groups implies H cannot be a proper subgroup. Hence $Gal(f)$ fixes each root of f and is therefore trivial. So we get f to be a linear polynomial. This proves the assertion.

A more general density theorem of chebotarev implies: For a monic, irreducible, integral polynomial f, there are infinitely many primes p modulo which f is irreducible if and only if, $Gal(f)$ contains an element of order n. More generally, suppose that f is an integral polynomial and let $f = f_1 f_2 \ldots f_r$ be its decomposition into irreducible, integral polynomials. For each $i \le r$, let θ_i be a fixed root of f_i and consider the field K_i obtained by adjoining θ_i. If G is the Galois group of f and G_i is the subgroup of all those permutations in G which fix θ_i, then look at the set $S = \bigcup_{i=1}^{r} G_i$. Finally, look at the product of the resultants[1] of f_i and f_i'; let $p_1^{b_1} \ldots p_r^{b_r}$ be the prime decomposition of this product. Then, we can prove similarly as above that:

f has a root modulo the single integer

$$p_1^{2b_1+1} \ldots p_r^{2b_r+1} \quad and \quad G = \bigcup_{g \in G} gSg^{-1}$$

if and only if f has a root modulo every integer.

1.6   Combinatorial applications - Polya's theory

Groups evidently help in counting problems. In fact, this goes a very long way as the so-called Polya's theory shows. Here we illustrate it by an example. Consider the problem of painting the faces of a cube either red or green. One wants to know how many such distinct coloured cubes one can make. Since the cube has 6 faces, and one has 2 colours to choose from, the total number of possible coloured cubes is 2^6. But, painting the top face red and all the other faces green produces the same result as painting the bottom face red and all the other faces green. The answer is not so obvious but turns out to be 10. Let us see how. To find the various possible colour patterns which are inequivalent, we shall exploit the fact that the rotational symmetries of the cube have the structure of a group. Let us explain the above in precise terms. Let D denote a set of objects to be coloured (in our case, the 6 faces of the cube) and R denote the range of colours (in the above case {red, green}). By a colouring of D, one means a mapping $\phi : D \to R$. Let X be the set of colourings. If G denotes a group of permutations of D, we can define a relation on the set of colourings as follows:

$\phi_1 \sim \phi_2$ if, and only if, there exists some $g \in G$ such that $\phi_1 g = \phi_2$.

By using the fact that G is a group, it is easy to prove that $\sim$ is an equivalence relation on X, and so it partitions X into disjoint equivalence

---

[1]Resultant of two polynomials $a_n \prod_{i=1}^{n}(x - \alpha_i)$ and $b_m \prod_{j=1}^{m}(x - \beta_j)$ is $a_n^m b_m^n \prod_{i,j}(\alpha_i - \beta_j)$

classes. Now for each $g \in G$, consider the map $\pi_g : X \to X$ defined as $\pi_g(\phi) = \phi g^{-1}$; it is a bijection from X to itself. In other words, for each $g \in G$, we have $\pi_g \in \text{Sym } X$, where Sym X = the group of all permutations on X. The map $f : G \to \text{Sym } X$ as $f(g) = \pi_g$ is a homomorphism; i.e., G can be regarded as a group of permutations of X.

It is clear that the orbits of the action described above are precisely the different colour patterns i.e., the equivalence classes under $\sim$. Therefore, we need to find the number of inequivalent colourings, i.e. the number of equivalence classes of $\sim$, i.e. the number of orbits of the action of G on X. Note that, like in the example of the cube we shall consider only finite sets D, R. The answer will be provided by a famous theorem of Polya. Polya's theorem was published first in a paper of J.H.Redfield in 1927 and, apparently no one understood this paper until it was explained by F. Harary in 1960. *Polya's theorem is considered one of the most significant papers in 20th-century mathematics. The article contained one theorem and 100 pages of applications.* For a group G of permutations on a set of n elements and variables $s_1, s_2, \ldots, s_n$, one defines a polynomial expression (called the cycle index) for each $g \in G$. If $g \in G$, let $\lambda_i(g)$ denote the number of i-cycles in the disjoint cycle decomposition of g. Then, the cycle index of G, denoted by $z(G; s_1, s_2, \ldots, s_n)$ is defined as the polynomial expression

$$z(G; s_1, s_2, \ldots, s_n) = \frac{1}{|G|} \sum_{g \in G} s_1^{\lambda_1(g)} s_2^{\lambda_2(g)} \ldots s_n^{\lambda_n(g)}.$$

For instance,

$$z(S_n; s_1, s_2, \ldots, s_n) = \sum_{\lambda_1 + 2\lambda_2 + \cdots + k\lambda_k = n} \frac{s_1^{\lambda_1} s_2^{\lambda_2} \ldots s_k^{\lambda_k}}{1^{\lambda_1} \lambda_1! 2^{\lambda_2} \lambda_2! \ldots k^{\lambda_k} \lambda_k!}$$

Polya's theorem asserts:

Suppose D is a set of m objects to be coloured using a range R of k colours. Let G be the group of symmetries of D. Then, the number of colour patterns $= \frac{1}{|G|} z(G; k, k, \ldots, k)$.

Our group of rotations of the cube consists of:

- 90 degree (clockwise or anti-clockwise) rotations about the axes joining the centres of the opposite faces - there are 6 such;

- 180 degree rotations about each of the above axes - there are 3 such;

- 120 degree (clockwise or anti-clockwise) rotations about the axes joining the opposite vertices - there are 8 such;

- 180 degree rotations about the axes joining the midpoints of the opposite edges and;

- the identity.

The cycle index of G turns out to be

$$z(G; s_1, \ldots, s_6) = \frac{1}{24} \left(6s_1^2 s_4 + 3s_1^2 s_2^2 + 8s_3^2 + 6s_2^3 + s_1^6 \right).$$

So, in our example of the cube, the number of distinct coloured cubes

$$= \frac{1}{24} \left[2^6 + 6 \cdot 2^3 + 8 \cdot 2^2 + 3 \cdot 2^2 \cdot 2^2 + 6 \cdot 2^2 \cdot 2 \right] = 10.$$

There are 10 distinct cubes in all.

2   Some truths about Lie groups

Groups are effective when they also come with additional analytic structure. The prime examples of this marriage of groups and analysis are the so-called Lie (pronounced as 'Lee') groups named so after the Norwegian mathematician Sophus Lie. Lie's idea was to develop a theory for differential equations analogous to what Galois had done for algebraic equations. His key observation was that most of the differential equations which had been integrated by older methods were invariant under continuous groups which could be easily constructed. In general, he considered differential equations which are invariant under some fixed continuous group. He was able to obtain simplifications in the equations using properties of the given group. Lie's work was fundamental in the subsequent formulation of a Galois theory of differential equations by Picard, Vessiot, Ritt and Kolchin. A typical illustration can be given as follows. Consider an n-th order linear differential equation with polynomial coefficients, viz.,

$$\frac{d^n f}{dz^n} + a_{n-1}(z) \frac{d^{n-1} f}{dz^{n-1}} + \cdots + a_0(z) f = 0.$$

Let $f_1, \ldots, f_n$ be **C**-linearly independent holomorphic solutions. Let K be the field obtained by adjoining the f_i's and their derivatives. Let G denote the group of all $\mathbf{C}(z)$-linear automorphisms of K which commute with taking derivatives. This G is called the Galois group of the differential equation. For $g \in G$, write $g.f_i = \sum_j g_{ji} f_j$ for some complex numbers g_{ji}. Thus, G is a group of matrices and defines what is called an algebraic group. It turns out that the solutions to such a linear homogeneous differential equation are '*elementary*' functions if and only if the corresponding Galois group is a so-called solvable group; this is a consequence of the Lie-Kolchin theorem - a result on triangularizability of connected solvable algebraic groups. As several groups of matrices naturally form Lie groups, it should not be surprising that Lie groups find applications in other subjects like physics and chemistry. In fact, there are applications to the real world too! For instance, in navigation, surveillance etc., one needs very sophisticated cameras (called *catadioptric*) to obtain a wide field of view, even greater than 180 degrees. Surprisingly, some of these involve finding dimensions of some quotient spaces of certain Lie groups like the Lorentz group $O(3, 1)$ (see Geyer and Daniilidis (2001)). Often, Lie groups arise as groups of real or complex matrices. They appear naturally in geometric ways - the group of rotations of 3-space is $O(3)$. Nowadays, the theory of Lie groups is regarded as a special case of the theory of algebraic groups over real or complex fields. Before discussing Lie groups (which involves a discussion of certain spaces called manifolds), we recall the notion of a topological group.

Given a group G which also happens to be a Hausdorff topological space, this is supposed to yield a *topological group* if the map $(x, y) \mapsto xy^{-1}$ from the product space $G \times G$ to G is a continuous map. Basically, the continuity of the group operations is a powerful tool which often yields group-theoretical consequences. A typical example is the absolute Galois group of the algebraic numbers over the rational numbers; this encodes all the information about all the algebraic number fields at one stroke. Notice that individual finite Galois groups do not have any meaningful topological group structure but only the discrete group structure. So, stringing them all together in this absolute Galois group makes it a topological group which is *totally disconnected* (that is, only the points are connected components) but compact as a space! This is an example of a profinite group (a *projective limit* of finite groups). In Lie theory which is a special case of a topological group, one will look at the other end of the spectrum where the spaces are connected.

Let us define smooth manifolds briefly. Basically, one has to keep in mind the examples of smooth regular surfaces in 3-space. A Hausdorff space M with a countable basis is said to be a *smooth (real) manifold of dimension n* if:

(i) there is an open cover $\{U_\alpha\}_{\alpha\in\Lambda}$ of M and homeomorphisms $\phi_\alpha : U_\alpha$ onto open sets in $\mathbf{R}^n$,

(ii) the homeomorphisms above are such that the 'transition functions' $\phi_\alpha \circ \phi_\beta^{-1} : \phi_\beta(U_\alpha \cap U_\beta) \to \phi_\alpha(U_\alpha \cap U_\beta)$ are smooth whenever $U_\alpha \cap U_\beta \neq \emptyset$ and

(iii) the collection $(U_\alpha, \phi_\alpha)_\alpha$ is a maximal family satisfying (i) and (ii).

Thus, the manifold is a topological space which is locally like $\mathbf{R}^n$ on which it makes sense to do differential calculus. Hence, one can define smooth functions between smooth manifolds. It turns out that a smooth manifold can always be 'embedded' in $\mathbf{R}^N$ for a large enough N. The unit sphere is a typical example where one may take 4 open sets to cover. If there is a curve (image of a smooth map α from $[0, 1]$ to M) on M, one has its tangent vector at a point. One may define the tangent space to M at a point as the set of tangent vectors to curves passing through that point. However, one must take some care because different curves may give the same tangent vector. It turns out that the tangent space at any point is a vector space of dimension n.

A *Lie group* is a group G which is a smooth manifold as well as a topological group and, the group multiplication and the inverse map are smooth. Some Lie groups are $\mathbf{R}^n, \mathbf{C}^*, S^1 \times \cdots S^1$, GL$(n, \mathbf{R})$, GL$(n, \mathbf{C})$, SL$(n, \mathbf{R})$, SL$(n, \mathbf{C})$, O$(n)$, SO$(n)$, U$(n)$, SU$(n)$. Also, every closed subgroup of GL$(n, \mathbf{R})$ is a Lie group. The tangent space to G at the identity plays a very important role. In particular, this vector space is equipped with a (non-associative) multiplication operation, the Lie bracket, that makes it into what is called a Lie algebra. The theory of Lie groups begun by Lie was continued by Killing and Elie Cartan. The big result of the classical theory was a classification of the complex simple (note the combination of these two words!) Lie algebras. Later, Hermann Weyl proved that every complex semisimple Lie algebra has a compact real form. This amounted to a classification of certain geometric objects called the real symmetric spaces. These are special Riemannian manifolds which possess certain distance-preserving maps inverting geodesics. We do not discuss Lie groups in more detail here but the interested reader may refer to Varadarajan (1984) and Warner (1971).

2.1 Algebraic groups - a(f)fine concept

Matrix groups are usually studied under the aegis of the so-called linear algebraic groups. These are more refined versions of Lie groups in the sense that they have more structure; in particular, they can be studied over any field. Here, the advantage is that one can use techniques from algebraic geometry. We do not go into the technical definitions here. Suffice it to say that these are subsets of n-tuples over a field k defined as zero sets of polynomials in n variables over k and have compatible group structures as well. Facts about polynomials play key roles. Some examples of algebraic groups over a field k are:

(i) The additive group of k.

(ii) The multiplicative group k^* identified with the set of zeroes of the polynomial $x_1 x_2 - 1$ in k^2.

   More generally, $G = T(n, k)$ is the group of $n \times n$ diagonal matrices over k with nonzero diagonal entries, viewed as

$$\left\{ (x_1, \ldots, x_{n+1}) \in k^{n+1} : x_1 x_2 \ldots x_{n+1} = 1 \right\}.$$

(iii) The set $G = GL(n, k)$ of all invertible matrices over k, viewed as the set

$$\left\{ (x_{ij}, y) \in k^{n^2+1} : det(x_{ij})y = 1 \right\}.$$

(iv) Let (V, q) be an n-dimensional vector space with a non-degenerate quadratic form $q : V \times V \to k$. The orthogonal group

$$O(V, q) = \{ g \in GL(V) : q(gv) = q(v) \}$$

can be viewed as an algebraic group as follows. Choose a basis $\{v_1, \ldots, v_n\}$ of V and write $B = (q(v_i, v_j))$ for the symmetric matrix associated to q. Note that for vectors $x = \sum_i x_i v_i, y = \sum_i y_i v_i \in V$, we have $q(x, y) = {}^t x B y$ where $t_x = (x_1, \ldots, x_n) \in k^n$. Then, $O(V, q)$ can be identified with

$$\left\{ g \in GL(n, k) : {}^t g B g = B \right\}.$$

(v) For any skew-symmetric matrix $\Omega \in GL(2n, k)$, the symplectic group

$$Sp(\Omega) = \left\{ g \in GL(2n, k) : {}^t g \Omega g = \Omega \right\}.$$

(vi) Let D be a division algebra with center k (its dimension as a k-vector space must be n^2 for some n). Let v_i; $1 \le i \le n^2$ be a k-basis of D. The right multiplication by v_i gives a linear transformation $R_{v_i} \in GL(n^2, \bar{k})$ for $i = 1, 2, \ldots, n^2$ where $\bar{k}$ is the algebraic closure of k. This is because when the scalars are extended to $\bar{k}$, the algebra D 'becomes' isomorphic to the algebra $M(n^2)$ of all $n^2 \times n^2$ matrices. The group

$$\left\{ g \in GL(n^2, \bar{k}) : gR_{v_i} = R_{v_i}g \; \forall \, i = 1, 2, \ldots, n^2 \right\}$$

is an algebraic group which can be identified over k with the group D^*.

We point out that applications to other subjects like differential geometry, number theory and finite geometries involve more general fields like the real field, finite fields and algebraic number fields.

2.2   Classification of Lie and algebraic groups

In an algebraic group over an algebraically closed field, one may "strip off" the maximal solvable normal algebraic subgroup and obtain the importance class of the so-called semisimple groups. These are classified in terms of certain associated finite reflection groups (as in § 1.4) called their Weyl groups. The same classification is valid for the complex Lie groups which are semisimple and simply connected. Over a general field, there may be different forms (or avataars) of the same semisimple algebraic group over the algebraic closure and, one uses Galois cohomology to classify and study them. Roughly, one obtains that the (non-exceptional or the) classical groups are the symmetriy groups symplectic, Hermitian or skew-Hermitian forms over division algebras over the field. Thus, their classification reduces to the classification of such forms over the underlying field. The classification over a finite field plays a key role in the classification of finite simple groups (see § 1.1). Looking at the classification of Lie groups or algebraic groups, there is a lot in common and this brings forth a unity in mathematics - although the objects are different, their classifications are very similar. The informed reader can refer to Carter (2005), Humphreys (1975), Springer (1981), Varadarajan (1984) for the theory of Lie algebras, algebraic groups and their classification.

3   Modular group - an arithmetic group

Often, classical number-theoretic problems can be re-phrased in terms of groups; the groups which arise in this context are called arithmetic groups. For instance, the classical Dirichlet unit theorem can be interpreted and generalized in this framework. A prime example of an arithmetic group is the modular group $SL(2, \mathbf{Z})$. It is a discrete subgroup of $SL(2, \mathbf{R})$ such that the quotient space $SL(2, \mathbf{Z}) \backslash SL(2, \mathbf{R})$ is non-compact, but has a finite $SL(2, \mathbf{R})$-invariant measure. One defines a discrete subgroup Γ of a Lie group G to be an arithmetic group if there exists an algebraic group H defined over $\mathbf{Q}$ such that the group $H(\mathbf{R})$ of real points is isomorphic to $G \times K$ for a compact group K and Γ is 'commensurable' with the image of $H(\mathbf{Z})$ in G. Here, commensurability of two groups means that their intersection is of finite index in both. Some examples of arithmetic groups are:

(i)   any finite group,

(ii) a free abelian group of finite rank,

(iii) the group of units in the ring of integers of an algebraic number field,

(iv) $GL(n, \mathbf{Z})$, $SL(n, \mathbf{Z})$, $Sp_{2n}(\mathbf{Z})$, $SL_n(\mathbf{Z}[i])$, $SO(Q)(\mathbf{Z}) := \{g \in SL_n(\mathbf{Z}) : {}^t gQg = Q\}$ where Q is a quadratic form in n variables over $\mathbf{Q}$.

When G is a semisimple Lie group, a classical theorem due to A.Borel and Harish-Chandra shows that an arithmetic subgroup Γ has 'finite volume'. This means that there is a G-invariant finite measure on the quotient space G/Γ. The possibility of this quotient space being compact is reflected group-theoretically in terms of existence of unipotent elements in Γ. In general, one has a nice "fundamental domain" for Γ in G and this is given by what is known as reduction theory of arithmetic groups. Reduction theory for $SL(2, \mathbf{Z})$ is (roughly) to find a complement to $SL(2, \mathbf{Z})$ in $SL(2, \mathbf{R})$; a 'nice' complement is called a fundamental domain. Viewing the upper half-plane $\mathbf{H}$ as the quotient space $SL(2, \mathbf{R})/SO(2)$, the subset

$$\left\{ z \in \mathbf{H} : \mathrm{Im}(z) \geq \sqrt{3}/2, |Re(z)| \leq 1/2 \right\}$$

is (the image in $\mathbf{H}$) of a fundamental domain (figure below):

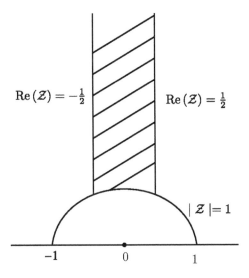

Fundamental domains can be very useful in many ways; for example, they give even a presentation for the arithmetic group. Indeed, the above fundamental domain gives the presentation $\langle x, y | x^2, y^3 \rangle$ for the group $PSL(2, \mathbf{Z}) = SL(2, \mathbf{Z})/\{\pm I\}$; that is, $PSL(2, \mathbf{Z})$ is a free product of cyclic groups of orders 2 and 3. The modular group $SL(2, \mathbf{Z})$ itself is thus an amalgamated free product of cyclic groups of orders 4 and 6 amalgamated along a subgroup of order 2. In this case, such a domain is written in terms of the Iwasawa decomposition of $SL(2, \mathbf{R})$. One has $SL(2, \mathbf{R}) = KAN$ by Gram-Schmidt process where $K = SO(2)$, $A = \{diag(a, a^{-1}) : a > 0\}$, $N = \left\{ \begin{pmatrix} 1 & x \\ 0 & 1 \end{pmatrix} : x \in R \right\}$. The, reduction theory for $SL(2, \mathbf{Z})$ says $SL(2, \mathbf{R}) = KA_{\frac{2}{\sqrt{3}}} N_{\frac{1}{2}} SL(2, \mathbf{Z})$. Here $A_t = \{diag(a_1, a_2) \in SL(2, \mathbf{R}) : a_i > 0 \text{ and } \frac{a_1}{a_2} \leq t\}$ and $N_u = \left\{ \begin{pmatrix} 1 & x \\ 0 & 1 \end{pmatrix} \in N : |x| \leq u \right\}$.

This is connected with the theory of quadratic forms as follows. We shall consider only positive definite, binary quadratic forms over $\mathbf{Z}$. Any such form looks like $f(x, y) = ax^2 + bxy + cy^2$ with $a, b, c \in \mathbf{Z}$; it takes only values > 0 except when $x = y = 0$. Two forms f and g are said to be equivalent (according to Gauss) if $\exists A = \begin{pmatrix} p & q \\ r & s \end{pmatrix} \in SL(2, \mathbf{Z})$ such that $f(x, y) = g(px + qy, rx + sy)$. Obviously, equivalent forms represent the same values. Indeed, this is the reason for the definition of equivalence.

One defines the discriminant of f to be $\text{disc}(f) = b^2 - 4ac$. Further, f is said to be primitive if $(a, b, c) = 1$. Note that if f is positive-definite, the discriminant D must be < 0 (because $4a(ax^2+bxy+cy^2) = (2ax+by)^2 - Dy^2$ represents positive as well as negative numbers if $D > 0$.) A primitive, +ve definite, binary quadratic form $f(x, y) = ax^2+bxy+cy^2$ is said to be reduced if $|b| \leq a \leq c$ and $b \geq 0$ if either $a = c$ or $|b| = a$. These easily imply

$$0 < a \leq \sqrt{\frac{|D|}{3}}.$$

For example, the only reduced form of discriminant $D = -4$ is $x^2 + y^2$. The only two reduced forms of discriminant $D = -20$ are $x^2 + 5y^2$ and $2x^2 + 2xy + 3y^2$.

$GL(2, \mathbf{R})$ acts on the space S of +ve-definite, binary quadratic forms as follows: Each $P \in S$ can be represented by a +ve-definite, symmetric matrix. For $g \in GL(2, \mathbf{R})$, ${}^t\!gPg \in S$. This action is transitive and the isotropy at $I \in S$ is $O(2)$. In other words, S can be identified with $GL(2, \mathbf{R})/O(2)$ i.e. $S = \{{}^t\!gg : g \in GL(2, \mathbf{R})\}$. In general, this works for +ve-definite quadratic forms in n variables.

It is easy to use the above identification and the reduction theory statement for $SL(2, \mathbf{Z})$ to show that each +ve definite, binary quadratic form is equivalent to a unique reduced form.

Indeed, writing $f = {}^t\!gg$ and $g = kan\gamma$, ${}^t\!gg = {}^t\!\gamma\,{}^t\!na^2 n\gamma$ with $n \in U_{1/2}$ and $a^2 \in A_{4/3}$; so ${}^t\!na^2 n$ is a reduced form equivalent to f.

To see how useful this is, let us prove a beautiful discovery of Fermat, viz., that any prime number $p \equiv 1 \bmod 4$ is expressible as a sum of two squares.

Since $(p - 1)! \equiv -1 \bmod p$ and since $(p - 1)/2$ is even, it follows that $(\frac{p-1}{2}!)^2 \equiv -1 \bmod p$ i.e.,

$$\left(\left(\frac{p - 1}{2}\right)!\right)^2 + 1 = pq$$

for some natural number q. Now the form $px^2 + 2(\frac{p-1}{2})!xy + qy^2$ is +ve definite and has discriminant -4. Now, the only reduced form of discriminant -4 is $x^2 + y^2$ as it is trivial to see. Since each form is equivalent to a reduced form (by reduction theory), the forms $px^2 + 2\frac{p-1}{2}!xy + qy^2$ and $x^2 + y^2$ must be equivalent. As the former form has p as the value at $(1, 0)$, the latter also takes the value p for some integers x, y.

In fact, reduction theory can also be used to show:

For any D < 0, there are only finitely many classes of primitive, positive-definite forms of discriminant D.

The number of classes alluded to is the class number $h(D)$ of the field $\mathbf{Q}(\sqrt{D})$; an isomorphism is obtained by sending $f(x, y)$ to the ideal $a\mathbf{Z} + \frac{-b + \sqrt{D}}{2}\mathbf{Z}$.

The idea is to use reduction theory to show that each form is equivalent to a unique 'reduced' form. 'Reduced' forms can be computed - there are even algorithms which can be implemented in a computer which can determine $h(D)$ and even the $h(D)$ reduced forms of discriminant D.

For more on arithmetic aspects, the initiated reader can refer to Margulis (1991), Platonov and Rapinchuk (1994) and Raghunathan (1972) but we end with the statement of a somewhat recent deep application of the theory of arithmetic groups to a classical problem of number theory. The *Oppenheim conjecture* asserts that if Q is a real, indefinite quadratic form in $n \geq 3$ variables which is not a multiple of a rational quadratic form, then its values at integer lattice points form a dense subset of **R**. This was reformulated in terms of Lie groups by Raghunathan and Dani following which Margulis solved it. The statement in terms of Lie groups is the following:

Let $G = SL(3, \mathbf{R}), \Gamma = SL(3, \mathbf{Z}), H = SO(X_1X_3 - X_2^2)$. Then, for any $g \in G$ such that $Hg\Gamma/\Gamma$ has compact closure, this space $Hg\Gamma/\Gamma$ is actually compact.

References

Carter, R. 1993. *Finite groups of Lie type, Conjugacy classes and complex characters*, Reprint of the 1985 original, Wiley Classics Library, John Wiley and Sons, Ltd., Chichester.

Carter, R. 2005. *Lie algebras of finite and affine type*, Cambridge Studies in Advanced Mathematics 96, Cambridge University Press.

Geyer, C. and Daniilidis, K. 2001. Catadioptric projective geometry, *International Journal of Computer vision*. **43**:223–243.

Humphreys, J. 1975. *Linear algebraic groups*, Graduate Texts in Mathematics 21, Springer.

Humphreys, J. 1997. *Reflection groups and Coxeter groups*, Cambridge University Press.

Margulis, G.A. 1991. *Discrete subgroups of semisimple Lie groups*, Ergebnisse der Mathematik und ihrer Grenzgebiete 17, Springer-Verlag.

Platonov, V.P. and Rapinchuk, A.S. 1994. *Algebraic groups and number theory*, Translated from the 1991 Russian original by Rachel Rowen, Pure and Applied Mathematics 139, Academic Press.

Raghunathan, M.S. 1972. *Discrete subgroups of Lie groups*, Ergebnisse der Mathematik und ihrer Grenzgebiete, Band 68, Springer-Verlag.

Schattschneider, D. 1978. The Plane Symmetry Groups: Their recognition and notation, *American Mathematical Monthly*, **85**:439–450.

Springer, T.A. 1981. *Linear algebraic groups*, Progress in Mathematics 9, Birkhauser.

Varadarajan, V.S. 1984. *Lie groups, Lie algebras and their representations*, Graduate Texts in Mathematics 102, Springer.

Warner, F. 1971. *Foundations of differentiable manifolds and Lie groups*, Scott, Foresman and Co., Glenview, Ill.-London.

CHAPTER 2

Splitting Primes

Chandan Singh Dalawat

Harish-Chandra Research Institute,
Chhatnag Road, Jhunsi,
Allahabad 211019, India.
e-mail: dalawat@gmail.com

We give an elementary introduction, through illustrative examples but without proofs, to one of the basic consequences of the Langlands programme, namely the law governing the primes modulo which a given irreducible integral polynomial splits completely. Some recent results, such as the modularity of elliptic curves over the rationals, or the proof of Serre's conjecture by Khare and Wintenberger, are also illustrated through examples.

Certainly the best times were when I was alone with mathematics, free of ambition and pretense, and indifferent to the world.
                                                                    — **Robert Langlands**.

Groups, rings, fields, polynomials, primes, functions: most mathematics students at the university have come across these concepts. It so happens that one of the basic problems of number theory can be formulated using nothing more than these notions; we shall try to explain Langlands' conjectural solution to this problem. Everything we say here has been said before, but some readers might not have access to the original sources. This is the pretext for putting together these observations; the aim is not so much to instruct as to convey the impression that these examples are part of a larger pattern. Proofs are altogether omitted.

Let us consider a monic polynomial $f = T^n + c_{n-1}T^{n-1} + \cdots + c_1 T + c_0$ of degree $n > 0$ with coefficients c_i in the ring $\mathbf{Z}$ of rational integers (the adjective *monic* means that the coefficient of T^n is 1). Suppose that the polynomial f is irreducible: it cannot be written as a product $f = gh$ of two polynomials $g, h \in \mathbf{Z}[T]$ of degree $< n$. Basic examples to be kept in mind are $f = T^2 + 1$ or $f = T^3 - T - 1$. (There are irreducible polynomials of every possible degree. For example, Selmer showed that $T^n - T - 1$ is irreducible for every $n > 1$.)

For every number $m > 0$, we have the finite ring $\mathbf{Z}/m\mathbf{Z}$ with m elements each of which is of the form $\bar{a}$ for some $a \in \mathbf{Z}$, with the relation $\bar{a} = \bar{b}$ if m divides $a - b$; this relation is also written $a \equiv b \pmod{m}$, and we say that $\bar{a}$ is the *reduction* of a modulo m. The group of invertible elements of this ring is denoted $(\mathbf{Z}/m\mathbf{Z})^\times$, so that $\bar{a} \in (\mathbf{Z}/m\mathbf{Z})^\times$ if and only if $\gcd(a, m) = 1$. The order of this group is denoted $\varphi(m)$. When the integer m is a prime number p, the ring $\mathbf{Z}/p\mathbf{Z}$ is a field denoted $\mathbf{F}_p$ and the group $\mathbf{F}_p^\times$ is cyclic of order $\varphi(p) = p - 1$.

For every prime p, the polynomial $f \in \mathbf{Z}[T]$ gives rise to a degree-n polynomial $\bar{f} = T^n + \bar{c}_{n-1} T^{n-1} + \cdots + \bar{c}_1 T + \bar{c}_0$ whose coefficients $\bar{c}_i \in \mathbf{F}_p$ are the reductions of c_i modulo p; this $\bar{f} \in \mathbf{F}_p[T]$ is called the *reduction* of f modulo p.

Now, although our polynomial f is irreducible by hypothesis, there may be some primes p modulo which its reduction $\bar{f}$ has n distinct roots. For example, if $f = T^2 + 1$ and $p = 5$, then $\bar{f} = (T + \bar{2})(T - \bar{2})$. More generally, we have $\bar{f} = (T + \bar{1})^2$ if $p = 2$, $\bar{f}$ has two distinct roots if $p \equiv 1 \pmod{4}$ and $\bar{f}$ remains irreducible if $p \equiv -1 \pmod{4}$. One sees this by remarking that, for odd p, a root $x \in \mathbf{F}_p$ of $\bar{f}$ is an order-4 element in the cyclic group $\mathbf{F}_p^\times$ of order $p - 1$, so that x exists if and only if $4 \mid p - 1$. This example goes back to Fermat.

Take $f = T^2 - T - 1$ as the next example. If $p = 5$, then $\bar{f} = (T + \bar{2})^2$. It can be checked that for $p \neq 5$, the polynomial $\bar{f}$ has two distinct roots if and only if $p \equiv \pm 1 \pmod{5}$, and that it is irreducible if and only if $p \equiv \pm 2 \pmod{5}$.

We notice that for these two quadratic polynomials f, the primes p for which $\bar{f}$ has two distinct roots are specified by "congruence conditions": by $p \equiv 1 \pmod{4}$ when $f = T^2 + 1$, by $p \equiv \pm 1 \pmod{5}$ when $f = T^2 - T - 1$.

It can be shown that for any (monic, irreducible) quadratic polynomial $f \in \mathbf{Z}[T]$, the primes p for which $\bar{f} \in \mathbf{F}_p[T]$ has two distinct roots are given by certain congruence conditions modulo some number D_f depending on f. This statement (with a precise expression for D_f which we omit) is implied by the law of quadratic reciprocity, first proved by Gauss. The law in question says that for any two odd primes $p \neq q$, we have

$$\left(\frac{p}{q}\right)\left(\frac{q}{p}\right) = (-1)^{\frac{p-1}{2}\frac{q-1}{2}} \tag{1}$$

where, by definition, $\left(\frac{p}{q}\right) = 1$ if $\bar{p} \in \mathbf{F}_q^\times$ is a square (if $\bar{p} = \bar{a}^2$ for some $\bar{a} \in \mathbf{F}_q^\times$), and $\left(\frac{p}{q}\right) = -1$ if $\bar{p} \in \mathbf{F}_q^\times$ is not a square.

With the help of this law, the reader should be able to check that the reduction $\bar{f}$ (modulo p) of $f = T^2 - 11$ has two distinct roots if and only if, modulo 44,

$$\bar{p} \in \left\{\bar{1}, \bar{5}, \bar{7}, \bar{9}, \overline{19}, \overline{25}, \overline{35}, \overline{37}, \overline{39}, \overline{43}\right\}.$$

(Incidentally, given any integer $m > 0$ and any $\bar{a} \in (\mathbf{Z}/m\mathbf{Z})^\times$, there are infinitely many primes $p \equiv a$ (mod. m); in a precise sense, the proportion of such primes is $1/\varphi(m)$, as shown by Dirichlet.)

The law of quadratic reciprocity can be formulated as an equality of two different kinds of *L-functions*, an "Artin *L*-function" and a "Hecke *L*-function", both functions of a complex variable s. The first one carries information about the primes p modulo which the polynomial $T^2 - q^*$ (where $q^* = (-1)^{(q-1)/2}q$, so that $q^* \equiv 1$ (mod. 4)) has two distinct roots, and is defined as

$$L_1(s) = \prod_p \frac{1}{1 - \left(\frac{q^*}{p}\right)p^{-s}}$$

where $\left(\frac{q^*}{2}\right) = (-1)^{(q^2-1)/8}$ and $\left(\frac{q^*}{q}\right) = 0$. The second one carries information about which primes p lie in which arithmetic progressions modulo q, and is defined as

$$L_2(s) = \prod_p \frac{1}{1 - \left(\frac{p}{q}\right)p^{-s}},$$

where we put $\left(\frac{q}{q}\right) = 0$. The law of quadratic reciprocity (1) is equivalent to the statement that $L_1(s) = L_2(s)$ (except possibly for the factor at the prime 2), as can be seen by comparing the coefficients of p^{-s} and noting that $\left(\frac{-1}{p}\right) = (-1)^{(p-1)/2}$.

What are the (monic, irreducible) degree-n polynomials f for which the primes p where the reduction $\bar{f}$ has n distinct roots in $\mathbf{F}_p$ can be specified by similar congruence conditions? Before characterising such polynomials, let us discuss one more example.

For every prime l, consider the (unique) monic polynomial $\Phi_l \in \mathbf{C}[T]$ whose roots are precisely the $l - 1$ primitive l-th roots $e^{2i\pi k/l}$ $(0 < k < l)$ of 1, namely $\Phi_l = T^{l-1} + T^{l-2} + \cdots + T + 1$; notice that $\Phi_l \in \mathbf{Z}[T]$. It can be shown (in much the same way as our discussion of $T^2 + 1$ above) that the reduction $\bar{\Phi}_l$ modulo some prime $p \neq l$ has $l - 1$ distinct roots in $\mathbf{F}_p$ if and only if $l \mid p - 1$, or equivalently $p \equiv 1 \pmod{l}$.

More generally, for every integer $m > 0$, the primitive m-th roots of 1 in $\mathbf{C}$ are ζ_m^k, where $\zeta_m = e^{2i\pi/m}$ and $\gcd(k, m) = 1$. If $\gcd(k', m) = 1$ and if $\bar{k}' = \bar{k}$ in $(\mathbf{Z}/m\mathbf{Z})^\times$, then $\zeta_m^{k'} = \zeta_m^k$ and conversely, so the number of such primitive roots is $\varphi(m)$, the order of the group $(\mathbf{Z}/m\mathbf{Z})^\times$. Let $\Phi_m \in \mathbf{C}[T]$ be the monic polynomial whose roots are precisely these $\varphi(m)$ numbers ζ_m^k, where $\gcd(k, m) = 1$, so that $\Phi_4 = T^2 + 1$. It can be shown that $\Phi_m \in \mathbf{Z}[T]$ and that it is irreducible; Φ_m is called *the cyclotomic polynomial* of level m.

For which primes p does the reduction $\bar{\Phi}_m$ have $\varphi(m)$ distinct roots in $\mathbf{F}_p$? Some work is required to answer this question, and it must have been first done by Kummer in the XIXth century, if not by Gauss before him. *It turns out that the reduction $\bar{\Phi}_m$ modulo p of the level-m cyclotomic polynomial Φ_m has $\varphi(m) = \deg \Phi_m$ distinct roots in $\mathbf{F}_p$ if and only if $p \equiv 1 \pmod{m}$.*

Now, for any monic irreducible polynomial f of degree $n > 0$, we can adjoin the n roots of f in $\mathbf{C}$ to $\mathbf{Q}$ to obtain a field K_f. The dimension of K_f as a vector space over $\mathbf{Q}$ is finite (and divides $n!$). For example, when $f = T^2 + 1$ and $i \in \mathbf{C}$ is a root of f, the field K_f consists of all complex numbers $x + iy$ such that $x, y \in \mathbf{Q}$. When $f = \Phi_l$ for some prime l, K_f consists of all $z \in \mathbf{C}$ which can be written $z = b_0 + b_1\zeta_l + \cdots + b_{l-2}\zeta_l^{l-2}$ for some $b_i \in \mathbf{Q}$.

The automorphisms of the field K_f form a group which we denote by Gal_f in honour of Galois. When f is quadratic, Gal_f is isomorphic to $\mathbf{Z}/2\mathbf{Z}$; when $f = \Phi_m$ is the cyclotomic polynomial of level $m > 0$, then Gal_f is canonically isomorphic to $(\mathbf{Z}/m\mathbf{Z})^\times$. The group Gal_f is commutative in these two cases; we say that the polynomial f is *abelian* (in honour of Abel) if the group Gal_f is commutative.

It has been shown that, $f \in \mathbf{Z}[T]$ being an irreducible monic polynomial of degree $n > 0$, the set of primes p modulo which $\bar{f}$ has n distinct roots can be characterised by congruence conditions modulo some number D_f depending only on f if and only if the group Gal_f is commutative (as is for example the case when $\deg f = 2$ or when $f = \Phi_m$ for some $m > 0$).

Visit `http://mathoverflow.net/questions/11688` for a discussion of this equivalence; the precise version of this statement is a consequence of the following beautiful theorem of Kronecker and Weber: *For every abelian polynomial f, there is some integer $m > 0$ such that $K_f \subset K_{\Phi_m}$: every root of f can be written as a linear combination of $1, \zeta_m, \ldots, \zeta_m^{\varphi(m)-1}$ with coefficients in* **Q**.

The generalisation of this theorem to all "global fields" constitutes "class field theory", which was developed in the first half of the XXth century by Hilbert, Furtwängler, Takagi, Artin, Chevalley, Tate, ... This theory is to abelian polynomials f what the quadratic reciprocity law is to quadratic polynomials over **Q**. As we saw above for the latter, class field theory can be formulated as the equality between an "Artin L-function" and a "Hecke L-function", the first one carrying information about the primes modulo which f splits completely and the second carrying information about which primes lie in which arithmetic progressions modulo D_f.

How about polynomials f which are not abelian, the ones for which the group Gal_f is not commutative? Such an f, namely $f = T^3 - 2$, was already considered by Gauss. He proved that this f splits completely modulo a prime $p \neq 2, 3$ if and only if $p = x^2 + 27y^2$ for some $x, y \in \mathbf{Z}$. For general f, how can we characterise the primes p modulo which $\bar{f}$ has $n = \deg f$ distinct roots? The Langlands programme, initiated in the late 60s, provides the only satisfactory, albeit conjectural, answer to this basic question (in addition to many other basic questions). This is what we would like to explain next.

Denote by $N_p(f)$ the number of roots of $\bar{f}$ in $\mathbf{F}_p$, so that for the abelian polynomial $f = T^2 - T - 1$ we have $N_5(f) = 1$ and

$$N_p(f) = \begin{cases} 2 & \text{if } p \equiv \pm 1 \ (\text{mod. } 5), \\ 0 & \text{if } p \equiv \pm 2 \ (\text{mod. } 5), \end{cases}$$

as we saw above. It can be checked that $N_p(f) = 1 + a_p$ for all primes p, where a_n is the coefficient of q^n in the formal power series

$$\frac{q - q^2 - q^3 + q^4}{1 - q^5} = q - q^2 - q^3 + q^4 + q^6 - q^7 - q^8 + q^9 + \cdots$$

in the indeterminate q. We can be said to have found a "formula" for $N_p(T^2 - T - 1)$. Notice that the a_n are *strongly multiplicative* in the sense that $a_{mm'} = a_m a_{m'}$ for all $m > 0, m' > 0$.

It follows from class field theory that there are similar "formulae" or "reciprocity laws" for $N_p(f)$ for every abelian polynomial f. What about polynomials f for which Gal_f is *not* abelian?

The fundamental insight of Langlands was that there is a "formula" for $N_p(f)$ for every polynomial f, abelian or not, and to say precisely what is meant by "formula". We cannot enter into the details but content ourselves with providing some illustrative examples (without proof). I learnt the first one, which perhaps goes back to Hecke, from Emerton's answer to one of my questions (visit http://mathoverflow.net/questions/12382) and also from a paper of Serre (Serre, 2003).

Consider the polynomial $f = T^3 - T - 1$ which is *not* abelian; the group Gal_f is isomorphic to the smallest group which is not commutative, namely the symmetric group $\mathbf{S}_3$ of all permutations of three letters. Modulo 23, $\bar{f}$ has a double root and a simple root, so $N_{23}(f) = 2$. If a prime $p \neq 23$ is a square in $\mathbf{F}_{23}^{\times}$, then p can be written either as $x^2 + xy + 6y^2$ or as $2x^2 + xy + 3y^2$ (but not both) for some $x, y \in \mathbf{Z}$; we have $N_p(f) = 3$ in the first case whereas $N_p(f) = 0$ in the second case. Finally, if a prime $p \neq 23$ is not a square in $\mathbf{F}_{23}^{\times}$, then $N_p(f) = 1$. The smallest p for which $N_p(f) = 3$ is $59 = 5^2 + 5.2 + 6.2^2$.

It can be checked that for all primes p, we have $N_p(f) = 1 + a_p$, where a_n is the coefficient of q^n in the formal power series $\eta_{1^1,23^1}$:

$$q \prod_{k=1}^{+\infty} (1 - q^k)(1 - q^{23k}) = q - q^2 - q^3 + q^6 + q^8 - q^{13} - q^{16} + q^{23} - q^{24} + \cdots$$

It follows that $a_p \in \{-1, 0, 2\}$, and a theorem of Chebotarev (generalising Dirichlet's theorem recalled above) implies that the proportion of p with $a_p = -1, 0, 2$ is $1/3, 1/2, 1/6$ respectively.

We thus have a "formula" for $N_p(T^3 - T - 1)$ similar in some sense to the one for $N_p(T^2 - T - 1)$ above. It is remarkable that $a_p = \frac{1}{2}(B_p - C_p)$ where B_n, C_n are defined by the identities

$$\sum_{j=0}^{+\infty} B_j q^j = \sum_{(x,y) \in \mathbf{Z}^2} q^{x^2 + xy + 6y^2}, \quad \sum_{j=0}^{+\infty} C_j q^j = \sum_{(x,y) \in \mathbf{Z}^2} q^{2x^2 + xy + 3y^2}.$$

Even more remarkable is the fact that if we define a function of a complex variable τ in $\mathbf{H} = \{x + iy \in \mathbf{C} \mid y > 0\}$ by $F(\tau) = \sum_{n=1}^{+\infty} a_n e^{2i\pi\tau.n}$, then F is analytic and "highly symmetric": for every matrix $\begin{pmatrix} a & b \\ c & d \end{pmatrix}$ with $a, b, c, d \in$ **Z**, $ad - bc = 1$, $c \equiv 0$ (mod. 23), so that in particular the imaginary part of $\dfrac{a\tau + b}{c\tau + d}$ is > 0, we have

$$F\left(\frac{a\tau + b}{c\tau + d}\right) = \left(\frac{d}{23}\right)(c\tau + d)\, F(\tau), \qquad \left(\frac{d}{23}\right) = \begin{cases} +1 & \text{if } \bar{d} \in \mathbf{F}_{23}^{\times 2} \\ -1 & \text{if } \bar{d} \notin \mathbf{F}_{23}^{\times 2}. \end{cases}$$

These symmetries of F, along with the fact that the coefficients a_n can be computed recursively from the a_p in a certain precise manner encoded in the equality

$$\sum_{n=1}^{+\infty} a_n n^{-s} = \prod_p \left(1 + a_p p^{-s} + \left(\frac{p}{23}\right) p^{-2s}\right)^{-1},$$

make it a "primitive eigenform of weight 1, level 23, and character $\left(\frac{\cdot}{23}\right)$": no mean achievement. This constitutes a "reciprocity law" for $T^3 - T - 1$.

There are "reciprocity laws" even for some polynomials $f \in \mathbf{Z}[S, T]$ in *two* indeterminates such as $f = S^2 + S - T^3 + T^2$: there are nice "formulae" for the number $N_p(f)$ of zeros $(s, t) \in \mathbf{F}_p^2$ of f (the number of pairs $(s, t) \in \mathbf{F}_p^2$ satisfying $s^2 + s - t^3 + t^2 = 0$), and the sequence of these numbers (or rather the numbers $a_p = p - N_p(f)$, which are in the interval $[-2\sqrt{p}, +2\sqrt{p}]$ by a theorem of Hasse) is "highly symmetric" as in the example of the polynomial $T^3 - T - 1$ above. Let me explain.

For our new $f = S^2 + S - T^3 + T^2$, defining B_n, C_n by the identities

$$\sum_{j=0}^{+\infty} B_j q^j = \sum_{(x,y,u,v)\in\mathbf{Z}^4} q^{x^2+xy+3y^2+u^2+uv+3v^2}$$

$$= 1q^0 + 4q^1 + 4q^2 + 8q^3 + 20q^4 + 16q^5 + 32q^6 + 16q^7 + \cdots$$
$$+ 4q^{11} + 64q^{12} + 40q^{13} + 64q^{14} + 56q^{15} + 68q^{16} + 40q^{17} + \cdots$$

$$\sum_{j=0}^{+\infty} C_j q^j = \sum_{(x,y,u,v)\in\mathbf{Z}^4} q^{2(x^2+y^2+u^2+v^2)+2xu+xv+yu-2yv}$$

$$= 1q^0 + 0q^1 + 12q^2 + 12q^3 + 12q^4 + 12q^5 + 24q^6 + 24q^7 + \cdots$$
$$+ 0q^{11} + 72q^{12} + 24q^{13} + 48q^{14} + 60q^{15} + 84q^{16} + 48q^{17} + \cdots$$

we have $a_p = \frac{1}{4}(B_p - C_p)$ for every prime $p \neq 11$. Amazing!

This "formula" may look somewhat different from the previous one but it is actually similar: $\frac{1}{4}(B_p - C_p) = c_p$ for every prime p (and hence $a_p = c_p$ for $p \neq 11$), where c_n is the coefficient of q^n in the formal product

$$\eta_{12,11^2} = q \prod_{k=1}^{+\infty}(1 - q^k)^2(1 - q^{11k})^2 = 0 + 1.q^1 + \sum_{n>1} c_n q^n.$$

We have already listed $c_p = \frac{1}{4}(B_p - C_p)$ for $p < 19$; here is another sample:

$p =$	19	23	29	31	37	41	$\cdots$	1987	1993	1997	1999
$c_p =$	0	−1	0	7	3	−8	$\cdots$	−22	−66	−72	−20

As in the case of $T^3 - T - 1$, if we put $q = e^{2i\pi\tau}$ (with $\tau = x + iy$ and $y > 0$), we get an analytic function $F(\tau) = \sum_{n=1}^{+\infty} c_n e^{2i\pi\tau.n}$ of $\tau \in \mathbf{H}$ which satisfies $F\left(\dfrac{a\tau + b}{c\tau + d}\right) = (c\tau + d)^2 F(\tau)$ for every matrix $\begin{pmatrix} a & b \\ c & d \end{pmatrix}$ $(a, b, c, d \in \mathbf{Z})$ with $ad - bc = 1$ and $c \equiv 0 \pmod{11}$. In addition, F has many other remarkable properties (such as $c_{mm'} = c_m c_{m'}$ if $\gcd(m, m') = 1$, $c_{p^r} = c_{p^{r-1}} c_p - p c_{p^{r-2}}$ for primes $p \neq 11$ and $r > 1$, and $c_{11^r} = c_{11}^r$ for $r > 0$) giving it the elevated status of "a primitive eigenform of weight 2 and level 11".

The analogy with the properties of the previous F is perfect, with $c \equiv 0 \pmod{11}$ replacing $c \equiv 0 \pmod{23}$ and $(c\tau + d)^2$ replacing $(\frac{d}{23})(c\tau + d)$ in the expression for the value at $(a\tau + b)/(c\tau + d)$. (There is a sense in which the prime 11 is "bad" for this $f = S^2 + S - T^3 + T^2$, much as 2 is bad for $T^2 + 1$, 5 is bad for $T^2 - T - 1$, the prime divisors of m are bad for Φ_m, and 23 is bad for $T^3 - T - 1$.)

The above example is taken from Langlands and Harder, but perhaps goes back to Eichler and implicitly to Hecke; what we have said about f (namely, $a_p = c_p$ for all primes $p \neq 11$) is summarised by saying that "the elliptic curve (of conductor 11) defined by $f = 0$ is *modular*". It can also be summarised by saying that the "Artin L-function"

$$L_1(s) = \prod_{p \neq 11} \frac{1}{1 - a_p.p^s + p.p^{-2s}}$$

(with an appropriate factor for the prime 11) is the same as the "Hecke L-function" $L_2(s) = \sum_{n>0} c_n n^{-s}$.

There is a similar equality (up to factors at finitely many primes) of the Artin L-function $\prod_p (1 - a_p . p^{-s} + p . p^{-2s})^{-1}$ of the polynomial $f = S^2 - T^3 + T$, defined using the number of zeros $N_p = p - a_p$ of f modulo various primes p, with the Hecke L-function $\sum_{n>0} c_n n^{-s}$ of

$$\eta_{4^2, 8^2} = q \prod_{k=1}^{+\infty} (1 - q^{4k})^2 (1 - q^{8k})^2 = 0 + 1.q^1 + \sum_{n>1} c_n q^n$$

the function $F(\tau) = \sum_{n>0} c_n e^{2i\pi\tau.n}$ corresponding to which enjoys the symmetries

$$F\left(\frac{a\tau + b}{c\tau + d}\right) = (c\tau + d)^2 F(\tau)$$

for every $\tau \in \mathbf{H}$ (so $\tau = x + iy$ with $x, y \in \mathbf{R}$, $y > 0$) and every matrix

$$\begin{pmatrix} a & b \\ c & d \end{pmatrix} \quad (a, b, c, d \in \mathbf{Z}), \quad ad - bc = 1, \quad c \equiv 0 \pmod{32}.$$

As a final example, consider $f = S^2 - T^3 - 1$ with $N_p = p - a_p$ points modulo p on the one hand, and

$$\eta_{6^4} = q \prod_{k>0} (1 - q^{6k})^4 = 0 + 1.q^1 + \sum_{n>1} c_n q^n$$

on the other, for which the function $F(\tau) = \sum_{n>0} c_n e^{2i\pi\tau.n}$ has the above symmetries with 32 replaced by 36. Each of these defines an L-function, the Artin L-function $\prod_p (1 - a_p . p^{-s} + p . p^{-2s})^{-1}$ and the Hecke L-function $\sum_{n>0} c_n n^{-s}$. Remarkably, they are the same (except for some adjustment at finitely many primes)!

Wiles and others (Taylor, Diamond, Conrad, Breuil) have proved that similar "formulae" and "symmetries" are valid — such equalities of Artin L-functions with Hecke L-functions are valid — for each of the infinitely many $f \in \mathbf{Z}[S, T]$ which define an "elliptic curve" (such as, among many others tabulated by Cremona,

$$f = S^2 + S - T^3 + T, \quad and \quad f = S^2 + S - T^3 - T^2 + T + 1,$$

for which the primes 37 and 101 are "bad" respectively), thereby settling a conjecture of Shimura, Taniyama and Weil, and providing the first proof of Fermat's Last Theorem in the bargain.

The precise version of this uninformative statement goes under the slogan "Every elliptic curve over Q is modular" and has been called "the theorem of the century"; it relies on the previous work of many mathematicians in the second half of the XXth century, among them Langlands, Tunnell, Serre, Deligne, Fontaine, Mazur, Ribet, ...

The reciprocity law for abelian polynomials (class field theory, 1900–1940), the reciprocity law for the polynomial $T^3 - T - 1$ of group S_3, the modulariy of the "elliptic curve" $S^2 + S - T^3 + T^2 = 0$ and more generally the modulariy of every elliptic curve defined over Q (1995–2000), and many other known reciprocity laws (Drinfeld, Laumon-Rapoport-Stuhler, Lafforgue, Harris-Taylor, Henniart) which we have not discussed, are all instances of the Langlands programme (1967–+∞?). At its most basic level, it is a search for patterns in the sequence $N_p(f)$ for varying primes p and a fixed but arbitrary polynomial f with coefficients in Q. Our examples exhibit the "pattern" for some specific f.

Confining ourselves to monic irreducible polynomials $f \in Z[T]$, Langlands predicts that there is a "reciprocity law" for f as soon as we give an embedding $\rho : \text{Gal}_f \to GL_d(C)$ of the group Gal_f into the group $GL_d(C)$ of invertible square matrices of size $d \times d$ with entries in C. (What was the embedding ρ for $f = T^3 - T - 1$? It came from the unique irreducible representation of S_3 on C^2.) Class field theory is basically the case $d = 1$ of this vast programme. Essentially the only other known instances of the precise conjecture are for $d = 2$ and either the group $\rho(\text{Gal}_f)$ is solvable (Langlands-Tunnell) or the image $\rho(c)$ of the element $c \in \text{Gal}_f$ which sends $x + iy$ to $x - iy$ is not of the form $\begin{pmatrix} a & 0 \\ 0 & a \end{pmatrix}$ for any $a \in C^\times$ (Khare-Wintenberger-Kisin).

So what is a reciprocity law? At the most basic level, it is the law governing the primes modulo which a given monic irreducible $f \in Z[T]$ factors completely, such as the quadratic reciprocity law when f has degree 2, or class field theory when f is abelian. We can also think of a reciprocity law as telling us how many solutions a given system of polynomial equations with coefficients in Q has over the various prime fields F_p. But we have seen that these laws can be expressed as an equality of an Artin L-function with a Hecke L-function, and that the modularity of elliptic curves over Q can also be expressed as a similar equality of L-functions. We may thus take the view that a reciprocity law is encapsulated in the equality of two L-functions initially defined in two quite different ways, Artin L-functions coming from *galoisian representations* and Hecke L-functions coming from *automorphic represetations*.

The rest of this Note requires greater mathematical maturity. We explain what kind of "reciprocity laws" Serre's conjecture provides, but pass in silence the recent proof of this conjecture by Khare and Wintenberger.

For any finite galoisian extension $K|\mathbf{Q}$ and any prime p which does not divide the discriminant of K, we have the "element" $\mathrm{Frob}_p \in \mathrm{Gal}(K|\mathbf{Q})$, only determined up to conjugation in that group. If f_p is the order of Frob_p, then p splits in K as a product of $g_p = [K : \mathbf{Q}]/f_p$ primes of K of residual degree f_p, so that p splits completely ($f_p = 1$) if and only if $\mathrm{Frob}_p = \mathrm{Id}_K$. The extension K is uniquely determined, among finite galoisian extensions of $\mathbf{Q}$, by the set $S(K)$ of primes which split completely in K. For example, if $S(K)$ is the set of $p \equiv 1 \pmod{m}$ for some given $m > 0$, then $K = \mathbf{Q}(\zeta_m)$, where ζ_m is a primitive m-th root of 1. Given a faithful representation of $\mathrm{Gal}(K|\mathbf{Q})$ in $\mathrm{GL}_d(\mathbf{C})$, Langlands' programme characterises the subset $S(K)$ in terms of certain "automorphic representations of GL_d of the adèles".

Serre's conjecture deals with faithful representations ρ of $\mathrm{Gal}(K|\mathbf{Q})$ into $\mathrm{GL}_2(\bar{\mathbf{F}}_l)$ which are odd in the sense that the image $\rho(c)$ of the "complex conjugation" $c \in \mathrm{Gal}(K|\mathbf{Q})$ has determinant -1. The oddity is automatic if $l = 2$; for odd l, it implies that K is totally imaginary.

A rich source of such representations is provided by elliptic curves E over $\mathbf{Q}$. For every prime l, we have the 2-dimensional $\mathbf{F}_l$-space $E[l] = {}_lE(\bar{\mathbf{Q}})$ of l-torsion points of E together with a continuous linear action of the profinite group $G_{\mathbf{Q}} = \mathrm{Gal}(\bar{\mathbf{Q}}|\mathbf{Q})$. Adjoining the l-torsion of E to $\mathbf{Q}$ we get a finite galoisian extension $\mathbf{Q}(E[l])$ which is unramified at every prime p prime to Nl, where N is the conductor of E, and comes with an inclusion of groups $\mathrm{Gal}(\mathbf{Q}(E[l])|\mathbf{Q}) \subset \mathrm{Aut}_{\mathbf{F}_l}(E[l])$ (which is often an equality). Choosing an $\mathbf{F}_l$-isomorphism $E[l] \to \mathbf{F}_l^2$, we obtain a faithful representation $\rho_{E,l} : \mathrm{Gal}(\mathbf{Q}(E[l])|\mathbf{Q}) \to \mathrm{GL}_2(\mathbf{F}_l)$ which is odd because $\det \rho_{E,l}$ is the mod-l cyclotomic character.

Given a finite galoisian extension K of $\mathbf{Q}$ and a faithful representation $\rho : \mathrm{Gal}(K|\mathbf{Q}) \to \mathrm{GL}_2(\bar{\mathbf{F}}_l)$ which is odd and irreducible (for the sake of simplicity), Serre defines two numbers k_ρ, N_ρ and conjectures that ρ "comes from a primitive eigenform of weight k_ρ and level N_ρ". Instead of trying to explain this, let me illustrate it with the example of the elliptic curve $E : y^2 + y = x^3 - x^2$ which we have already encountered.

This curve has good reduction at every prime $p \neq 11$. Serre shows that the representation $\rho_{E,l} : \text{Gal}(\mathbf{Q}(E[l])|\mathbf{Q}) \to \text{GL}_2(\mathbf{F}_l)$ is an isomorphism for every $l \neq 5$. For every $p \neq 11, l$, the characteristic polynomial of $\rho_{E,l}(\text{Frob}_p) \in \text{GL}_2(\mathbf{F}_l)$ is

$$X^2 - \bar{c}_p X + \bar{p} \in \mathbf{F}_l[X]$$

where c_n is the coefficient of q^n in $\eta_{1^2,11^2} = q \prod_{k>0}(1 - q^k)^2(1 - q^{11k})^2$ as above. Recall that p splits completely in $\mathbf{Q}(E[l])$ if and only if

$$\rho_{E,l}(\text{Frob}_p) = \begin{pmatrix} 1 & 0 \\ 0 & 1 \end{pmatrix},$$

which implies, but is not equivalent to, $p, c_p \equiv 1, 2 \pmod{l}$. For $l = 7$, the first ten p satisfying these congruences are

$$113, 379, 701, 1051, 2437, 2521, 2731, 2857, 3221, 3613;$$

none of them splits completely in $\mathbf{Q}(E[7])$. Nevertheless, the representation $\rho_{E,l}$ is explicit enough to determine the splitting of rational primes in $\mathbf{Q}(E[l])$; the first ten such primes for $l = 7$ are

$$4831, 22051, 78583, 125441, 129641, 147617, 153287, 173573, 195581, 199501,$$

as obligingly computed by Tim Dokchitser at my request.

In summary, we have the following "reciprocity law" for $\mathbf{Q}(E[l])$:

"$p \neq 11, l$ splits completely in $\mathbf{Q}(E[l])$" $\Leftrightarrow$ "$E_p[l] \subset E_p(\mathbf{F}_p)$",

where E_p is the reduction of E modulo p. Indeed, reduction modulo p identifies $E[l]$ with $E_p[l]$ and the action of Frob_p on the former $\mathbf{F}_l$-space with the action of the canonical generator $\varphi_p \in \text{Gal}(\bar{\mathbf{F}}_p|\mathbf{F}_p)$ on the latter $\mathbf{F}_l$-space. To say that φ_p acts trivially on $E_p[l]$ is the same as saying that $E_p[l]$ is contained in the $\mathbf{F}_p$-rational points of E_p. The analogy with the multiplicative group μ is perfect:

"$p \neq l$ splits completely in $\mathbf{Q}(\mu[l])$" $\Leftrightarrow$ "$\mu_p[l] \subset \mu_p(\mathbf{F}_p)$"

($\Leftrightarrow l \mid p - 1 \Leftrightarrow p \equiv 1 \pmod{l}$), where μ_p is *not* the p-torsion of μ — that would be $\mu[p]$ — but the reduction of μ modulo p, namely the multiplicative group over $\mathbf{F}_p$.

With the proof by Khare and Wintenberger (2006–2009) of Serre's modularity conjecture (1973–1987), we can write down such reciprocity laws for every finite galoisian extension $K|\mathbf{Q}$ as soon as a faithful odd irreducible representation $\rho : \mathrm{Gal}(K|\mathbf{Q}) \to \mathrm{GL}_2(\bar{\mathbf{F}}_l)$ is given.

Sources. The reciprocity laws for quadratic and cyclotomic polynomials can be found in most textbooks on number theory; they are also treated at some length by Wyman (Wyman, 1973). An advanced introduction to class field theory is contained in the book *Algebraic Number Theory* edited by Cassels and Fröhlich (Cassels and Fröhlich, 2010) (recently reissued by the London Mathematical Society with a long list of corrections, available on Buzzard's website).

The polynomial $T^3 - 2$ is discussed by Cox (Cox, 1989). I learnt the example $T^3 - T - 1$ from Emerton and later in a paper by Serre (Serre, 2003). It is also discussed in an earlier paper (Serre, 1977), as is the polynomial $T^3 + T - 1$, for which the corresponding modular form is

$$\frac{1}{2}\left(\sum_{(x,y)\in\mathbf{Z}^2} q^{x^2+xy+8y^2} - \sum_{(x,y)\in\mathbf{Z}^2} q^{2x^2+xy+4y^2} \right).$$

He also mentions a different kind of example, given by the dihedral extension $\mathbf{Q}(\sqrt[4]{1}, \sqrt[4]{12})$ of $\mathbf{Q}$, for which the corresponding modular form is

$$\eta_{12^2} = q\prod_{k=1}^{+\infty}(1 - q^{12k})^2 = \sum_{x,y\equiv 1,0(3),x+y\equiv 0(2)} (-1)^y q^{x^2+y^2}.$$

The elliptic curve $E : y^2 + y = x^3 - x^2$ and its associated modular form $\eta_{1^2,11^2}$ are mentioned by Langlands (*A little bit of number theory*, unpublished but available on the website of the Institute for Advanced Study), and discussed in his talk (Langlands, 1990) at the Gibbs symposium (also available on the same website). It is also discussed by Harder (Harder, 2008) and available on the website of the International Centre for Theoretical Physics), and by Zagier (Zagier, 1985).

This is the curve (with a different defining equation) used by Shimura (Shimura, 1966) to study reciprocity laws for $\mathbf{Q}(E[l])$ for primes $l \in [7, 97]$. Serre showed (Serre, 1972) that $\rho_{E,l}$ is surjective for every $l \neq 5$; cf. the online notes on Serre's conjecture by Ribet and Stein. The other two elliptic curves, namely $y^2 = x^3 - x$ and $y^2 = x^3 + 1$, of conductor 32 and 36 respectively, are taken from a paper by Honda (Honda, 1973).

Resources. In addition to the sources cited above, a number of articles providing an introduction to the Langlands Programme are freely available online. There is a very accessible account by R. Taylor (Taylor). There are introductory accounts by Gelbart (Gelbart, 1984), Arthur (Arthur, 1981), (Arthur, 2003), Knapp (Knapp, 1977), and the more informal essays by Arthur, Knapp and Rogawski in the *Notices* of the American Mathematical Society. Knapp also has a number of other related introductory texts on his website. For the more advanced reader, there is a book by Bump, Cogdell, de Shalit, Gaitsgory, Kowalski, and Kudla *An introduction to the Langlands program*, edited by Bernstein and Gelbart, Birkhäuser (2003).

Applications. Reciprocity laws are a source of intense pleasure for the mathematician; they are also immensely useful within mathematics. For the rest, Simon Singh says that "There are also important implications for the applied sciences and engineering. Whether it is modelling of the interactions between colliding quarks or discovering the most efficient way to organise a telecommunications network, often the key to the problem is performing a mathematical calculation. In some areas of science and technology the complexity of the calculations is so immense that progress in the subject is severly hindered. If only mathematicians could prove the linking conjectures of the Langlands programme, then there would be shortcuts to solving real-world problems, as well as abstract ones" (*Fermat's Last Theorem*, p. 214).

A final example. Let me end with another striking application of the Langlands programme: it allows us to prove some arithmetical statements which have a fairly elementary formulation but for which no other proof is available, elementary or otherwise.

First recall that the proportion of primes p for which $T^2 + 1$ has no roots (resp. two distinct roots) in $\mathbf{F}_p$ is $1/2$ (resp. $1/2$), and that the proportion of p for which $T^3 - T - 1$ has no roots (resp. exactly one root, resp. three distinct roots) in $\mathbf{F}_p$ is $1/3$ (resp. $1/2$, resp. $1/6$).

What is the analogue of the foregoing for the number of roots $N_p(f)$ of $f = S^2 + S - T^3 + T^2$ in $\mathbf{F}_p$? We have seen that $a_p = p - N_p(f)$ lies in the interval $[-2\sqrt{p}, +2\sqrt{p}]$, so $a_p/2\sqrt{p}$ lies in $[-1, +1]$. What is the proportion of primes p for which $a_p/2\sqrt{p}$ lies in a given interval $I \subset [-1, +1]$? It was predicted by Sato (on numerical grounds) and Tate (on theoretical grounds), not just for this f but for all $f \in \mathbf{Z}[S, T]$ defining an "elliptic curve without complex multiplications", that the proportion of such p is equal to the area

$$\frac{2}{\pi} \int_I \sqrt{1 - x^2}\, dx.$$

of portion of the unit semicircle projecting onto I. The Sato-Tate conjecture for elliptic curves over $\mathbf{Q}$ was settled in 2008 by Clozel, Harris, Shepherd-Barron and Taylor.

There is an analogue for "higher weights". Let c_n ($n > 0$) be the coefficient of q^n in the formal product

$$\eta_{12^4} = q \prod_{k=1}^{+\infty} (1 - q^k)^{24} = 0 + 1.q^1 + \sum_{n>1} c_n q^n.$$

In 1916, Ramanujan had made some deep conjectures about these c_n; some of them, such as $c_{mm'} = c_m c_{m'}$ if $\gcd(m, m') = 1$ and

$$c_{p^r} = c_{p^{r-1}} c_p - p^{11} c_{p^{r-2}}$$

for $r > 1$ and primes p, which can be more succintly expressed as the identity

$$\sum_{n>0} c_n n^{-s} = \prod_p \frac{1}{1 - c_p.p^{-s} + p^{11}.p^{-2s}}$$

when the real part of s is $> (12 + 1)/2$, were proved by Mordell in 1917. The last of Ramanujan's conjectures was proved by Deligne only in the 70s: for every prime p, the number $t_p = c_p/2p^{11/2}$ lies in the interval $[-1, +1]$.

All these properties of the c_n follow from the fact that the corresponding function $F(\tau) = \sum_{n>0} c_n e^{2i\pi\tau.n}$ of $\tau = x + iy$ ($y > 0$) in $\mathbf{H}$ is a "primitive eigenform of weight 12 and level 1" (which basically amounts to the identity $F(-1/\tau) = \tau^{12} F(\tau)$). Here are the first few c_p (usually denoted $\tau(p)$, but our τ is in $\mathbf{H}$) computed by Ramanujan:

$p =$	2	3	5	7	11	13	17 $\cdots$
$c_p =$	−24	252	4830	−16744	534612	−577738	−6905934 $\cdots$

(Incidentally, Ramanujan had also conjectured some congruences satisfied by the c_p modulo 2^{11}, 3^7, 5^3, 7, 23 and 691, such as $c_p \equiv 1 +$

p^{11} (mod. 691) for every prime p; they were at the origin of Serre's conjecture mentioned above.)

We may therefore ask how these $t_p = c_p/2p^{11/2}$ are distributed: for example are there as many primes p with $t_p \in [-1,0]$ as with $t_p \in [0,+1]$? Sato and Tate predicted in the 60s that the precise proportion of primes p for which $t_p \in [a,b]$, for given $a < b$ in $[-1,+1]$, is

$$\frac{2}{\pi} \int_a^b \sqrt{1 - x^2} \, dx.$$

This is expressed by saying that the $t_p = c_p/2p^{11/2}$ are *equidistributed* in the interval $[-1,+1]$ with respect to the measure $(2/\pi) \sqrt{1 - x^2} \, dx$. Recently Barnet-Lamb, Geraghty, Harris and Taylor have proved that such is indeed the case.

Their main theorem implies many such equidistribution results, including the one recalled above for the elliptic curve $S^2 + S - T^3 + T^2 = 0$; for an introduction to such *density theorems*, see (Taylor).

Acknowledgments

I heartily thank Tim Dokchitser for computing the first ten primes which split completely in $\mathbf{Q}(E[7])$, and my colleague Rajesh Gopakumar for a careful reading of the text.

References

Arthur, J. 1981. *Automorphic representations and number theory*, in CMS Conf. Proc., 1, Amer. Math. Soc., Providence.

Arthur, J. 2003. The principle of functoriality, *Bull. Amer. Math. Soc. (N.S.)* **40** 1, 39–53.

Cassels, J. and Fröhlich, A. 2010. *Algebraic number theory*. Proceedings of the instructional conference held at the University of Sussex, Brighton, September 1–17, 1965, Corrected reprint, London Mathematical Society.

Cox, D. 1989. Primes of the form $x^2 + ny^2$. *Fermat, class field theory and complex multiplication*. John Wiley & Sons, New York.

Gelbart, S. 1984. An elementary introduction to the Langlands program, *Bull. Amer. Math. Soc. (N.S.)* **10** 2, 177–219.

Harder, G. 2008. *The Langlands program (an overview)*, School on Automorphic Forms on GL(*n*), 207–235, ICTP Lect. Notes, 21, Abdus Salam Int. Cent. Theoret. Phys., Trieste.

Honda, T. 1973. Invariant differentials and *L*-functions. Reciprocity law for quadratic fields and elliptic curves over **Q**, *Rend. Sem. Mat. Univ. Padova* **49**, 323–335.

Knapp, A. 1977. Introduction to the Langlands program, in *Proc. Sympos. Pure Math.*, 61, Amer. Math. Soc., Providence.

Langlands, R. 1990. Representation theory: its rise and its role in number theory, *Proceedings of the Gibbs Symposium* (New Haven, 1989), 181–210, Amer. Math. Soc., Providence, RI.

Serre, J-P. 1972. *Propriétés galoisiennes des points d'ordre fini des courbes elliptiques*, Invent. Math. **15**, 259–331.

Serre, J-P. 1977. Modular forms of weight one and Galois representations, in *Proc. Sympos.* (Durham, 1975), 193–268, Academic Press, London.

Serre, J-P. 2003. On a theorem of Jordan, *Bull. Amer. Math. Soc. (N.S.)* **40**, no. 4, 429–440.

Shimura, G. 1966. A reciprocity law in non-solvable extensions, *J. Reine Angew. Math.* **221**, 209–220.

Taylor, R. *Reciprocity laws and density theorems*, on the author's website.

Wyman, B. 1973. What is a reciprocity law? *Amer. Math. Monthly* **79** (1972), 571–586; *correction*, ibid. **80**, 281.

Zagier, D. 1985. *Modular points, modular curves, modular surfaces and modular forms*, Lecture Notes in Math., 1111, Springer, Berlin.

CHAPTER 3

Elliptic Curves and Number Theory

R. Sujatha

School of Maths, TIFR,
Mumbai 400 005, India.
e-mail: sujatha@math.tifr.res.in

The aim of this article is to provide a brief introduction to elliptic curves along with some deep, ancient and still largely unsolved related problems in arithmetic. It is fair to say that the study of elliptic curves stretches over a vast mathematical canvas that straddles diverse areas such as complex analyis, topology, algebraic geometry, cryptography and even mathematical physics. However, we shall focus on specific aspects of this study related to number theory, where some of the deepest arithmetic problems occur. As usual, $\mathbf{Z}$ denotes the ring of integers and $\mathbf{Q}$ the field of rational numbers.

1 Curves and equations

Diophantus of Alexandria (200–284 A.D.) was a Greek mathematician, best known for his work '*Arithmetica*', a treatise on the study of numbers. This book sets out the first systematic study of equations with integer coefficients (however, see the article (Sridharan, 2011) for the much earlier work on related numerical examples). For this reason, polynomial equations in integers or rational numbers are called *Diophantine equations*.

D I O P H A N T I
ALEXANDRINI

Rerum Arithmeticarum
Libri fex,
quorū primi duo adiecta habent Scholia,
Maximi (ut coniectura eft)
Planvdis.

Item Liber de nvmeris Polygonis
feu Multiangulis,

Opus incomparabile,uera Arithmetica Logiftica perfectio-
nem continens,paucis adhuc uifum,

A' Gvil. Xylandro Auguftano incredibili labore
Latinè redditum, & Commentariis ex-
planatum, inq; lucem editum.
A D
Illuftrif. Principe Lvdovicvm *Vuirtembergenfem.*

B A S I L E AE
Per Evsebivm Episcopivm,
& Nicolai Fr. hæredes.
M D LXXV.

Historically, solving degree three polynomial equations ("cubic equations") in one variable, viz.

$$f(x) := ax^3 + bx^2 + cx + d = 0,$$

where the coefficients a, b, c, d are in $\mathbf{Z}$ or $\mathbf{Q}$, presented the first serious challenge in the study of equations in one variable. A complete solution to this problem was achieved only in the 16th century, thanks to the efforts of several mathematicians (see (Van der Waerden, 1991) or any book on Modern Algebra). Simultaneously, mathematicians like Fermat, Euler, Lagrange were beginning to uncover the deep problems arising from the study of equations over $\mathbf{Z}$ and $\mathbf{Q}$ in small degrees in two variables.

The simplest among these are the *linear* equations

$$aX + bY = c, \qquad a \text{ or } b \neq 0,$$

where the coefficients are in $\mathbf{Z}$ or $\mathbf{Q}$. Clearly, any such equation always has solutions over $\mathbf{Q}$ and has integer solutions when a, b, c are in $\mathbf{Z}$, if and only if the greatest common divisor of a and b divides c. *Quadratic equations* are of the form

$$g(X, Y) = aX^2 + bXY + cY^2 + dX + eY + f = 0,$$

(a) Fermat(1601–1655)   (b) Euler(1707–1783)

Color images of these figures appears in the color plate section at the end of the book.

with a, b, or $c \neq 0$, and $a, \ldots, f \in \mathbf{Z}$. These describe conic sections and specific examples are the ellipse, hyperbola and parabola. One has a fairly good understanding of the solutions (over $\mathbf{Z}$ or $\mathbf{Q}$) of such equations, and non-trivial arithmetic ideas (eg. "Quadratic reciprocity", "Hasse-Minkowski principle") are already apparent in the Diophantine analysis of quadratic equations.

1.1 Affine and Projective curves

Let $f(X, Y)$ be a polynomial in the variables X, Y with coefficients in $\mathbf{Q}$. Clearly, an equation $f(X, Y) = 0$ over $\mathbf{Q}$ can be viewed as an equation over the field $\mathbf{R}$ or the field $\mathbf{C}$ of complex numbers. The solution set over the reals gives *plane curves*, i.e. curves in the Euclidean plane of which the circle and other conic sections are examples. This also ushers in a geometric view point. Modern algebraic geometry enhances this basic philosophy with additonal structures, and encapsulates it in the study of *algebraic curves* and *algebraic varieties* over an arbitrary base field (see Fulton (1989)). To keep the discussion simple, we shall mainly work with the base field $\mathbf{Q}$ in this article. A polynomial $f(X, Y)$ over $\mathbf{Q}$ naturally gives rise to an *affine algebraic curve* over $\mathbf{Q}$ by considering the solutions of the equation $f(X, Y) = 0$, and *projective algebraic curves* correspond to studying the solutions of the corresponding homogeneous polynomial equation where $f(X, Y)$ is made homogeneous by introducing an extra variable Z. For instance, if $f(X, Y)$ is the 'affine' polynomial $X^2 + 3Y + 2X - 5$, then the corresponding homogenisation is $F(X, Y, Z) = X^2 + 3YZ + 2XZ - 5Z^2$. The solutions (X, Y, Z) of $F(X, Y, Z) = 0$ in $\mathbf{Q}$ are studied in *projective coordinates* i.e. (X, Y, Z) such that not all the entries are zero, and where two solutions $P = (a, b, c)$ and $Q = (a', b', c')$ are deemed equivalent if one is a scalar multiple of the other i.e. $P = \lambda.Q$ for a nonzero element λ of $\mathbf{Q}$. To obtain the affine polynomial from its homogenisation, one simply puts $Z = 1$ in the homogenised polynomial. Thus there is a bijective map $(X, Y) \mapsto (X, Y, 1)$ from the set of solutions of $f(X, Y) = 0$ into the set of solutions of $F(X, Y, Z) = 0$ with projective coordinates (X, Y, Z) and such that $Z \neq 0$. A projective algebraic curve given by a homogeneous equation $F(X, Y, Z) = 0$, is *nonsingular* if the polynomials F and its partial derivatives $\partial F/\partial X, \partial F/\partial Y, \partial F/\partial Z$ do not all have a common zero (in projective coordinates) over $\mathbf{C}$. A common solution of this set of homogeneous polynomials is a *singularity* or a *singular point*. For example, $X^2 + Y^2 + Z^2 = 0$ is nonsingular, while the point $(0, 0, 1)$ is a singular point of the curve $Y^2Z - X^3 = 0$. The *degree* of the curve defined by the homogeneous equation $F(X, Y, Z) = 0$ is the degree of F. A basic invariant associated to algebraic curves is the *genus*. If F has degree d, and is nonsingular, then the genus is given by $(d - 1)(d - 2)/2$. We shall adopt the geometric view point in the definition and study of elliptic curves.

1.2 Elliptic curves

For the purposes of this article, an *elliptic curve* over $\mathbf{Q}$ will mean a nonsingular algebraic curve of genus one, defined by an affine equation

$$E : Y^2 = f(X), \tag{1}$$

where $f(X) \in \mathbf{Q}[X]$ is a cubic polynomial with distinct roots. Examples are given by

$$Y^2 = X^3 - X$$
$$Y^2 = X^3 + 17. \tag{2}$$

More generally, any equation of the form (1) over $\mathbf{Q}$ can be brought, by a simple change of variables, to the form

$$E : Y^2 = X^3 + AX + B, \tag{3}$$

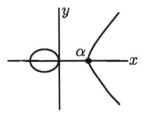

A Cubic Curve with Two Real Components

with A, $B \in \mathbf{Q}$, and its homogenization is

$$E : Y^2Z = X^3 + AXZ^2 + BZ^3. \tag{4}$$

The "obvious" solution $(0, 1, 0)$ (or equivalently, $(0, \lambda, 0)$ for any non-zero λ in $\mathbf{Q}$) is called the *point at infinity*. The *discriminant* Δ of E is defined by

$$\Delta = -16(4A^3 + 27B^2).$$

The nonvanishing of Δ is equivalent to the polynomial $f(X)$ having distinct roots. Another important (but more technical to define) invariant is the *conductor* of E, which essentially has the same set of prime divisors as the discriminant. For instance, the first curve in (2) has discriminant 64 and conductor 32, while the second has discriminant -124848 and conductor 10404. The interested reader can consult (Tate and Silverman, 1992), (Silverman, 2009) for more details. We denote the set of solutions of E with coordinates in $\mathbf{Q}$, together with the point at infinity, by $E(\mathbf{Q})$. As we shall see below, studying $E(\mathbf{Q})$ takes on a completely different nature to that of the study of the solutions of cubic equations in one variable.

2 Elliptic curves and the Birch and Swinnerton-Dyer conjecture

The central conjecture in the study of the arithmetic of elliptic curves is the Birch and Swinnerton-Dyer conjecture. This astounding conjecture relates two quite different invariants that arise in the study of elliptic curves, one arithmetic in nature and the other analytic, and we describe them below.

2.1   Group structure and the algebraic rank

A remarkable fact is that $E(\mathbf{Q})$ comes equipped with a nonevident *abelian group law*, which we denote by $\oplus$. The group operation is defined using basic facts from geometry (specifically, *Bezout's theorem*, in the particular form that states that any straight line intersects E in three projective points), along with elementary coordinate geometry. Let P_1, P_2 be two points of $E(\mathbf{Q})$ with coordinates $P_1 = (x_1, y_1)$ and $P_2 = (x_2, y_2)$, with x_i, y_i in $\mathbf{Q}$. For simplicity, assume that $x_1 \neq x_2$. Then the coordinates (x_3, y_3) for $P_1 \oplus P_2$ are given by

$$x_3 = \lambda^2 - x_1 - x_2, \qquad y_3 = -\lambda x_3 - \nu,$$

where $y = \lambda x + \nu$ is the equation for the straight line through P_1 and P_2. Also, the point with projective coordinates $(0, 1, 0)$ plays the role of the zero element for this group structure. A celebrated result of Mordell asserts that $E(\mathbf{Q})$ is a finitely generated abelian group. Recall that any finitely generated abelian group G is isomorphic to a direct sum $G_f \oplus G_t$, where G_f is a *free* abelian group isomorphic to $\mathbf{Z}^r$ (r copies of the group of integers), and G_t is a *torsion* finite abelian group. This integer r is defined to be the *rank* of G. Applying this to $E(\mathbf{Q})$ enables us to define g_E, the *algebraic rank* of E, as the rank of the finitely generated abelian group $E(\mathbf{Q})$. The curves in (2) have algebraic ranks 0 and 2 respectively; for the second curve, the points $(-2, 3)$ and $(2, 5)$ give two independent generators of $E(\mathbf{Q})$ modulo torsion.

A general effective method to compute the rank of an elliptic curve remains unknown, but the torsion subgroup is relatively easy to compute both theoretically and in numerical examples. Though the rank of a randomly chosen elliptic curve over $\mathbf{Q}$ tends to be quite small, it is conjectured that there exist elliptic curves of arbitrarily large rank. Answers to other related questions like the numbers which arise as ranks of elliptic curves, the distribution of ranks, remain largely unknown. The largest known rank to date is ≥ 28, discovered by N. Elkies in 2006, while the previous record was that of a curve of rank at least 24 due to Martin-McMillen in 2000. Sometimes the "simplest" points of infinite order on an elliptic curve can be very complicated. For example, the curve

$$Y^2 = X^3 - (157)^2 X$$

has rank one and the simplest point of infinite order on it is given by

$$(X, Y) = \left(\frac{27748777873292446321169121}{60976025066561516725072} , \frac{22826630568289716631287654159126420}{47614438250616355400538204422449067} \right). \qquad (5)$$

2.2   *L*-functions and analytic rank

The *Hasse-Weil L*-function of E, denoted $L(E, s)$, is a function of the complex variable s and is an ingenious generalisation of the classical Riemann zeta function

$$\zeta(s) = \sum_{n=1}^{\infty} 1/n^s = \prod_{p \ prime} (1 - p^{-s})^{-1}.$$

To define $L(E, s)$, we have to introduce the *global minimal Weierstrass equation* for E. This is an equation of the form

$$Y^2 + a_1 XY + a_3 Y = X^3 + a_2 X^2 + a_4 X + a_6$$

with a_i's in **Z**, and such that its discriminant Δ_E is minimal amongst all such equations (see (Silverman, 2009) for the full technical details). For example, if E is given by $Y^2 = X^3 + 16$, then the substitution $X = 4X_1$ and $Y = 8Y_1 + 4$ gives the global minimal Weierstrass equation $Y_1^2 + Y_1 = X_1^3$. Obviously, global minimal Weierstrass equations exist for elliptic curves over **Q**. For each prime p, reducing the coefficients of the global minimal form modulo p, gives a polynomial $\bar{f}_p(X, Y)$ with coefficients in $\mathbf{F}_p$, the finite field of p elements. We denote the corresponding curve over $\mathbf{F}_p$ defined by $\bar{f}_p(X, Y) = 0$, by $\tilde{E}_p$. A prime p is a *prime of good reduction* for E, if the reduced curve $\tilde{E}_p$ is still nonsingular (note that the criterion for nonsingularity defined in §1 makes sense for polynomials f in $k[X, Y]$, where k is an arbitrary base field, when we replace **C** by the algebraic closure of k). These correspond to primes p not dividing the discriminant of E. In this case, the integers a_p are defined as

$$a_p := 1 + p - \# \tilde{E}_p(\mathbf{F}_p), \tag{6}$$

where $\# \tilde{E}_p(\mathbf{F}_p)$ denotes the cardinality of solutions of the reduced equation in $\mathbf{F}_p$. The numbers a_p are interesting in their own right and can be computed by hand for small primes p. The *L*-series $L(E, s)$ is defined by an infinite product

$$L(E, s) = \prod_{p \nmid \Delta_E} \left(1 - a_p p^{-s} + p^{1-2s}\right)^{-1} \prod_{q | \Delta_E} (1 - a_q a^{-s})^{-1},$$

where a_p is as in (6) for $p \nmid \Delta_E$, and $a_q = 0, +1$ or -1 for $q | \Delta_E$ depending on the 'singularity' of the reduced equation $\tilde{E}_q$ (see (Silverman, 2009,

Chap. VII)). The *L*-function also has an expression by what is called a
Dirichlet series

$$L(E, s) = \sum_{n=1}^{\infty} a_n/n^s.$$

It is a classical result that the series converges in the complex plane when
the real part of s is strictly greater than 3/2. It is however a recent and
deep result (proved around a decade ago) that $L(E, s)$ has an analytic con-
tinuation to the entire complex plane. This is a consequence of a profound
theorem due to Wiles (see (Wiles, 1995), (Taylor and Wiles, 1995), (Breuil
et al., 2001)) about modular forms. This 'modularity result' linking elliptic
curves to modular forms plays a crucial role in the proof of Fermat's last
theorem (see Cornell *et al.* (1995)). Thus for any point s_0 in **C**, one has the
power series expansion

$$L(E, s_0) = \sum_{k \geq 0} c_k(s - s_0)^k.$$

The least integer k for which $c_k \neq 0$ is called the *order* of $L(E, s)$ at $s = s_0$.
The *analytic rank* of E, denoted r_E is defined to be the order of vanishing
of $L(E, s)$ at $s = 1$. We remark that $s = 1$ is a special point because there is
a functional equation relating $L(E, s)$ and $L(E, 2 - s)$ in general, so $s = 1$
is the "centre of the critical strip".

2.3 Birch and Swinnerton-Dyer conjecture

In the early 1960's, Birch and Swinnerton-Dyer, two mathematicians at
Cambridge University, carried out extensive computations for the analytic
rank and algebraic rank for certain families of elliptic curves. Based on
the numerical evidence, they proposed a revolutionary conjecture which is
named after them, and which we shall abbreviate to BSD in this article. In
its weak form, it is the following assertion:

Conjecture 1. *(BSD) For any elliptic curve E over* **Q***, the invariants* g_E
and r_E *are equal.*

In addition, the full form of the BSD conjecture gives an exact formula for the leading term (i.e. the first nonzero coefficient c_k, with $k \geq 1$),
of the Taylor expansion of $L(E, s)$ at $s = 1$. This exact formula involves
other invariants associated to elliptic curves and is technical to explain

here. However, we remark that it also involves a mysterious group called the *Tate-Shafarevich group* of E, which is conjectured to be always finite, and of which, very little is known even today. This group has a purely algebraic definition but also a geometric interpretation. We refer to (Silverman, 2009) for the technical details.

3   Theory of Complex multiplication

In this section we shall work with the field $\mathbf{C}$ of complex numbers as the base field. Any elliptic curve E defined over $\mathbf{Q}$ can also be viewed as an elliptic curve over the complex numbers, as $\mathbf{Q}$ is a subfield of $\mathbf{C}$. We denote by $E(\mathbf{C})$ the set of solutions in the field $\mathbf{C}$ for the equation defining E. It is also an abelian group under $\oplus$.

3.1   Lattices and Elliptic functions

Recall that a lattice $\mathbf{L}$ is an additive subgroup of $\mathbf{C}$ which is generated by two complex numbers w_1 and w_2 that are linearly independent over $\mathbf{R}$. For example, the additive subgroup generated by $\langle 1, i \rangle$ is a lattice and consists of all elements of the form $\{m + ni \mid m, n \in \mathbf{Z}\}$. An *elliptic function* (relative to the lattice $\mathbf{L}$) is a meromorphic function $f(z)$ on $\mathbf{C}$ that satisfies

$$f(z + w) = f(z), \quad \text{for all } z \in \mathbf{C} \text{ and all } w \in \mathbf{L}.$$

A consequence of this periodicity property is that a nonconstant elliptic function necessarily has both poles and zeros. The prototype elliptic function is the Weierstrass $\wp$-function, denoted by $\wp(z, \mathbf{L})$, and defined by the series

$$\wp(z, \mathbf{L}) = 1/z^2 + \sum_{\substack{w \in \mathbf{L} \\ w \neq 0}} \left(1/(z - w)^2 - 1/w^2 \right),$$

which is absolutely convergent. The derivative $\wp'(z, \mathbf{L})$ is again an elliptic function and we have the fundamental identity

$$\wp'(z, \mathbf{L})^2 = 4\wp(z, \mathbf{L})^3 - g_2(\mathbf{L})\wp(z) - g_3(\mathbf{L}), \tag{7}$$

where

$$g_2(\mathbf{L}) = 60 \sum_{w \in \mathbf{L} \backslash 0} 1/w^4, \quad g_3(\mathbf{L}) = 140 \sum_{w \in \mathbf{L} \backslash 0} 1/w^6, \tag{8}$$

and both the series are absolutely convergent. The alert reader would have noticed that putting $Y = \wp'(z, \mathbf{L})/2$ and $X = \wp(z, \mathbf{L})$ in (7) defines an elliptic curve. Any nonzero elliptic function relative to the lattice $\mathbf{L}$ is in fact a rational function over $\mathbf{C}$ in $\wp(z, \mathbf{L})$ and $\wp'(z, \mathbf{L})$ (i.e. a ratio of two nonzero polynomials in $\wp(z, \mathbf{L})$ and $\wp'(z, \mathbf{L})$.)

Two lattices $\mathbf{L}$ and $\mathbf{L}'$ are said to be *homothetic* if $\mathbf{L} = \lambda \mathbf{L}'$ where λ is a non-zero complex number. Homothety defines an equivalence relation on the set of lattices in $\mathbf{C}$. If $f(z)$ is an elliptic function relative to the lattice $\mathbf{L}$, then $f(\lambda^{-1}z)$ is an elliptic function for the lattice $\lambda \mathbf{L}$, and further, we have

$$\lambda^2 \wp(z, \mathbf{L}) = \wp(\lambda z, \lambda \mathbf{L}).$$

The *j-invariant* of a lattice $\mathbf{L}$ is defined by

$$j(\mathbf{L}) = 1728 \frac{g_2(\mathbf{L})^3}{\Delta(\mathbf{L})},$$

where

$$\Delta(\mathbf{L}) = g_2(\mathbf{L})^3 - 27g_3(\mathbf{L})^2,$$

with $g_2(\mathbf{L})$ and $g_3(\mathbf{L})$ as in (8). The *j-invariant* classifies lattices up to homothety, and $\Delta(\mathbf{L})$ is called the *discriminant* of the lattice. The group $E(\mathbf{C})$ is related to lattices by the following theorem:

Theorem 1 (Uniformization Theorem). *Let A and B be complex numbers satisfying $4A^3 - 27B^2 \neq 0$. Then there exists a unique lattice $\mathbf{L} \subset \mathbf{C}$ such that $g_2(\mathbf{L}) = A$, $g_3(\mathbf{L}) = B$. If $E/\mathbf{C}$ is an elliptic curve, then there exists a lattice $\mathbf{L}$ unique upto homothety, and an analytic isomorphism*

$$\mathbf{C}/\mathbf{L} \simeq E(\mathbf{C})$$
$$z \mapsto [\wp(z, \mathbf{L}), \wp'(z, \mathbf{L})].$$

The *j*-invariant of an elliptic curve E is defined as the *j*-invariant of the corresponding lattice, given by the Uniformization Theorem. This invariant classifies elliptic curves over $\mathbf{C}$ up to isomorphism between algebraic curves.

3.2   Elliptic curves with Complex Multiplication

Given a lattice $\mathbf{L}$, put $S(\mathbf{L}) = \{\alpha \in \mathbf{C} \mid \alpha \mathbf{L} \subset \mathbf{L}\}$. Clearly $\mathbf{Z} \subset S(\mathbf{L})$, as $\mathbf{L}$ is an additive subgroup of $\mathbf{C}$.

Definition 1. If $S(\mathbf{L})$ is strictly larger than $\mathbf{Z}$, then the lattice $\mathbf{L}$ is said to have complex multiplication.

If $\alpha \in S(\mathbf{L})$ is not in $\mathbf{Z}$, then the field $\mathbf{Q}(\alpha)$ generated by $\mathbf{Q}$ and α in $\mathbf{C}$ is a quadratic imaginary field extension of $\mathbf{Q}$, i.e. $\mathbf{Q}(\alpha) = \mathbf{Q}(\sqrt{-D})$ for some positive square free integer D.

This definition has a natural extension to elliptic curves. Let E be an elliptic curve over $\mathbf{C}$ and let $\mathbf{L}$ be a lattice such that $E(\mathbf{C}) \simeq \mathbf{C}/\mathbf{L}$. Recall that an endomorphism of a group is a homomorphism of the group into itself. Multiplication by a nonzero complex number α defines an endomorphism of the group $\mathbf{C}/\mathbf{L}$ precisely when $\alpha\mathbf{L} \subset \mathbf{L}$. The group of endomorphisms $\mathrm{End}(\mathbf{C}/\mathbf{L})$ clearly contains the integers. If it is strictly larger than $\mathbf{Z}$, it is a subring of the ring of integers of an imaginary quadratic extension of $\mathbf{Q}$, which is strictly bigger than $\mathbf{Z}$ (recall that if $\mathbf{Q}(\sqrt{-D})$ is an imaginary quadratic extension of $\mathbf{Q}$, then its ring of integers consists of all elements α such that α satisfies a polynomial equation with coefficients in $\mathbf{Z}$.) The elliptic curve E is said to have *complex multiplication* if $\mathrm{End}(E)$ is strictly larger than $\mathbf{Z}$.

Consider the family of elliptic curves $Y^2 = X^3 - DX$, defined over $\mathbf{Q}$, with D a squarefree integer. As $(X, Y) \mapsto (-X, iY)$ (where $i^2 = -1$) is an endomorphism not in $\mathbf{Z}$, these curves have complex multiplication with $\mathrm{End}(E)$ isomorphic to the ring of Gaussian integers $\mathbf{Z}[i] = \{m + ni \,|\, m, n \in \mathbf{Z}\}$. The theory of complex multiplication has interesting connections with *class field theory*, where one studies field extensions of $\mathbf{Q}$. Finally, all deep phenomena that occur in the study of the arithmetic of elliptic curves are believed to already occur in the class of elliptic curves with complex multiplication. For elliptic curves over $\mathbf{Q}$ with complex multiplication, it is a classical result that their L-functions are analytic, which does not need the deep techniques of Wiles, *et al.*

4   Further Vistas

In this final section, we briefly discuss other areas of mathematics that are related to elliptic curves, starting with an ancient problem that is over a thousand years old.

4.1 Congruent Numbers

Definition 2. A natural number $N > 1$ is said to be *congruent* if there exists a right angled triangle all of whose sides have lengths which are rational numbers and whose area is N.

In other words, there should exist rational numbers a, b, c such that $a^2 + b^2 = c^2$ and $(ab)/2 = N$. Ancient Arab manuscripts from around the ninth century A.D. have long lists of congruent numbers. It is clear from the article (Sridharan, 2011) that Indian mathematicians would also have been interested in congruent numbers from even earlier times. If N is congruent, it can be shown that there are always infinitely many right angled triangles with rational sides whose area is N. Fermat showed that 1 is not a congruent number, introducing his celebrated *infinite descent* argument (see (Coates, 2005) for a discussion on congruent numbers). The smallest congruent number is 5. An example of a large prime number that is congruent is 1,119,543,881 (see (Chahal, 2006) for a corresponding right angled triangle). The problem of determining whether a number N is congruent is easily seen to be equivalent to deciding whether the elliptic curve

$$E_N : Y^2 = X^3 - N^2 X \tag{9}$$

has a solution (X, Y) over $\mathbf{Q}$ with $Y \neq 0$; such a solution is a point of infinite order on E_N (see (5) for the case $N = 157$). In fact, no algorithm for settling whether such a point exists in general has ever been proven. Note that the curves E_N have complex multiplication by $\mathbf{Z}[i]$.

If the BSD conjecture were true, then it follows that N is congruent if and only if the L-function $L(E_N, s)$ does vanish at $s = 1$. The beauty of being able to work with L-functions is that they are highly computable. Using two different connections between elliptic curves and modular forms Koblitz (1993), one can show that if the BSD conjecture were true, then there is even an explicit algorithm consisting of a finite number of steps enabling one to determine whether a given integer N is congruent or not. We also mention that a folklore conjecture based on vast numerical evidence asserts that any number N which leaves a remainder of 5, 6 or 7 modulo 8 is always congruent. Further, it follows from the theory of L-functions that if N is congruent to 5, 6, or 7 modulo 8, then $L(E_N, s)$ always has a zero of odd order at $s = 1$, thereby lending credence to the conjecture above. The best result on this to date is the following theorem whose proof involves techniques from Iwasawa theory (see Sujatha (2010)).

Theorem 2. *Let N be an integer congruent to 5, 6, or 7 modulo 8 and E_N the elliptic curve as in (9). If the p-primary torsion subgroup of the Tate-Shafarevich subgroup of E_N is finite for some prime p, then N is congruent.*

Of course, there are congruent numbers N that are not congruent to 5, 6 or 7 modulo 8, for example $N = 34$.

4.2   Computational Aspects

The deep arithmetic theory of elliptic curves has given rise to a rapidly developing field of computational work on elliptic curves and algorithms. In one direction, there are algorithms related to the connections between modular forms and elliptic curves, the computation of L-series, and for computing the group of points on a given elliptic curve. The interested reader is referred to Cremona's tables (Cremona, 1977), PARI (PARI/GP, 2005) and SAGE (Stein, 2010). In another direction, some of these algorithms have proved especially useful in cryptography. The main applications of elliptic curves here are in the fields of primality testing and factorization methods, both of which lie at the heart of devising secure cryptosystems. The group law on the points of an elliptic curve forms the basis for certain cryptosystems in public key cryptography. Moreover, the rich theory of elliptic curves over different base fields provides a wide array of structures and operations which are amenable to computation and efficient algorithmic implementation in the form of computer programs. These are ideal for designing cryptosystems as many of them yield two way functions f such that f is easy to compute, while the inverse f^{-1} is extremely hard to compute in principle, but becomes computable when an additional piece of information is made available. Hendrik Lenstra has developed a factorization algorithm that uses elliptic curves, which today is among the fastest known algorithms for factoring large integers. The interested reader may refer to (Koblitz, 1994) for applications of number theory and elliptic curves in cryptography.

4.3   Arithmetic Geometry

Many of the phenomena studied for elliptic curves turn out to be extendable to a much wider context in algebraic geometry and Galois theory. This large body of work is called *Arithmetic Geometry* where the interplay between geometry and arithmetic comes to the fore. It has spawned a host of other deep conjectures, many of which are inspired or guided by problems occurring in the study of elliptic curves. It is beyond the scope of this article to enter into a detailed discussion of these.

The proof of Fermat's Last Theorem by Wiles (1995) masterfully brought together areas such as modular forms, Galois representations and elliptic curves. Another example that is more recent, is the proof by Richard Taylor of the Sato-Tate conjecture (see Taylor (2008), Kumar Murty and Ram Murty (2007)) which gives a precise prediction of how the integers a_p vary as the prime p varies. This proof builds on fundamental work due to Clozel *et al.* (2008) in the area of automorphic forms.

We end this article with a discussion on the state of our knowledge on the Birch and Swinnerton-Dyer conjecture. The first breakthrough was the result of Coates and Wiles (1977) where they showed that if E is an elliptic curve over $\mathbf{Q}$ with complex multiplication and if the algebraic rank g_E is positive, then $L(E, 1) = 0$. The first examples of elliptic curves over $\mathbf{Q}$ with finite Tate-Shafarevich group were given by Rubin (1987) and Kolyvagin independently. The most general result known in this direction is the following theorem due to Kolyvagin (1988):

Theorem 3 (Kolyvagin). *Let E be an elliptic curve over $\mathbf{Q}$. If the analytic rank $r_E \leq 1$, then we have $g_E = r_E$ and the Tate-Shafarevich group of E is finite.*

Note that this result in particular implies that when $g_E \geq 2$ we have $r_E \geq 2$. Earlier work of Gross and Zagier (1986) proved the existence of a point of infinite order on E when $r_E = 1$. If the analytic rank of E is strictly greater than one, the BSD conjecture remains a mystery! Here is a concrete example however. Take E to be the curve over $\mathbf{Q}$ defined by the equation

$$Y^2 = X^3 - 82X.$$

As remarked earlier, this curve has complex multiplication by $\mathbf{Z}[i]$, and its algebraic rank is three. The points $(-9, 3)$, $(-8, 12)$, $(-1, 9)$ generate the group $E(\mathbf{Q})$ modulo torsion. By Kolyvagin's result, the analytic rank is at least three and a numerical computation shows that it is exactly three, thereby establishing BSD for this curve. The conjecture of Birch and Swinnerton-Dyer predicts that the Tate-Shafarevich group of E is zero but we do not even know if it is finite. However, it can be shown (Coates *et al.*, 2010) by using subtle theoretical results about elliptic curves, along with delicate computations that the p-primary part of the Tate- Shafarevich group is finite for a large number of primes p congruent to 1 modulo four.

Acknowledgements

We would like to thank John Coates and Sudhanshu Shekhar for helpful comments.

References

Breuil, C., Conrad, B., Diamond, F., Taylor, R. 2001. On the modularity of elliptic curves over **Q**: wild 3-adic exercises, *J. Amer. Math. Soc.* **14**:843–939.

Chahal, J.S. 2006. Congruent numbers and elliptic curves, *Amer. Math. Monthly.* **113**:308–317.

Clozel, L., Harris, M. and Taylor, R. 2008. Automorphy for some l-adic lifts of automorphic mod l representations, *Pub. Math. IHES.* **108**:1–181.

Coates, J. 2005. Congruent number problem, *Q. J. Pure Appl. Math.* **1**:14–27.

Coates, J. and Wiles, A. 1977. On the conjecture of Birch and Swinnerton-Dyer, *Invent. Math.* **39**:223–251.

Coates, J., Liang, Z. and Sujatha, R. 2010. The Tate-Shafarevich group for elliptic curves with complex multiplication, *Milan Journal of Math.* **78**:395–416.

Cornell, G., Silverman, J. and Stevens, G. 1995. *Modular forms and Fermat's last theorem.* Papers from the Instructional Conference on Number Theory and Arithmetic Geometry held at Boston University, Boston, MA, August 9–18.

Cremona, J. 1977. *Algorithms for modular elliptic curves*, Second ed. Cambridge University Press, Cambridge.

Fulton, W. 1989. *Algebraic curves: An introduction to algebraic geometry*, Notes written with the collaboration of Richard Weiss. Reprint of 1969 original. Advanced Book Classics. Addison-Wesley Publishing Company, Advanced Book Program, Redwood City, CA.

Gross, B. and Zagier, D. 1986. Heegner points and derivatives of L-series, *Invent. Math.* **84**:225–320.

Koblitz, N. 1993. *Introduction to elliptic curves and modular forms*, Second ed. Graduate Texts in Mathematics 97, Springer-Verlag, New York.

Koblitz, N. 1994. *A course in Number Theory and Cryptography*, Second Ed. Graduate Texts in Mathematics 114, Springer-Verlag, New York.

Kolyvagin, V. 1988. Finiteness of $E(\mathbf{Q})$ and $\text{Ш}(E/\mathbf{Q})$ for a class of Weil curves, *Izv. Akad. Nauk SSSR* 52.

Murty, M. Kumar and Murty, M. Ram. 2007. The Sato-Tate conjecture and generalizations, *Current Trends in Science*, Platinum Jubilee volume of the Indian Academy of Sciences. 635–646.

PARI/GP. 2005. http://pari.math.u-bordeaux.fr/

Rubin, K. 1987. Tate-Shafarevich groups and L-functions of elliptic curves with complex multiplication, *Invent. Math.* **89:**527–560.

Silverman, J. 2009. *The Arithmetic of Elliptic Curves*, Second ed. Graduate Texts in Mathematics, 106 Springer, Dordrecht.

Sridharan, R. 2011. *Rational Quadrilaterals from Brahmagupta to Kummer*, this volume.

Stein, W. 2010. The modular forms database, http://modular.fas.harvard.edu/Tables

Sujatha, R. 2010. *Arithmetic of elliptic curves through the ages*, European Women in Mathematics, Proceedings of the 13th General Meeting, ed. Catherine Hobbs, Sylvie Pacha, World Scientific Press. 71–89.

Tate, J. and Silverman, J. 1992. *Rational points on elliptic curves*, Undergraduate Texts in Mathematics. Springer-Verlag, New York.

Taylor, R. 2008. Automorphy for some l-adic lifts of automorphic mod l representations. II, *Pub. Math. IHES.* **108:**183–239.

Taylor, R. and Wiles, A. 1995. Ring-theoretic properties of certain Hecke algebras, *Ann. of Math.* **141:**553–572.

Waerden, B.L. Van der. 1991. *Algebra* Vols I and II, Springer-Verlag, New York.

Wiles, A. 1995. Modular elliptic curves and Fermat's last theorem, *Ann. of Math.* **141:**443–551.

CHAPTER 4

Curvature and Relativity

Siddhartha Gadgil

Department of Mathematics,
Indian Institute of Science,
Bangalore 560012, India.
e-mail: gadgil@math.iisc.ernet.in

Roads in mountains are curved while those in deserts are usually straight. Thus, we all understand curvature in some sense – indeed it would not be possible to ride a bicycle if we could not see a road curving. It is not difficult to formalise this into a measure of curvature of curves in the plane.

The case of two dimensional objects is a little more subtle. A football is clearly curved while the top of a table is not. The case of cylinder is a little more complicated – it is curved in one direction and not curved in another. The surface of an egg is even more complicated – while it is clearly curved at all points in all directions, how curved it is varies with both points and directions. Things are even more subtle in the case of surfaces with saddles – like the bridge of your nose (where one's spectacles rest). Indeed a straight line can spin to give the illusion of a curved surface (as one may see, for instance, at the Visvesvaraya Museum in Bangalore).

Nevertheless, one can capture all this complexity in a simple way, as discovered by Euler. All the situations we have considered, however, correspond to the more primitive notion of curvature, so called *extrinsic curvature*. There is a more mysterious notion, that of *intrinsic curvature*. This was understood through the work of Gauss and Riemann.

Intrinsic curvature is a much harder concept to understand as it is alien to our everyday experience. However, it is this form of curvature that is central to Einstein's General theory of Relativity. It is through the natural development of mathematics that intrinsic curvature was understood and developed to the point where it was ready to be put to use by Einstein. Thus, this story illustrates how mathematics, developing according to its own rules and values, gives insights into the natural world.

We shall sketch the major conceptual steps leading to Riemann's notion of intrinsic curvatures, allowing the development of General relativity. We shall also mention some more recent developments.

1 Curvature of a curve in the plane

Let us begin with the simplest case – smooth curves in the plane. Clearly lines are not curved, while circles are curved. We would like to say more. Namely, we would like to associate to a point on a curve a number that measures how curved it is.

We build on the simplest cases. A line clearly has zero curvature. As a circle is symmetric, its curvature at all points on it should be the same. Furthermore, the curvature of a circle clearly decreases as its radius increases. Thus, it is natural to take the curvature of a circle of radius r to be the reciprocal $1/r$ of its radius.

Observe that if we consider circles C_r through a fixed point P and with a fixed tangent line l at P, and let the radius of the circles go to infinity, then the circles C_r approach the line l. Indeed one can define a precise notion of limits so that the circles have limit the line l. We urge students to try to find such a definition (the answer is subtle, but one can learn a lot from the attempt).

Thus, it will be convenient for us to regard lines as circles with radius $r = \infty$ and hence curvature $1/r = 0$. Given three points P, Q and R in the plane, basic geometry tells us that either all three points lie on a line or there is a unique circle passing through all of them. Thus, there is a unique circle or line through P, Q and R. We shall use this to define the curvature of a curve.

We define the curvature at a point P on a curve C in the plane just like tangents are defined in Calculus. Recall that to define the tangent line at the point P, we take a point Q on C that is close to P and consider the secant line through P and Q. As Q approaches P, the secant lines approach the tangent line at P.

To define the curvature at P, we take two points Q and R on the curve C close to P. We have seen that there is a unique circle or line through these points. Consider the reciprocal $1/r(P, Q, R)$ of the radius $r(P, Q, R)$ of the circle through P, Q and R, where we take $1/r(P, Q, R) = 0$ if the points P, Q, and R are contained in a line. As Q and R approach P, we get a limit, which we define to be the curvature.

We can view this in terms of calculus. Recall that the tangent line to a graph $y = f(x)$ at the point $(a, f(a))$ is the unique line passing through $(a, f(a))$ with first derivative at a the same as that of $f(x)$. Similarly, we can find an arc of a circle (or a line) with the same first and second derivatives at a as $f(x)$. We take the curvature to be the reciprocal of the radius of this arc of a circle.

2 Curvature of surfaces

The case of surfaces is a little more complicated. While a sphere is clearly curved by the same amount at all points and directions, a cylinder is curved in one direction but not the other. Thus, the curvature of a surface depends not only on a point on the surface but also a (tangent) direction at that point. Hence one seems to have infinitely many numbers measuring the curvature in different directions.

Firstly, we need to understand what we mean by the curvature in a given direction. Given a point P on a smooth surface S, we can speak of the tangent plane T to S at the point P. This can be defined by considering all smooth curves in S passing through the point P. The set of tangents at P of all such curves is the tangent plane P. A unit normal vector n at P is a vector of unit length that is perpendicular to T.

Suppose we are given a direction v in the tangent plane at P. Consider the unique plane π containing v and the unit normal n. The intersection $S \cap \pi$ of the surface S and the plane π is a smooth curve in the plane π. We can thus define its curvature as we did for curves in the plane. This gives the curvature of S in the direction v.

Euler observed that the curvatures in the various directions are determined by two numbers – the *principal curvatures*. This is analogous to an ellipse being determined by its major and minor axes – indeed, both cases are governed by a quadratic function, which can be simplified by completing squares. This can be seen by considering the graph of a function giving a piece of the surface.

By rotating, and translating, we can ensure that the point P on the surface S is the origin $(0,0,0)$ and the tangent plane T at P is the xy-plane. A piece of the surface S can then be described as the graph of a function $z = f(x,y)$, with $f(x,y)$ having a Taylor expansion with no constant or linear terms, i.e.,

$$f(x,y) = ax^2 + 2bxy + cy^2 + \cdots$$

As in the study of conic sections, we can complete squares. This means that we can perform a rotation so that S becomes the graph of a function

$$g(x,y) = \kappa_1 x^2 + \kappa_2 y^2$$

just as we can rotate an ellipse so that its major and minor axis become the x-axis and y-axis.

Both the principal curvatures at any point of a sphere of radius r are $1/r$ while for a cylinder of radius r the principal curvatures are $1/r$ and 0. The principal curvatures of a plane are 0.

3   Gauss maps the earth

Geometric properties of a surface can be classified as *extrinsic* or *intrinsic*. An intrinsic property of a surface is one that depends only on distances measured along the surface, i.e., is unchanged on bending the surface. Thus, a square can be bent into a piece of a cylinder, so intrinsic properties of the square and the corresponding piece of the cylinder must coincide.

Intrinsic Differential Geometry, the study of intrinsic properties, was born with the work of Gauss, who was involved in surveys to make maps of regions large enough that the curvature of the earth mattered. While the curvature of the earth can be understood in terms of how it lies in space – for instance, by the vanishing of ships beyond a horizon, while conducting a survey one actually makes measurements only along the surface of the earth. So one needs to understand the relation between the geometry of the earth and measurements made along the earth.

Gauss observed that we cannot make a map of a region of a sphere without distorting distances. Thus, while we can wrap a piece of paper around a cylinder smoothly, we cannot do so on a sphere without introducing wrinkles.

The fundamental discovery of Gauss was that the product of the principal curvatures, which we now call the Gaussian curvature, depended only on distances measured along the surface, i.e., is an intrinsic property of the surface (i.e., a *bending invariant*). For example, as one of the principal curvatures of a cylinder is 0, the Gaussian curvature is 0. This should be the case as a piece of the cylinder can be obtained by bending a square.

Note that the principal curvatures themselves are not intrinsic (consider for example a square and a piece of the cylinder). Hence, while the Gaussian curvature is intrinsic, it is defined in terms of quantities that are not. This mathematically unsatisfactory situation was remedied by Riemann, who developed a genuinely intrinsic framework for Differential geometry, i.e., one where we consider just a surface and the distances between points on it obtained by measuring along the surface. In the process, Riemann defined intrinsic curvatures for what we call Riemannian manifolds, which are the natural higher dimensional analogues of surfaces with distances coming from space.

It is not difficult to see that a sphere is intrinsically curved. One way to see this is to consider the area of discs with small radii on different surfaces. In the case of the plane, we know that the disc of area r has area πr^2. On the other hand, the area in the sphere of radius r is less than πr^2. Note that the area of a disc in a surface does not change if we bend the surface, so by measuring areas we obtain a bending invariant, i.e., an intrinsic property.

More precisely, we consider the area $A(r)$ of a disc or radius r around P on the given surface S. We consider the Taylor expansion of $A(r)$ near $r = 0$. This is of the form

$$A(r) = \pi r^2 - \frac{\kappa}{12} r^4.$$

The (Gaussian) curvature is then κ.

While the above allows us to measure curvature, it does not tell us how to study the geometry of curved space. In particular, we need substitutes for straight lines and parallel vectors. To understand these, we take a different view of curvature.

4   Wind-vanes on spheres

Consider a frictionless wind-vane in vacuum, as in figure 1. Assume that the wind-vane is initially stationary. Then if we rotate the base of the wind-vane, the arrow does not move (as there is no friction).

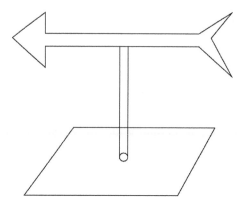

Figure 1: *The Wind-vane.*

Now assume that the wind-vane is placed flat on a plane. Let us move the base of the wind-vane smoothly along a curve in the plane. Then the wind-vane points in a constant direction, i.e., the direction of the wind-vane gives *parallel* vectors along the curve. This gives a physical meaning to vectors at different points being parallel.

Something more interesting happens if we perform the same experiment on the surface of a sphere (see figure 2). Suppose the wind-vane is initially placed at the point *A* which we take to be the north pole, and points in the direction of *B* on the equator. Let *C* be another point on the equator. We shall move the wind-vane along the longitude *AB*, then along the equator from *B* to *C* and finally back to *A* along a longitude.

As we move along *AB*, the wind-vane is always pointing in the direction of motion. We can see this by symmetry – as the two sides of the longitude of a sphere look the same, there is no reason for the wind-vane to turn right rather than left (or vice versa). When we reach *B*, the wind-vane is perpendicular to the equator, so that it is pointing due south. Now as we move along the equator to *C* (by symmetry) it continues to point south. Hence when we reach *C*, we are pointing along the longitude through *C* towards the south pole. Finally, as we move the wind-vane back along the longitude *AC*, it continues to point towards *C*.

Notice that something remarkable has happened. The wind-vane which was pointing towards *B* is now pointing towards *C*. Thus if we look at *parallel* vectors along the loop *ABCA*, we end up with a different vector from the one we had started with. Thus *parallel transport* around a loop on a

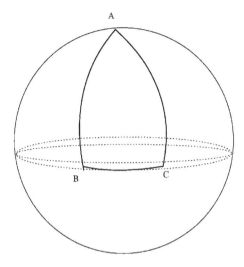

Figure 2: *A path with holonomy.*

sphere leads to a rotation, in technical language the *holonomy* of the parallel transport. This happens because the sphere is curved, and the amount by which we rotate depends on the *curvature*. By contrast, in the plane the holonomy is always the identity, i.e., each vector is taken to itself.

5 Parallel transport and curvature

We shall now take a more geometric look at *parallel transport* and another view of *curvature*. Suppose we have a curve α on a surface S from a point P to a point Q and a vector V at P that is tangent to S. Then we can *parallel transport* the vector V along α to get a parallel vector field along α. This corresponds to moving our wind-vane.

Let us see what properties such a parallel vector field should satisfy. Firstly, the vectors of the vector field have the same length. Further, the angle between two parallel vector fields is constant. Properties similar to this will hold for parallel transport in a more general setting. These properties amount to saying that the holonomy about a loop is a rotation.

The third property is more subtle. If we move in a straight line in the plane at constant speed, then the *velocity vector* clearly gives a parallel vector field along the line. We have a similar property for more general surfaces. The *shortest* curve on a surface S joining two points P and Q is a *geodesic*. If we move along a geodesic at constant speed, then the velocity vectors are parallel. This condition is called *torsion freeness* and will play a different role from the first two conditions in the general setting.

It turns out that these conditions essentially determine parallel transport on a surface (though not in higher dimensions). Observe that both these conditions are *intrinsic*.

The *curvature* of a surface can be viewed as measuring the holonomy. More precisely, let us take a small closed curve, say a parallelogram beginning and ending at a point P. We measure the angle by which parallel transport along the curve rotates vectors. We divide this by the area enclosed by the curve. As we take smaller and smaller curves, this ratio approaches the (Gaussian) curvature κ at P.

6 Higher dimensions and intrinsic curvature

Curvature and parallel transport in higher dimensions are more subtle than for surfaces. For instance, if we know the length of a vector V and the angle it subtends with another vector W, then V is determined up to a reflection. This together with continuity allows parallel transport to be determined by simple conditions as in the previous section. However, in 3-dimensions there are in general infinitely many vectors V with a given length and subtending a given angle with a vector W. Hence we must have a more complicated definition of parallel transport.

Similarly, for surfaces, the holomony of parallel transport along a curve can be measured by the angle formed by the final vector with the initial vector. For the same reason as above, the curvature, which measures parallel transport, must be a far more complicated object than in the case of surfaces.

Explaining the construction of parallel transport and curvatures in higher dimensions is beyond our present scope. We merely remark that it is the natural outgrowth of an attempt to understand better the work of Gauss and put in a fully intrinsic form. Such a goal is natural in mathematics but not especially so if one is only concerned with practical calculations.

7   Curvature and relativity

In Einstein's general relativity, gravity is due to the intrinsic curvature of space. While the curvature of space seems to most people to be a bizarre notion, there is no better reason for space to be flat (i.e., not curved) than for the Earth to be flat (as was once believed to be the case), especially once one accepts that there is no absolute medium called space but only distances between events.

Merely accepting that space could be intrinsically curved is of little use in developing relativity – one needs a precise and quantitative concept of intrinsic curvature. It is this that developed through the work of Riemann and others who built on it.

Mechanics can be formulated in terms of the principle of least action (generalising the principle that light travels by the fastest path). Einstein formulated general relativity in terms of an action that depended on curvature – specifically Ricci curvature of a Lorentzian manifold. Such a formulation made use of not only Riemann's curvature tensor, but the work of several mathematicians that went beyond this.

8   To the present

Our story so far has been concerned with events from long ago. But, as relativity is based on Riemannian Geometry, one should expect that advances in Riemannian Geometry, from within mathematics, should have an impact on relativity. This is indeed the case, as we illustrate with a recent result of Michael Anderson.

We consider vacuum space-times, i.e., space-times with no matter. The end state of the evolution of such a space-time is expected to be stationary, i.e., time independent. Anderson showed that the only possible vacuum stationary space time is ordinary space-time. This result used Cheeger-Gromov theory, a recent far-reaching development in Riemannian geometry.

CHAPTER 5

Generating Functions

S. Ramanan

Department of Mathematics,
Chennai Mathematical Institute,
Siruseri 603103, India.
e-mail: sramanan@cmi.ac.in

Power series

The use of generating functions is one way of book-keeping. Let $a : \mathbf{Z}^+ \to \mathbf{C}$ be any function, in other words, a sequence a_n of complex numbers. Then one associates to it the *power series* $\sum a_n x^n$. The power series may not converge anywhere except $x = 0$, that is to say, its radius of convergence may be 0. If b is another such sequence, then the product of the two power series is associated to the function $n \mapsto \sum_{i=0}^{i=n} a_i b_{n-i}$. This is the *Cauchy product* of two sequences and so already we notice that associating the power series to the sequence has some advantage.

But if the power series has a positive radius of convergence, say R, then it represents a (holomorphic) function in the open set $\{|x| < R\}$.

Take the simplest such function $n \mapsto 1$ for all n. The associated power series is $\sum x^n$. We know that this power series has 1 for its radius of convergence and represents the function $\frac{1}{1-x}$ in the open set $|x| < 1$. Also we know that the series $\sum(-1)^n$ is not convergent and indeed oscillates between 1 and 0. However, we can substitute the value of the function $\frac{1}{1-x}$ at the point $x = -1$ and get $\frac{1}{2}$ and pretend that the series $\sum(-1)^n$, although not convergent, represents $\frac{1}{2}$ in some sense.

This is called the *summability method* of Abel. To be precise, if a_n is a sequence and the power series $\sum a_n x^n$ has R as its radius of convergence R, then the power series represents a function in $\{|x| < R\}$. If λ is such that

$|\lambda| = R$, then the function may have a continuous extension to λ as well. If it does, then one defines the *Abel sum* of the series $\sum a_n \lambda^n$ to be the value of the function at λ. This is of course consistent with the limit within the radius of convergence.

The sequence $a_n = \frac{1}{n!}$ can be kept track of by the power series $\sum(\frac{1}{n!})x^n$ which is the exponential function $exp(x)$. We know that the exponential function satisfies $exp(x + y) = exp(x)exp(y)$. In particular, we have the formula $exp(2x) = (exp(x))^2$. By computing the coefficients of x^n on both sides we get $\sum \binom{n}{r} = 2^n$. Thus the equality $exp(2x) = exp(x)^2$ is a succinct way of providing the set of formulas (for each n) given above.

One can actually get identities which are not so straightforward by using generating functions. Consider the power series $1 - exp(-x)$. This is a power series without constant term which is convergent everywhere. Hence the series $(1 - exp(-x))^r$ starts with the term x^r with coefficient 1. On the other hand, it can be expanded by binomial theorem as $\sum(-1)^p\binom{r}{p}$ $(exp(-x))^p = \sum(-1)^p\binom{r}{p}exp(-px)$. Hence the coefficient of x^q in the power series $(1 - exp(-x))^r$ is $\frac{(-1)^q}{q!}\sum_{p=0}^{p=q}(-1)^p\binom{r}{p}p^q$. We conclude that the following identities hold for all r and $q < r$.

$$\sum_{p=0}^{p=q}(-1)^p\binom{r}{p}p^q = 0$$

and

$$\sum_{p=0}^{p=r}(-1)^p\binom{r}{p}p^r = (-1)^r r!.$$

The power series $\frac{1-exp(-x)}{x}$ is invertible and its inverse $\frac{x}{1-exp(-x)}$ plays an important role in analysis as well as the modern algebraic geometric theorem, called the Riemann Roch theorem. The above identities are fundamental to Patodi's approach via the so-called *heat equation method* to the Riemann-Roch theorem.

Arithmetic Functions

A sequence $n \mapsto a_n$ of integers is called an *arithmetic function*. We may reverse the above procedure and start with a power series (with integral coefficients) and study the properties of the arithmetical function it gives rise to. One famous example of this genre is the *Ramanujan function*. Consider the infinite product $x\Pi(1 + x^r)^{24}$. Expand it in power series $\sum \tau(n)x^n$. The arithmetic function $n \mapsto \tau(n)$ is called the *Ramanujan function*. Ramanujan studied this function and proved or conjectured many of its properties. We will mention one below. But some of the congruence properties of this function that Ramanujan suggested are mysterious. To give an example, let $\sigma_k(n)$ be the sum of the k-th powers of all the divisors of n. Then $\tau(n) - \sigma_{11}(n)$ is divisible by 691! (not factorial, but exclamation!).

We will now explain its connection with algebraic geometry. Substitute for x in the power series $\sum \tau(n)x^n$ the term $q = e^{2\pi i z}$. We get then a function on the upper half plane $\{z \in \mathbf{C} : Imz > 0\}$. This function has invariance properties with respect to the action of the multiplicative group of $(2, 2)$ matrices with determinant 1 and integral entries, otherwise known as $SL(2, \mathbf{Z})$. The action is given by

$$\begin{pmatrix} a & b \\ c & d \end{pmatrix}(z) = \frac{az + b}{cz + d}.$$

With respect to this action, the function $f(q) = \sum \tau(n)q^n$ satisfies the equation

$$f\left(\frac{az + b}{cz + d}\right) = (cz + d)^{12} f(z).$$

Functions on the upper half plane which have invariance properties with respect to the above action of $SL(2, \mathbf{Z})$ are related to the description of all lattices in $\mathbf{C}$. A *lattice L* is the subgroup generated by two complex numbers ω, ω' which are linearly independent over $\mathbf{R}$. The quotient of $\mathbf{C}$ by this lattice is then a torus of real dimension 2 on which one can talk of meromorphic functions, called *elliptic* or *doubly periodic* functions on $\mathbf{C}$. One such function is the famous Weierstrass' $\wp$ function, which is defined by $\wp(z) = \frac{1}{z^2} + \sum_{\lambda \in L\setminus\{0\}} \left(\frac{1}{(z+\lambda)^2} - \frac{1}{\lambda^2}\right)$. Of course then the function $\wp'(z) = (d/dz)(\wp(z))$ is also an elliptic function. The two are related by the equation

$$\wp'^2 = 4(\wp)^3 + a\wp + b,$$

where a, b are complex constants determined by L.

The map $z \mapsto (\wp(z), \wp'(z))$ identifies the (complement of $\{0\}$ in the) torus with the cubic curve $y^2 = 4x^3 + ax + b$ in $\mathbf{C}^2$. Indeed, the map $z \to (\wp(z), \wp'(z), 1)$ of $\mathbf{C}$ into the complex projective plane $\mathbf{CP}^2$ identifies the torus with the algebraic curve in $\mathbf{CP}^2$ given by the homogeneous equation

$y^2z = 4x^3 + axz^2 + bz^3$, where (x, y, z) are homogeneous coordinates in the projective plane.

Two lattices L, L' are *equivalent* if there is a transformation $z \rightarrow pz, p \neq 0$ which takes one lattice to another. This is a natural equivalence, if we are interested in elliptic functions, since the field of elliptic functions defined by two lattices are 'the same' if and only if the lattices are equivalent. The set of equivalence classes of lattices is bijective to the upper half plane, modulo the action of $SL(2, \mathbf{Z})$ given above. Thus the function f which we considered above is closely related to the 'moduli' of all lattices in $\mathbf{C}$, or what is the same the moduli of elliptic curves or moduli of cubic curves. Thus algebraic geometry enters the picture in a decisive way.

Ramanujan made another conjecture regarding the τ-function, namely that $\tau(p) < 2p^{11/2}$ for all primes p This was proved by Pierre Deligne in the 70's. Deligne made full use of the arsenal of algebraic geometry and applied it to the moduli of elliptic curves in establishing this result.

Dirichlet series

There are other ways of book-keeping which are useful as well. One can associate for example the *Dirichlet series* $\sum \frac{a_n}{n^s}$ to the sequence a_n. Here s is a complex variable. Again we will take the simplest example, namely $a_n = 1$ for all n. The associated Dirichlet series is $\zeta(s) = \sum \frac{1}{n^s}$. As we are all familiar with, this series converges whenever $Re(s) > 1$. The series represents a holomorphic function in this right half plane, called *Riemann's zeta function*. Indeed, there is a (unique) holomorphic function ζ in the complement of 1 in $\mathbf{C}$, which restricts to the above function in the half plane. Now we can take its value at -1 as we did in the case of the power series and define the *Dirichlet sum* of $\sum n$ to be $\zeta(-1)$. One can actually compute it to be $-\frac{1}{12}$.

If I may add a personal note here, the above fact is partly responsible for my taking mathematical research as a career! In my seventh standard at school (then called the second form), there was prescribed for our non-detailed study, a Tamil book entitled *Moonru Indiyap Periyorgal – 'Three great men of India'*. These great achievers were Srinivasa Ramanujan, P.C. Ray and Jagadish Chandra Bose. In it was reproduced a letter that Ramanujan had written to G.H. Hardy saying, "the sum $\sum n$ can be construed in some sense to be $-1/12$. When you hear this, you may think that

I ought to be in a mental asylum but I can justify my claim". Of course, this was intriguing and I kept asking the mathematics teachers and other senior colleagues what this meant. No one could throw light on this, and I told myself that I would some day understand the meaning of his remark! This partly played a part in my choice of career.

Although the zeta function is known by the name of Riemann, the mathematician who first conceived of using this kind of series in number theory was Euler. He used it for example to show that there are infinitely many prime numbers. Fundamental to this approach is the *Euler product*. Let us assume that there are only finitely many primes. Notice that since every natural number is the product of positive primes in a unique way, we have $\sum \frac{1}{n} = \Pi(\sum \frac{1}{p^i})$, where the product extends over all primes. Since each of the factors in the product on the right side is convergent and this is a finite product under the assumption, the rearranged series $\sum \frac{1}{n}$ would converge, which of course is a contradiction. Indeed, a closer look at this proof actually gives the stonger assertion that $\sum \frac{1}{p}$ is divergent, for otherwise $\Pi(1 - 1/p)$ would converge, along with $\Pi(1 - 1/p)^{-1}$ which is just $\sum \frac{1}{n}$. In other words, prime numbers are not merely infinite, but are numerous enough to ensure that $\sum \frac{1}{p}$ is divergent. This gives one a hint that a closer study of the zeta function might throw light on the distribution of prime numbers. Thus the zeta function became the centre piece of activity among the *analytic number theorists*. One of the celebrated conjectures in this regard is the following. It is known that the zeta function has zeros at $-2m$, for all natural numbers m. But it has other zeros elsewhere as well. The question if all these remaining zeros lie on the line $Re(z) = \frac{1}{2}$ is an outstanding unsolved problem, known as the *Riemann hypothesis*.

An arithmetic function a is said to be *multiplicative* if $a(mn) = a(m) \cdot a(n)$ whenever m and n are coprime. The associated Dirichlet series $\sum a(n)/n^s$ has in that case, an Euler product expansion $\sum \frac{a(n)}{n^s} = \Pi(\sum \frac{a(p^i)}{(p^i)^s})$.

One such multiplicative function is the *Möbius function* μ defined as follows. $\mu(n) = (-1)^r$, where r is the number of prime factors of n, if n is square free, and is 0 otherwise. The associated Dirichlet series is therefore $\sum \frac{\mu(n)}{n^s} = \Pi(1 - p^s)$ and this is the inverse of the zeta function (which is associated to the function $u(n) = 1$ for all n).

If a and b are two arithmetic functions, then one can define their *Dirichlet product a * b* as the arithmetic function whose Dirichlet series is the product of the Dirichlet series of a and b. It is easy to see that this Dirichlet product is the arithmetic function $n \mapsto \sum a(d)b(n/d)$ where the sum extends over all divisors of n. It is obvious that this product is commutative and associative so that $(a * u) * \mu = a * (u * \mu) = a$. In other words, given an arithmetic function a, the function $a * u$ is $\sum a(d)$, the sum being over all divisors of n, and a can be recovered from $a * u$ by taking the Dirichlet product with the Möbius function. This is called the *inversion formula*. Incidentally, one of the conjectures of Ramanujan regarding the τ-function is that it is multiplicative. This was later proved by Mordell.

Poincaré Polynomials and Power Series in Topology

Associating the generating function to an arithmetic function is useful in other contexts as well. Let X be a topological space. Then to keep track of its geometric particularity of X in each dimension, one can associate a number b_i called the *i-th Betti number*, after the Italian mathematician, Enrico Betti. This was later formalised by the French mathematician Henri Poincaré. It was explained still later by Emmy Noether that Betti numbers are best defined by associating to X an abelian group $H^i(X)$ in each dimension i and defining b_i to be the rank of this group. The group $H^i(X)$ is itself called the *i-th cohomology group* of X.

The sequence of Betti numbers b_i can be kept track of by introducing the power series $\sum b_i t^i$. If X is of finite dimension n, the Betti numbers b_i are 0 for all $i > n$, so that in that case this power series is actually a polynomial called the *Poincaré polynomial*. It is interesting to note that the Poincaré polynomial of the product of two topological spaces is the product of those of the component spaces.

For the sphere of n dimension, b_i is 0 for $i \neq 0$, or n, and $b_0 = b_n = 1$. In other words its Poincaré polynomial is $1 + t^n$. In particular, the circle S^1 has the Poincaré polynomial $1 + t$.

Another example is that of the complex projective space $\mathbf{CP}^n$. Its real dimension is $2n$ and its Betti numbers are 1 for even $i \leq 2n$ and 0 otherwise. Again, this only means that the Poincaré polynomial of $\mathbf{CP}^n$ is

$$\sum_{i=0}^{i=n} t^{2i} = \frac{1 - t^{2n+2}}{1 - t^2}.$$

If we take the infinite dimensional projective space the Poincaré power series is $\frac{1}{1-t^2}$.

Actually the Poincaré power series of S^1 and that of $\mathbf{CP}^\infty$ are related in an organic fashion. Note that S^1 is a group as well. If G is any compact connected Lie group, its Poincaré polynomial is of the form $\Pi_{i=1}^{i=l}(1 + t^{2m_i-1})$. The integer l is called the *rank* of G. For S^1 we have $l = 1$ and $m_1 = 1$. One can associate to any such group G a topological space called its *classifying space* B_G. Although this space is not unique or canonical, its Poincaré power series is unique and is in fact $\Pi \frac{1}{1-t^{2m_i}}$. This fact plays a fundamental role in the theory of *characteristic classes*.

Diophantine Equations

Algebraic Geometry deals with loci defined by polynomial equations. If the polynomials in question have integral coefficients, then one may look for integral solutions which may or may not exist. Such problems were first handled by the Greek mathematician Diophantus. He was born in Alexandria and wrote a book called 'Arithmetic'. This was translated in many languages at that time. The Indian mathematician Brahmagupta of ancient times also made many important contributions to this area.

We will first give an example which is popularly known.

Consider the equation $x^2 + y^2 = z^2$. We can ask for integral solutions for this equation. There are infinitely many solutions and are known as *Pythagorian triples* since they are potentially sides of integral length of a right angled triangle. Indeed all solutions can be explicitly written down as $(p^2 - q^2)/2, pq, (p^2 + q^2)/2$, for any pair of integers p, q, both odd or both even.

The corresponding equation of the third degree, namely $x^3 + y^3 = z^3$ was considered by Fermat, who proved that apart from trivial solutions in which one of x, y, z is zero, there are none. He also claimed that he had a proof that there are no non-trivial solutions of equations $x^n + y^n = z^n$ for all $n > 3$ as well. It is generally believed that he did not actually prove it. In any case, his claim was correct as was settled by Andrew Wiles in the 90's. Again Wiles had to use all the modern tools of mathematics in settling the question.

Weil Conjectures

Suppose we have homogeneous polynomial equations over rational numbers. We consider the set of solutions of these equations but identify two solutions which are obtained, one from the other by multiplying by a (nonzero) scalar. In other words, we look for solutions in the projective space. We may actually interpret this as finding solutions in integers for a set of polynomial equations with integral coefficients, by simply clearing the denominators.

Consider the equation

$$Y^2 = X^3 - X + 2.$$

This does not have a solution as we see by reading it modulo 3. In $\mathbf{F}_3$, all elements x satisfy $x^3 - x = 0$, while 2 is not a square in $\mathbf{F}_3$.

Whenever we have an equation with integral coefficients as above, it is therefore fruitful to ask whether there are solutions when all coefficients are read modulo a prime p. In other words, we will then have an equation over the *field* $\mathbf{F}_p$. If there is no solution there, then there cannot be a solution in integers either. Indeed, by extending the field to any finite field $\mathbf{F}_{p^n}$ we can look for solutions there also.

So we start with homogeneous polynomial equations with coefficients in any finite field $\mathbf{F}_q$. We seek solutions in $\mathbf{F}_q\mathbf{P}^r$ as before. By extending the field to $\mathbf{F}_{q^n}$, we look for solutions there as well. In any case, the solutions are contained in $\mathbf{F}_{q^n}^{r+1}$ and so are finite in number. Let v_n be the number of solutions over $\mathbf{F}_{q^n}$. Then we consider the generating function of this sequence. One may thus associate to this sequence the generating series $\sum v_n t^n$. But in practice one actually takes the series $\sum \frac{v_n}{n} t^n$. Moreover one exponentiates this and defines

$$\zeta(V, t) = exp\left(\sum \frac{v_n}{n} t^n\right)$$

Weil made many conjectures about the zeta function of smooth projective varieties. The first is that it is a rational function. This was first proved to be true by Dwork.

So one may write this function of t as $\frac{\Pi(1-t\alpha_j)}{\Pi(1-t\beta_k)}$. The numbers α_j (resp. β_k) are all algebraic integers and their moduli are of the form $q^{r/2}$ where r is a positive odd (resp. even) integer $\leq 2n$. He also conjectured a functional equation for the function.

The functional equation is the following:

$$\zeta(1/(q^n.t)) = q^{n.\chi/2} t^\chi \zeta(t)$$

where χ is $deg(den) - deg(num)$. It is natural to group all the terms $1 - t\alpha_j$ with $|\alpha_j| = q^{r/2}$ and define $f_r = \prod(1 - t\alpha_j)$ with those factors. Thus $\zeta = (f_1 \cdot f_3 \cdots f_{2n-1})/(f_0 \cdot f_2 \cdots f_{2n})$.

Then Weil also proposed a general scheme in which the problem may be tackled and all his conjectures proved. To this end he proposed that there ought to be a theory of cohomology of algebraic varieties, over algebraically closed fields of characteristic $p > 0$ with properties analogous to those over complex numbers. If it also has properties like Fixed Point theorems, which are true for the classical theory, then his conjectures should be a consequence.

What kind of topology can one use on complex varieties which would make sense for varieties defined over fields of characteristic $p > 0$ as well? Zariski introduced, more as a sort of language, a topology on algebraic varieties. The idea is to have just enough open sets for polynomials to be continuous. Formally, one defines a subset of a variety V in $\mathbf{C}^n$ to be open if its complement is the set of zeros of a set of polynomial functions on V. For any variety in general, one declares a set to be open if its intersection with any affine subvariety is open. Although this topology is very weak, it has at least the merit of codifying what used to be called the *principle of irrelevance of algebraic inequalities* For example, if one wishes to prove the Cayley-Hamilton theorem, namely, any (n, n) matrix satisfies its characteristic polynomial, one may assume that the characteristic polynomial has no multiple roots (a condition which is defined by algebraic inequalities) and therefore diagonalisable. For diagonal marices, the assertion is obvious!

But then the Zariski topology seems to be very weak to be of much serious use. However Serre showed that this topology is good enough to be used in what is known as the *theory of cohomology of coherent sheaves*. So the question arises if there is a modification of this topology which can be used for defining cohomology of the variety *á la* Poincaré.

Consider the simplest case, namely when the variety is $\mathbf{C}$. Proper Zariski open sets are complements of finite sets. So neigbourhoods in this sense are hardly ever simply connected. Good cohomology theories on topological spaces require that locally the spaces be contractible! Notice however that the universal covering spaces of the complement of finite sets are indeed contractible. Is it possible then that the lack of simple connectivity is the only fly in the ointment? If so, how can one redress this defect?

If coverings of open subsets are also considered to be open sets, then Grothendieck surmised that Betti numbers can be defined and computed.

The way this was achieved by Grothendieck is amazing. To this end he re-defined the notion of topology itself. Roughly speaking, he decreed that open sets of a topological space need not be subsets! With a slight alteration, *étale topology* was born. The key thing to check is that if a smooth projective variety is defined by equations with integral coefficients, then the traditional Betti numbers of the complex variety defined by the same equations, is the same as those obtained via étale topology after taking the equations modulo prime p, and passing to its algebraic closure, at any rate, for most primes.

The final result relating to the absolute value of the roots of numerator and the denominator and the degree of the decomposition of the numerator and the denominator being the Betti numbers was settled by Deligne.

I mentioned earlier that Deligne proved the Ramanujan conjecture. Actually what he did was to apply the Weil conjecture to a suitable object and later proved the Weil conjecture itself. For this great achievement, he was awarded the Fields medal.

This short account is intended to highlight how different pieces of mathematics come together and why the study of mathematics as a unified discipline is so much more enriching. While one specialises in certain branches of mathematics, one ought to keep the perspective of the whole of mathematics as a single pursuit.

CHAPTER 6

Approximating Continuous Functions by Polynomials

Gautam Bharali

Department of Mathematics, Indian Institute of Science,
Bangalore 560 012, India.
e-mail: bharali@math.iisc.ernet.in

The aim of this article is to discuss some ideas relating to the following basic question: *How well can we approximate a continuous function on an interval of the real line by simpler functions?* This question has given rise to a flourishing area of research. Perhaps the best-known answer to this question (and there is *more than one answer* to this question depending on the precision of the answer one seeks and on the application that motivated the question) is the Weierstrass Approximation Theorem. It says:

Theorem 1. *(Weierstrass, 1885) Let f be a continuous real-valued function defined on a closed interval $[a, b]$, and let $\varepsilon > 0$ be given. Then, there exists a polynomial p with real coefficients such that $|f(x) - p(x)| < \varepsilon$ for every $x \in [a, b]$.*

Polynomials are among the simplest smooth functions that we work with. Yet, given a function f as described in Theorem 1—regardless of how non-smooth f is—Weierstrass' theorem says that there is a polynomial whose values are as close to the values of f as we choose. Expressed this way, Theorem 1 is a powerful result that requires a careful proof. However, apart from giving a proof, we shall treat Weierstrass' theorem as a backdrop for the following discussions:

- "natural" approaches to approximation of continuous functions by polynomials;

- techniques involved in the proof of Weierstrass' theorem that are of great utility in many other fields of mathematics;

- the notion of a "good approximation" in practical applications.

The last item on this list subsumes the question: *Why do we care about approximation by polynomials?* Our discussions on this issue will take us just beyond the threshold of a vast arena of new ideas. These represent a mix of late-20th-century Real Analysis and insights from science and engineering. Our discussion in that section (i.e., Section 4) will be motivational; the interested reader should look up the articles cited in Section 4. However, all the earlier sections will feature complete proofs.

1   Lagrange's interpolating polynomials

Let us consider a continuous real-valued function f defined on the interval $[a, b]$. A naive strategy for producing a polynomial p that can replace f in extensive computations—provided we are in a position to tolerate a certain level of error—is to select $n + 1$ distinct points $x_0, x_1, \ldots, x_n \in [a, b]$ and demand that

$$p(x_j) = f(x_j), \quad j = 0, \ldots, n. \tag{1}$$

Producing such a polynomial is quite easy. For instance the polynomial

$$P_f(x; x_0, \ldots, x_n) := \sum_{j=0}^{n} f(x_j) \prod_{k \neq j} \frac{x - x_k}{x_j - x_k} \tag{2}$$

clearly satisfies the above requirement. Furthermore, it is the *unique* polynomial p of degree $\leq n$ satisfying (1). To see this: notice that if p is a polynomial of degree $\leq n$ that satisfies (1), then $(P_f(\cdot; x_0, \ldots, x_n) - p)$ is a polynomial of degree $\leq n$ that vanishes at $n + 1$ *distinct* points. This means that $p = P_f(\cdot; x_0, \ldots, x_n)$. The polynomial $P_f(\cdot; x_0, \ldots, x_n)$ is called **Lagrange's interpolating polynomial for f for the points $x_0, \ldots, x_n$.**

If a polynomial p satisfies (1), we say that **p interpolates f at $x_0, \ldots, x_n$.**

A naive justification of Theorem 1 might go as follows: As $P_f(\cdot\,; x_0,$
$\dots, x_n)$ *coincides with* f *at* $x_0, x_1, \dots, x_n \in [a, b]$, *we just need to to consider more points at which* f *is interpolated, and* $P_f(\cdot\,; x_0, \dots, x_n) \longrightarrow$
f *as* $n \to \infty$. One issue that is most obviously *unclear* in this argument is whether the sequence $\{P_f(\cdot\,; x_0, \dots, x_n)\}_{n \in \mathbb{N}}$—for some choice of
$x_0, x_1, x_2 \dots$—converges *uniformly* to f. It is thus pertinent to ask: *How accurate is the heuristic reasoning given above?*

To answer this question, we need some notation. Given f and $x_0, \dots, x_n$
as above, we define the **kth-order divided difference**, $0 \le k \le n$, by the recurrence relation

$$f[x_0, \dots, x_k] := \begin{cases} f(x_0), & \text{if } k = 0, \\ \dfrac{f[x_1, \dots, x_k] - f[x_0, \dots, x_{k-1}]}{x_k - x_0}, & \text{if } k > 0. \end{cases}$$

(Recall that the points $x_0, \dots, x_n$ are distinct.) Let us now explore the connection between the kth-order divided differences and Lagrange's interpolating polynomials. Expressing Lagrange's interpolation polynomial for x_0 and x_1 in terms of divided differences is simple:

$$P_f(x; x_0, x_1) = f(x_0)\frac{x - x_1}{x_0 - x_1} + f(x_1)\frac{x - x_0}{x_1 - x_0}$$
$$= f[x_0] + f[x_0, x_1](x - x_0).$$

Next, referring to the property (1), we realise that the polynomial

$$P_f(\cdot\,; x_0, \dots, x_{k+1}) - P_f(\cdot\,; x_0, \dots, x_k)$$

vanishes at the points $x_0, \dots, x_k, k \ge 0$. However, since

$$\deg(P_f(\cdot\,; x_0, \dots, x_{k+1}) - P_f(\cdot\,; x_0, \dots, x_k)) \le k + 1,$$

we conclude that there is a constant $A_k \in \mathbb{R}$ such that

$$P_f(x; x_0, \dots, x_{k+1}) - P_f(x; x_0, \dots, x_k) = A_k(x - x_0)\dots(x - x_k) \quad \forall x \in \mathbb{R}.$$
$$(3)$$

The above identity brings us to a point where we can pose the following puzzle:

Exercise 1. Use mathematical induction and the identity (3) to show that

$$P_f(x; x_0, \ldots, x_n) = f[x_0] + \sum_{j=1}^{n} f[x_0, \ldots, x_j] \prod_{k=0}^{j-1} (x - x_k) \quad \forall x \in \mathbb{R}, \ n \geq 1.$$

(4)

Use the above identity to argue that if p is any polynomial with $\deg(p) \leq (n - 1)$, then $p[x_0, \ldots, x_n] = 0$ for any choice of distinct points $x_0, \ldots, x_n \in \mathbb{R}$.

Note the equivalence

$$|f(x) - p(x)| < \varepsilon \text{ for every } x \in [a, b] \iff \sup_{x \in [a,b]} |f(x) - p(x)| < \varepsilon.$$

This motivates the following terminology. Denote

$$\mathcal{P}_n := \text{the set of all real polynomials of degree } \leq n.$$

Then, a polynomial $P \in \mathcal{P}_n$—if such a polynomial exists—is called **the best uniform approximation of f in $\mathcal{P}_n$** if

$$\sup_{x \in [a,b]} |f(x) - p(x)| \geq \sup_{x \in [a,b]} |f(x) - P(x)| \quad \forall p \in \mathcal{P}_n. \quad (5)$$

We state for the record the following fact:

Theorem 2 (Tchebyshev). *Let f be a continuous real-valued function defined on a closed interval $[a, b]$. For each $n \in \mathbb{N}$, there is a unique best uniform approximation of f in $\mathcal{P}_n$.*

We shall omit the proof of the above theorem, which is quite technical. The reader will find a proof of Theorem 2 in (Rivlin, 2003, Section 1.2). We know that the class of continuous real-valued functions on a $[a, b]$ is metrized by the function $d(f, g) := \sup_{x \in [a,b]} |f(x) - g(x)|$. Thus, the number on the right-hand side of (5) is just the distance of f from the subset $\mathcal{P}_n$. We thus denote this number by $\mathrm{dist}_{[a,b]}(f, \mathcal{P}_n)$.

What now follows in this section is largely the work of Tchebyshev, as is Theorem 2. Note that, in view of the second part of Exercise 1, for any $p \in \mathcal{P}_n$, we know that $p[x_0, \ldots, x_{n+1}] = 0$ for any choice of distinct points $x_0, \ldots, x_{n+1} \in [a, b]$. Thus,

$$f[x_0, \ldots, x_{n+1}] = f[x_0, \ldots, x_{n+1}] - p[x_0, \ldots, x_{n+1}]$$

$$= (f - p)[x_0, \ldots, x_{n+1}]$$

$$= \sum_{j=0}^{n+1} \frac{f(x_j) - p(x_j)}{w'_{n+1}(x_j)},$$

where $w_{n+1}(x) := (x - x_0)(x - x_1) \cdots (x - x_{n+1})$. The last equality follows from algebraic manipulations. The above gives us the simple inequality:

$$\left| f[x_0, \ldots, x_{n+1}] \right| \leq \sup_{x \in [a,b]} |f(x) - p(x)| \cdot W(x_0, \ldots, x_{n+1}), \qquad (6)$$

where $W(x_0, \ldots, w_{n+1}) = \sum_{j=0}^{n+1} 1/|w'_{n+1}(x_j)|$. Since (6) holds true for *any* arbitrary $p \in \mathcal{P}_n$, it must hold true for the unique best uniform approximation of f in $\mathcal{P}_n$. But that tells us:

$$\left| f[x_0, \ldots, x_{n+1}] \right| \leq \mathrm{dist}_{[a,b]}(f, \mathcal{P}_n) \cdot W(x_0, \ldots, x_{n+1}). \qquad (7)$$

Let us now fix n distinct points $x_0, x_1, \ldots, x_n \in [a, b]$. We now appeal to the identity (4) with:

- $n + 1$ in place of n in that identity;
- the $(n + 1)$st point being any *arbitrarily chosen point* $x \in [a, b]$ different from $x_0, x_1, \ldots, x_n$.

Making these substitutions in (4), we get

$$f(x) - P_f(x; x_0, \ldots, x_n) = P_f(x; x_0, \ldots, x_n, x) - P_f(x; x_0, \ldots, x_n)$$
$$= f[x_0, \ldots, x_n, x](x - x_0) \cdots (x - x_n). \qquad (8)$$

Combining the above with (7), we get

$$|f(x) - P_f(x; x_0, \ldots, x_n)| \leq |x - x_0| \cdots |x - x_n| \cdot W(x_0, \ldots, x_n, x)$$
$$\cdot \mathrm{dist}_{[a,b]}(f, \mathcal{P}_n)$$

$$= \left\{ 1 + \sum_{j=0}^{n} \prod_{k \neq j} \left| \frac{x - x_k}{x_j - x_k} \right| \right\} \mathrm{dist}_{[a,b]}(f, \mathcal{P}_n)$$

$$\equiv (1 + \Lambda(x; x_0, \ldots, x_n)) \cdot \mathrm{dist}_{[a,b]}(f, \mathcal{P}_n)$$

for all $x \in [a, b]$. It now follows that

$$\sup_{x\in[a,b]} |f(x) - P_f(x; x_0,\ldots,x_n)| \le \left(1 + \sup_{x\in[a,b]} \Lambda(x; x_0,\ldots,x_n)\right)$$
$$\cdot \operatorname{dist}_{[a,b]}(f,\mathcal{P}_n). \tag{9}$$

The above estimate is very revealing from the viewpoint of the question we had raised earlier in this section.

The behaviour of the function $\Lambda(\cdot; x_0,\ldots,x_n)$ depends on the choice of the points $x_0,\ldots,x_n$, but $\sup_{x\in[a,b]} \Lambda(x; x_0,\ldots,x_n)$ increases with n, irrespective of the choice of of the points $x_0, x_1,\ldots,x_n$. We could try to understand how $\Lambda(\cdot; x_0,\ldots,x_n)$ behaves for n equally spaced points, i.e., when

$$x_j = a + j\left(\frac{b-a}{n}\right) =: x_j^{eq}, \quad j = 0,\ldots,n.$$

Rivlin has shown in (Rivlin, 2003, Section 1.2) that, in fact

$$\sup_{x\in[a,b]} \Lambda(x; x_0^{eq},\ldots,x_n^{eq}) \ge e^{n/2} \quad \forall n \in \mathbb{Z}_+. \tag{10}$$

This growth is *extremely rapid*, and casts doubt on whether Lagrange's interpolating polynomials provide uniform polynomial approximations in general! At this point, however, this is merely a suspicion. Let us, instead, list what is rigorously known at this point:

a) In the estimate (9), $\lim_{n\to\infty} \sup_{x\in[a,b]} \Lambda(x; x_0,\ldots,x_n) = +\infty$. (We have not yet proved this fact, but we shall see why this is true when we return to Lagrange's interpolating polynomials in Section 4.)

b) In view of (a), if we were to prove that Lagrange's interpolating polynomials uniformly approximate f by *just appealing to the estimate* (9), then we would necessarily have to show that

$$\lim_{n\to\infty} \operatorname{dist}_{[a,b]}(f,\mathcal{P}_n) = 0. \tag{11}$$

Now note that, in view of Tchebyshev's theorem (i.e., Theorem 2), the limit (11) is equivalent to Weierstrass' theorem. In other words, to answer the question raised at the beginning of this section, it does not seem promising that we can use Lagrange's interpolating polynomials to provide a constructive proof for Weierstrass' theorem.

For the sake of completeness, we must mention that Runge showed in (Runge, 1901) that Lagrange's interpolating polynomials—with equally spaced points—do *not* uniformly converge to the function $f(x) = 1/(1 + x^2)$,

$x \in [-5, 5]$. A very readable account of this phenomenon is given by Epperson (1987).

Despite this negative result, our discussion on Lagrange's interpolating polynomials must not be considered a waste. In this section, we used very elementary arguments to arrive at rather sophisticated estimates. We will revisit these, the estimate (9) in particular, in Section 4. In fact, *Lagrange's interpolating polynomials are widely used in many types of practical applications* where computations are made simpler by replacing a complicated continuous function f by suitable polynomials. This is because Lagrange's polynomials are so easy to construct. But, at a more quantitative level: why are they, in spite of Runge's counter-example, considered good approximating polynomials? We shall take up this question in Section 4, after examining a proof of Weierstrass' theorem.

2   The versatile convolution

The concluding discussion in Section 1 reveals that we will need additional ideas in order to prove Weierstrass's theorem. Since the appearance of Weierstrass' proof in Weierstrass (1885), many mathematicians have provided alternative proofs. A survey of these newer proofs can be found in the historical-mathematical survey by Pinkus (2000). A majority of these proofs bear a technical resemblance to Weierstrass' proof in that they rely upon the convolution integral.

Students who are familiar with the Lebesgue integral will quickly be able to re-cast the following discussion in a more general setting. However, *to appeal to a wider audience, this section will rely on the familiar Riemann integral*. Our first task is to clarify what we mean by the integral

$$I := \int_{\mathbb{R}} f(x) \, dx,$$

where f is a continuous real-valued function on the real line $\mathbb{R}$. Recall that the Riemann integral is classically defined only on *a closed, bounded interval*. Let us first consider the case when f is a non-negative continuous function. In this case, we say that the **Riemann integral of f exists** if the limit

$$\lim_{R \to +\infty} \int_{-R}^{R} f(x) \, dx$$

exists *and is finite*. This limit is denoted as $\int_{\mathbb{R}} f(x)\,dx$. The integral on the right-hand side above is the ordinary Riemann integral. Since, classically, the Riemann integral is defined only on closed, bounded intervals, the above integral is sometimes called the *improper Riemann integral of f*. This is the terminology Rudin uses in his textbook (Rudin, 1976, Chapter 6, Exercise 8). Now, for a general f, we define:

$$f^+(x) := \begin{cases} f(x), & \text{if } f(x) \ge 0, \\ 0, & \text{if } f(x) \le 0, \end{cases} \qquad f^-(x) := \begin{cases} 0, & \text{if } f(x) \ge 0, \\ -f(x), & \text{if } f(x) \le 0, \end{cases}$$

i.e., $f^\pm$ are the positive/negative parts of f.

Definition 1. Let f be a continuous real-valued function on $\mathbb{R}$. We say that the **Riemann integral of f exists** or, alternatively, that f **is Riemann integrable on** $\mathbb{R}$ if both the Riemann integrals

$$\int_{\mathbb{R}} f^+(x)\,dx \quad \text{and} \quad \int_{\mathbb{R}} f^-(x)\,dx$$

exist. In that case, the value of the Riemann integral of f is given as:

$$\int_{\mathbb{R}} f(x)\,dx := \int_{\mathbb{R}} f^+(x)\,dx - \int_{\mathbb{R}} f^-(x)\,dx.$$

One further point: *we can, in analogy with the above definition, define the existence of the Riemann integral on a semi-infinite subinterval of* $\mathbb{R}$ *too*; we leave the reader to carry out this exercise. We now pose a simple exercise that involves checking the above definition.

Exercise 2. Let f and g be two continuous real-valued functions on $\mathbb{R}$, and assume that the Riemann integrals of f and g exist. Assume further that g is a bounded function. Show that, for each $x \in \mathbb{R}$, the Riemann integral of the function $H_x(y) := f(x-y)g(y)$ exists. Then show that

$$\int_{\mathbb{R}} f(x-y)g(y)\,dy = \int_{\mathbb{R}} f(y)g(x-y)\,dy \quad \forall x \in \mathbb{R}.$$

The conclusions of the above exercise allow us to define the convolution.

Definition 2. Let f and g be two continuous real-valued functions on $\mathbb{R}$ that are Riemann integrable on $\mathbb{R}$, and assume that one of them is a bounded function. The **convolution** of f and g, denoted by $f * g$, is the function defined as

$$f * g(x) := \int_{\mathbb{R}} f(x - y)g(y)\, dy \quad \forall x \in \mathbb{R}.$$

In the remainder of this article, for any subset $S \subset \mathbb{R}$, $C(S)$ will denote the class of real-valued continuous functions on S. We wish to introduce notion that, at first sight, has no connection with the convolution. This notion has proved to be *extremely* useful throughout modern Real Analysis, and its usefulness can be appreciated by looking at the statements of the next two theorems. But first: consider a sequence of functions $\{K_n\}_{n \in \mathbb{Z}_+} \subset C(\mathbb{R})$. We say that $\{K_n\}_{n \in \mathbb{Z}_+}$ is an **approximation of the identity** if

(I) Each K_n is a non-negative bounded function;

(II) The Riemann integral of each K_n exists and

$$\int_{\mathbb{R}} K_n(x)\, dx = 1 \quad \forall n \in \mathbb{Z}_+;$$

(III) For any $\delta > 0$,

$$\lim_{n \to +\infty} \int_{|x| \geq \delta} K_n(x)\, dx = 0.$$

We refer to the italicized comment following Definition 1 to give meaning to the integral occurring in (III).

Remark 1. Very often in the literature, the functions K_n are generated by a single function. By this we mean that we consider a continuous function $\chi : \mathbb{R} \longrightarrow \mathbb{R}$ that satisfies conditions (I) and (II) above with χ replacing K_n. For each $\varepsilon > 0$, we set

$$\chi_\varepsilon(x) := \varepsilon^{-1}\chi(x/\varepsilon) \quad \forall x \in \mathbb{R}.$$

Then, defining $K_n := \chi_{1/n}$, $n \in \mathbb{Z}_+$, we can easily see that the sequence $\{K_n\}_{n \in \mathbb{Z}_+}$ thus defined has the properties (I)–(III).

Let us try to imagine what the graphs of the K_n's look like. Property (III) suggests that there is less and less area under the "tails" of the graphs of the K_n's as $n \to +\infty$. However, in view of (I), since the area under

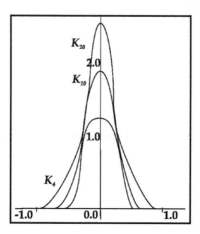

Figure 1: *Graphs of a few Landau functions*

the graph of each K_n equals 1, we deduce that the graphs of the K_n's have successively sharper and higher peaks around 0. A drawing of the graphs of some of the K_n's, with

$$K_n(x) := \begin{cases} C_n(1 - x^2)^n, & \text{if } x \in [-1, 1], \\ 0, & \text{if } x \notin [-1, 1], \end{cases}$$

where $C_n > 0$ are so defined that condition (II) is satisfied, is given in Figure 1. The importance of an approximation of the identity is partly sensed from the following theorem.

Theorem 3. *Let f be a bounded, continuous real-valued function on $\mathbb{R}$ that is Riemann integrable on $\mathbb{R}$, and let $\{K_n\}_{n \in \mathbb{Z}_+} \subset C(\mathbb{R})$ be an approximation of the identity. Given any a, b such that $-\infty < a < b < +\infty$, $K_n * f$ converges uniformly to f on $[a, b]$ as $n \to +\infty$.*

Proof. For any fixed $x \in [a, b]$, property (II) tells us that

$$K_n * f(x) - f(x) = \int_{\mathbb{R}} (f(x - y) - f(x)) K_n(y) \, dy \quad \forall n \in \mathbb{Z}_+. \tag{12}$$

Suppose we have been given $\varepsilon > 0$. We have to show that there exists an $N \in \mathbb{Z}_+$ such that

$$|K_n * f(x) - f(x)| < \varepsilon \quad \forall x \in [a, b] \quad \text{whenever } n \geq N. \tag{13}$$

Since the interval $[a - 1, b + 1]$ is a compact set, $f|_{[a-1,b+1]}$ is uniformly continuous. Hence, there exists a $\delta \in (0, 1/2)$—that is *independent of x and y*—such that

$$|f(x - y) - f(x)| < \varepsilon/2 \quad \forall x \in [a, b], \; \forall y : |y| \leq \delta. \tag{14}$$

By hypothesis, there is a number $M \in (0, +\infty)$ such that $|f(t)| < M \; \forall t \in \mathbb{R}$. We now interpret condition (III) quantitatively: using the $\delta > 0$ occurring in (14), we can find an $N \in \mathbb{Z}_+$ such that

$$\int_{|y| \geq \delta} K_n(y) \, dx < \frac{\varepsilon}{4M} \quad \forall n \geq N. \tag{15}$$

In view of (12), we have

$$|K_n * f(x) - f(x)| \leq \int_{\mathbb{R}} |f(x - y) - f(x)| K_n(y) \, dy$$

$$= \int_{|y| \leq \delta} |f(x - y) - f(x)| K_n(y) \, dy$$

$$+ \int_{|y| \geq \delta} |f(x - y) - f(x)| K_n(y) \, dy.$$

Applying (14) and (15) to the above inequality, we get

$$|K_n * f(x) - f(x)| < \frac{\varepsilon}{2} \int_{|y| \leq \delta} K_n(y) \, dy + \int_{|y| \geq \delta} (|f(x - y)| + |f(x)|) \, K_n(y) \, dy$$

$$\leq \frac{\varepsilon}{2} + \int_{|y| \geq \delta} 2M K_n(y) \, dy$$

$$< \varepsilon \quad \text{whenever } n \geq N.$$

We have just established (13), which is what was to be proved. □

Given a function $f : \mathbb{R} \longrightarrow \mathbb{R}$, we define the **support of f** as

$$\text{supp}(f) := \{x \in \mathbb{R} : f(x) \neq 0\}.$$

One of the many properties of the convolution that make it so useful in many areas of mathematics is that if f is a continuous function that is Riemann integrable on $\mathbb{R}$, and g is a *continuously differentiable* function of compact support, *then f * g is also continuously differentiable*, regardless of whether or not f is itself differentiable. In particular

$$\frac{d(f * g)}{dx}(x) = \int_{\mathbb{R}} f(y)g'(x - y)\, dy \quad \forall x \in \mathbb{R}. \tag{16}$$

Note that our conditions on g ensure that g' is a bounded, continuous function having compact support (and is hence Riemann integrable on $\mathbb{R}$). Thus, the right-hand side of (16) is well-defined. The reason for (16) is actually quite elementary. We appeal to the second part of Exercise 2 to simplify the following difference quotient:

$$\frac{f * g(x + h) - f * g(x)}{h} = \int_{\mathbb{R}} f(y)\frac{g(x + h - y) - g(x - y)}{h}\, dy$$

$$= \int_{\mathbb{R}} f(y)g'(x - y + \xi)\, dy \tag{17}$$

for some ξ between 0 *and h*, for any *fixed* $x \in \mathbb{R}$ and $h \neq 0$. The second equality follows from the fact that our conditions on g allow us to invoke the Mean Value Theorem. Let us now keep x fixed and let h approach 0. Suppose supp(g) $\subset [-R, R]$ for some $R > 0$. Then,

- for any h such that $0 < |h| < 1$, the second integral on the right-hand side above is just the classical Riemannn integral on the the interval $[-1 - R + x, 1 + R + x]$ (x fixed);

- owing to the continuity of g', the integrand of the second integral above converges uniformly to $f(y)g'(x - y)$ as $h \to 0$.

The last two remarks enable us to apply the familiar Uniform Convergence Theorem to (17), which gives us

$$\lim_{h \to 0} \frac{f * g(x + h) - f * g(x)}{h} = \int_{\mathbb{R}} f(y)g'(x - y)\, dy,$$

which is just the identity (16).

Let us, in Theorem 3, now pick an approximation of the identity $\{K_n\}_{n \in \mathbb{Z}_+}$ in which each K_n has compact support and is *infinitely differentiable*. Armed with the above argument, we merely have to apply mathematical induction to get:

Theorem 4. *Let f be a bounded, continuous real-valued function on* $\mathbb{R}$ *that is Riemann integrable on* $\mathbb{R}$, *and let a, b such that* $-\infty < a < b < +\infty$.

1) *Let* $\{K_n\}_{n\in\mathbb{Z}_+}$ *be an approximation of the identity in which each* K_n *is compactly supported and infinitely differentiable. Then, each* $K_n * f$ *is also infinitely differentiable.*

2) *Let* $\varepsilon > 0$ *be given. Then, there exists an infinitely differentiable function* Ψ *such that* $|f(x) - \Psi(x)| < \varepsilon$ *for every* $x \in [a, b]$.

The reader is urged to compare the statement of Theorem 1 with Part (2) of Theorem 4. The latter theorem says that—restricted to a compact interval — we can find a smooth function whose values are as close to the values of f as we choose. Loosely speaking, we "almost" have Weierstrass' theorem at this juncture!

What we have just demonstrated represents one of the powers of the convolution integral. There is a general philosophy here: *convolution with an appropriate approximation of the identity produces smooth approximinants.* In fact, there is no obstacle to generalising all of the above arguments to $\mathbb{R}^n$, $n \geq 1$. But one can do a *lot* more. For readers who are familiar with the Lebesgue integral, here is a sample of what the convolution can help to prove:

Theorem 5. *Let* f *be function on* $\mathbb{R}^n$ *of class* $\mathbb{L}^p(\mathbb{R}^n)$, $1 \leq p < \infty$.

1) *Let* $\{K_n\}_{n\in\mathbb{Z}_+}$ *be any approximation of the identity. Then,* $K_n * f \longrightarrow f$ *in* $\mathbb{L}^p$-*norm as* $n \to +\infty$.

2) *The class* $C_c^\infty(\mathbb{R}^n)$ *of infinitely differentiable, compactly supported functions is dense in* $\mathbb{L}^p(\mathbb{R}^n)$, $1 \leq p < \infty$.

Primarily because of its ability to produce smooth approximinants, the convolution is encounterd in a wide range of disciplines: not just in Real Analysis, but also in Complex Analysis, Electrical Engineering and Partial Differential Equations. Based on slightly different motivations (which we have not discussed here), there are discrete analogues of the convolution that are important in Arithmetic and Number Theory.

3   The proof of Weierstrass' theorem

In this section, we shall provide a proof of Theorem 1. As mentioned earlier, many mathematicians provided alternative proofs of Theorem 1 after the appearance of Weierstrass' proof in (Weierstrass, 1885). The proof that we are about to provide was given by Landau (1908). We choose this proof because it is very close to Weierstrass' original proof in spirit, but features a number of simplifications. Most notably, *absolutely no effort is required to see that the convolutions constructed are polynomials* in Landau's proof.

The proof of Weierstrass' theorem: This proof will simply be an application of Theorem 3. We first note that it suffices to prove the theorem for $[a, b] = [0, 1]$ and for a function f satisfying $f(0) = f(1)$. This is because, if we have proved this special case, we would then have—for the function

$$F(y) := f((b - a)y + a) - f(a) - y(f(b) - f(a)), \quad y \in [0, 1],$$

and for a given $\varepsilon > 0$—a polynomial P with real coefficients such that $|F(y) - P(y)| < \varepsilon$ for every $y \in [0, 1]$. Simple algebra reveals that for the new polynomial

$$p(x) := P\left(\frac{x - a}{b - a}\right) + f(0) + (f(b) - f(a))\left(\frac{x - a}{b - a}\right), \quad x \in [a, b],$$

we have $|f(x) - p(x)| < \varepsilon$ for every $x \in [a, b]$, which is precisely the desired conclusion. Hence, the above simplifying assumptions will be in effect for the remainder of the proof.

We extend the function f to $\mathbb{R}$—calling that extension $\widetilde{f}$—as follows

$$\widetilde{f}(x) := \begin{cases} f(x), & \text{if } x \in [0, 1], \\ 0, & \text{if } x \notin [0, 1]. \end{cases}$$

It is easy to see that $\widetilde{f}$ is continuous on $\mathbb{R}$ and that its Riemann integral exists ($\widetilde{f}$ is compactly supported!). Next, we propose a candidate for an approximation of the identity:

$$K_n(x) := \begin{cases} C_n(1 - x^2)^n, & \text{if } x \in [-1, 1], \\ 0, & \text{if } x \notin [-1, 1], \end{cases}$$

where $C_n > 0$ are so defined that condition (II) is satisfied. These functions are called **Landau functions**. Figure 1 shows the graphs of a few of the Landau functions.

Observe that, simply by construction, the family $\{K_n\}_{n \in \mathbb{Z}_+}$ satisfies conditions (I) and (II) for being an approximation of the identity. Figure 1 on the reveals that the K_n's peak at 0 in just the way one would expect from an approximation of the identity. However, we need to establish property (III) rigorously; appealing to a picture is not rigorous enough to be a proof!

To establish that $\{K_n\}_{n\in\mathbb{Z}_+}$ satisfies condition (III), we need an estimate of the size of C_n in terms of n. Since

$$\int_{\mathbb{R}} (1-x^2)^n\, dx = 2 \int_0^1 (1-x^2)^n\, dx$$

$$\geq 2 \int_0^{1/\sqrt{n}} (1-x^2)^n\, dx$$

$$\geq 2 \int_0^{1/\sqrt{n}} (1-nx^2)\, dx = \frac{4}{3\sqrt{n}},$$

it follows by definition that

$$C_n < \sqrt{n} \qquad \forall n \in \mathbb{Z}_+. \tag{18}$$

Let us choose a $\delta > 0$. Note that for $\delta \geq 1$, the integrals in (III) are already 0. Hence, let $\delta \in (0,1)$. Then

$$\int_{|x|\geq\delta} K_n(x)\, dx = \int_{-1}^{-\delta} C_n(1-x^2)^n\, dx + \int_\delta^1 C_n(1-x^2)^n\, dx$$

$$< 2\sqrt{n} \int_\delta^1 (1-x^2)^n\, dx$$

$$\leq 2(1-\delta)\sqrt{n}(1-\delta^2)^n$$

$$= 2(1-\delta)\sqrt{n}\exp\left(-|\log(1-\delta^2)|n\right) \longrightarrow 0 \quad as\ n \to +\infty.$$

The first inequality above is a consequence of (18), while the limit above is zero because the exponential decays to zero faster than $\sqrt{n}$ (or, for that matter, any positive power of n) diverges to $+\infty$. This establishes that $\{K_n\}_{n\in\mathbb{Z}_+}$ satisfies condition (III).

Finally, let us define $p_n(x) := K_n * \widetilde{f}(x) \forall x \in \mathbb{R}$. Note that the properties of K_n and $\widetilde{f}$ ensure that the convolution is well-defined. By Theorem 3, $K_n * \widetilde{f}$ converges uniformly to f on $[0,1]$ as $n \to +\infty$. The conclusion of Weierstrass' theorem is just a quantitative restatement of the above fact. Let us write out explicitly:

$$p_n(x) = \int_{\mathbb{R}} \widetilde{f}(x-y)K_n(y)\, dy = \int_{x-1}^x f(y-x)K_n(y)\, dy.$$

Making the change of variable $u := x - y$ in the above equation, and then relabelling each u as y, we get

$$p_n(x) = C_n \int_0^1 f(y)(1-(y-x)^2)^n\, dy \qquad \forall x \in \mathbb{R}.$$

We can expand out the integrand of the right-hand side in the above equation using the Binomial Theorem, and it becomes obvious that each p_n is a polynomial. This completes the proof. □

Before we close this section, let us establish a connection between this section and Section 1. Recall that, in view of Tchebyshev's theorem, Weierstrass' theorem is equivalent to the following statement:

$$\lim_{n\to\infty} \text{dist}_{[a,b]}(f, \mathcal{P}_n) = 0, \tag{19}$$

where f is exactly as in the hypothesis of Weierstrass' theorem. It is natural to ask: *How rapidly, in terms of n, does* $\text{dist}_{[a,b]}(f, \mathcal{P}_n)$ *decay?* This question is not just an exercise in idle puzzle-solving. Quantitative versions of Theorem 1 have practical consequences for the design of approximation algorithms. For instance, we will see in the next section an impact of this question on Lagrange's interpolation polynomials. To answer this question, we need a definition.

Definition 3. Let f be a continuous real-valued function defined on the closed interval $[a, b]$. The **modulus of continuity of** f is the function $\omega_f :$ $(0, +\infty) \to [0, +\infty)$ that, for each $\delta > 0$, measures the greatest possible separation between $f(x)$ and $f(y)$ among all $(x, y) \in [a, b] \times [a, b]$ satisfying $|x - y| \leq \delta$. Specifically

$$\omega_f(\delta) := \sup\{|f(x) - f(y)| : x, y \in [a, b], |x - y| \leq \delta\}.$$

Jackson provided the following answer to the question above:

Theorem 6 (Jackson). *There exists a constant $C > 0$ that depends only on a and b such that for any f that is continuous on the closed interval $[a, b]$,*

$$\text{dist}_{[a,b]}(f, \mathcal{P}_n) \leq C\omega_f\left(\frac{b-a}{n+1}\right). \tag{20}$$

The proof of Jackson's theorem uses a number of technical facts. However, the interested reader can find a comparatively streamlined proof of Theorem 6 in (Bojanic and DeVore, 1969).

4 Polynomial approximations in practice

We return to some of the issues introduced in Section 1. We had asserted, at the end of that section, that *despite Runge's counterexample*, Lagrange's interpolating polynomials (in some form) are still widely used. The simplicity of their construction is one reason for their popularity. We shall now examine this assertion more rigorously. We shall bring two different perspectives to bear upon this examination.

The first perspective is the more theoretical one. We refer to the estimate (9) and the observation about the function $\Lambda(\cdot\,; x_0, \ldots, x_n)$ that immediately follows it. We are motivated to ask: *Given $n \in \mathbb{Z}_+$, is there an arrangement of the points $x_0, x_1, \ldots, x_n \in [a, b]$ so that the quantity* $\sup_{x \in [a,b]} \Lambda(x\,; x_0, \ldots, x_n)$ *is minimised?*. This question has a very elegant answer. To understand the answer, we first consider the interval $[-1, 1]$. The **Tchebyshev polynomial of degree n**, denoted by T_n, is determined globally by defining $T_n|_{[-1,1]}$ by the rule

$$T_n(\cos \theta) = cos(n\theta).$$

The first few Tchebyshev polynomials are:

$$T_0(x) = 1, \quad T_1(x) = x,$$
$$T_2(x) = 2x^2 - 1, \quad T_3(x) = 4x^3 - 3x,$$
$$T_4(x) = 8x^4 - 8x^2 + 1, \quad T_5(x) = 16x^5 - 20x^3 + 5x.$$

It is easy to see, by the rule above, that all the zeros of T_n, $n \geq 1$, lie in $[-1, 1]$. It turns out that $\sup_{x \in [-1,1]} \Lambda(x\,; x_0, \ldots, x_n)$ is minimised when $x_0, x_1, \ldots, x_n$ are the zeros of the Tchebyshev polynomial of degree $n + 1$, i.e., when

$$x_j = \cos\left(\frac{2j+1}{2n+2}\pi\right), \quad j = 0, \ldots, n.$$

One way to deduce this fact is to use the method of Lagrange multipliers. We omit the long, but essentially elementary, calculations leading up to this fact. The relevant details can be found in (Rivlin, 2003, Section 1.2). For a general interval, the **Tchebyshev points of order n** adjusted for $[a, b]$, i.e.,

$$x_j := \frac{1}{2}\left[(a + b) + (a - b)\cos\left(\frac{2j+1}{2n+2}\pi\right)\right] =: x_j^T, \quad j = 0, \ldots, n,$$

give a $\Lambda(x\,; x_0, \ldots, x_n)$ whose supremun on $[a, b]$ is very close to the least possible (and which is, of course, the least possible when $[a, b] = [-1, 1]$).

In order to understand our next result, we need a definition.

Definition 4. Let $\alpha > 0$. A real-valued function f defined on the interval $[a, b]$ is said to be **Hölder continuous of order** α if there exists a number $K \in (0, +\infty)$ such that $|f(x) - f(y)| \leq K|x - y|^\alpha$ for every $x, y \in [a, b]$.

It is quite obvious that any f satisfying the Hölderian condition for some exponent $\alpha > 0$ is continuous. The functions $f_M(x) := |x|^{1/M}$, $x \in [-1, 1]$, where $M > 1$, are examples of Hölder continuous functions. The above definition is meant to account for non-differentiable functions whose graphs have cusps resembling the cusps of the graphs of the above f_M's at $x = 0$. The Hölder continuous functions that are interesting are those that are Hölder of order $\alpha \leq 1$. To see why, we pose a small puzzle:

Exercise 3. Let $[a, b]$ be some closed interval of the real line. Show that the only Hölder continuous functions of order α, for $\alpha > 1$, are the constant functions.

The connection between the Hölder continuous functions and Lagrange's interpolating polynomials is the following theorem.

Theorem 7. *Let f be a real-valued function defined on a closed interval $[a, b]$, and assume that f is Hölder continuous of order α for some $\alpha \in (0, 1)$. For each n, let $\{x_0^T, x_1^T, \ldots, x_n^T\}$ denote the Tchebyshev points of order n adjusted for $[a, b]$. The Lagrange interpolation polynomials $\{P_f(\cdot\,; x_0^T, \ldots, x_n^T)\}_{n \in \mathbb{Z}_+}$ converge to f uniformly on $[a, b]$.*

Proof. Recall the observation that $\sup_{x \in [a,b]} \Lambda(x\,; x_0, \ldots, x_n)$ increases to $+\infty$ with n, irrespective of the choice of of the points $x_0, x_1, \ldots, x_n$. This theorem hinges on the fact that $\sup_{x \in [a,b]} \Lambda(x\,; x_0^T, \ldots, x_n^T)$ grows very moderately. To be precise: there exists a universal constant $\beta > 0$ such that

$$\sup_{x \in [a,b]} \Lambda\left(x\,; x_0^T, \ldots, x_n^T\right) \leq \beta \log n \quad \forall n \in \mathbb{Z}_+; \tag{21}$$

see (Conte and de Boor, 1980, Section 6.1) for this and other properties of interpolation at the Tchebyshev points.

Let us denote by K_f the Hölder constant of f; i.e., $|f(x) - f(y)| \leq K_f|x - y|^\alpha \; \forall x, y \in [a, b]$. It is therefore clear that the modulus of continuity of f,

$$\omega_f(\delta) = K_f \delta^\alpha \quad \forall \delta > 0. \tag{22}$$

We now make use of the important estimate (9). Combining that estimate with Jackson's theorem, we get

$$\sup_{x\in[a,b]}\left|f(x) - P_f(x\,;x_0^T,\ldots,x_n^T)\right| \le \left(1 + \sup_{x\in[a,b]}\Lambda(x\,;x_0^T,\ldots,x_n^T)\right)$$
$$\cdot \operatorname{dist}_{[a,b]}(f,\mathcal{P}_n)$$
$$\le C\left(1 + \sup_{x\in[a,b]}\Lambda\left(x\,;x_0^T,\ldots,x_n^T\right)\right)\cdot\omega_f\left(\frac{b-a}{n+1}\right).$$

We now apply the inequalities (21) and (22) to the above calculation to get

$$\sup_{x\in[a,b]}\left|f(x) - P_f(x\,;x_0^T,\ldots,x_n^T)\right| \le C(1 + \beta\log n)$$
$$\cdot K_f|a-b|^\alpha(n+1)^{-\alpha} \qquad (23)$$
$$\longrightarrow 0 \quad as\ n\to+\infty.$$

The above limit is precisely the desired result. □

This result suggests a solution to a practical problem in approximation theory. The commonest version of this problem is to generate, using available computer resources, approximate values for some special function f in a given range $[a,b]$. One solution is to previously compute *and permanently store* tables of values of standard functions at the Tchebyshev points of order n, for various n, adjusted for the interval $[a,b]$. Then, depending on the desired accuracy, the appropriate table can be looked up and the formula (4) applied to compute a good approximate value of $f(x)$. That this possible to high accuracy is guaranteed by Theorem 7 (provided f is Hölder continuous of some order α).

We return to the central question: *Why do we care about approximation by polynomials?* The answer to this question is that polynomials are simple to manipulate. From the computational perspective, polynomials can be evaluated just by performing multiplications and additions, *which are the only arithmetic operations that computers know how to perform.* This makes polynomial approximinants especially valuable to designers of computational software meant for numerical manipulation of special functions. Unfortunately, despite the arithmetic simplicity of polynomials:

- multiplication is *expensive* in terms of the computer resources it requires;

- it is *a priori* unclear, because of the abstract nature of the estimate (9), as to how many points are required for $P_f(\cdot\,;x_0^T,\ldots,x_n^T)$ to be an approximation of the desired accuracy.

The reader might now wonder whether, given a real-valued function f defined on some $[a, b]$, it is ever possible to obtain a polynomial approximinant that: *i)* is explicitly known and easy to compute; and *ii)* provides an approximation of the desired accuracy! This is where the second—less rigorously justifiable, but nevertheless important—perspective must also be considered. This perspective is that even for a hard-to-compute function f, there is sufficient *coarse information* about f to be able to overcome the two hurdles just described. In this article, we have not focused on the first hurdle: i.e., minimising computational complexity posed by a large number of multiplications. However, the following problem-cum-strategy will serve well to illustrate how one overcomes the second hurdle:

Problem. *We need to work with a function f on the interval $[0, 100]$ that is Hölder continuous of known order α. Find $P_f(\cdot\,; x_0, \ldots, x_n)$, a Lagrange polynomial, such that $|f(x) - P_f(x\,; x_0, \ldots, x_n)| < 10^{-5} \; \forall x \in [1, 100]$.*

The strategy to tackle this problem rests upon the the asymptotic behaviour

$$\sup_{x \in [a,b]} \left| f(x) - P_f(x\,; x_0^T, \ldots, x_n^T) \right| \lesssim (\log n)(n + 1)^{-\alpha},$$

which we can infer from (23). Upper bounds for the constants $C, \beta > 0$—following the notation in (23)— are actually known (but the mathematics is somewhat beyond the scope of the present article). Enough coarse information about f is (generally) known for a crude estimate of K_f. In short: a *very conservative* upper bound κ for the product $\beta \cdot C \cdot K_f$ is generally available. We then pick an n^* such that $100^{\alpha}(2\kappa)(\log n^*)(n^* + 1)^{-\alpha} < 10^{-5}$. Then, by (23), $P_f(\cdot\,; x_0, \ldots, x_{n^*})$ works.

A lot of the modern innovations in the field of polynomial - approximation theory combine strategies for keeping computational complexity to a minimum (which involve arguments of a rather different flavour from the Real Analysis that we have focused on) with the idea of approximating using polynomials. The most successful insight that combines both these concerns is to approximate a given f over $[a, b]$ by *piecewise polynomial functions*. Such an approximinant is called a **spline**. It is generally accepted that the first mathematical references to a spline—meaning a piecewise polynomial approximinant—are the 1946 papers by Schoenberg (1946a,b). Another pioneering work on splines is the paper Birkhoff and de Boor (1964). The questions analysed in this paper are the ones we focused on in Section 1. Of course, the analysis in (Birkhoff and de Boor, 1964) is considerably more detailed than what we have seen. However, the ideas that we discussed here in the simpler setting of Weierstrass' theorem and Lagrange's interpolation polynomials touch upon all the analytical issues that one must worry about in studying approximation by splines. This aspect of approximation theory is a flourishing area of research today.

While the emphasis of this somewhat descriptive section has been on the issues encountered in the practical implementation of polynomial interpolation, we have seen—in the form of Theorem 7 and the illustration following it—that *one cannot do without a considerable amount of abstract reasoning when addressing these practical issues.* We conclude with a closing remark on the direction in which a more abstract appreciation of Theorem 1 has evolved. The pioneering work in this context is the *Stone–Weierstrass Theorem*, proved by Marshall Stone in (Stone, 1948). In this remarkable result, Stone abstracts the essential algebraic features of polynomials that would continue to be meaningful on a topological space, and proves an approximation theorem for continuous functions defined on a compact Hausdorff space. The strategies in (Stone, 1948), that combine algebra with analysis, have been very influential since they permit generalizations to abstract function algebras.

References

Birkhoff, G. and de Boor, C. 1964. Error bounds for spline interpolation, *J. Math. Mech.* **13:**827–835.

Bojanic, R. and DeVore, R. 1969. A proof of Jackson's theorem, *Bull. Amer. Math. Soc.* **75:**364–367.

Conte, S.D. and de Boor, C. 1980. *Elementary Numerical Analysis*, 3rd edition, McGraw-Hill.

Epperson, J.F. 1987. On the Runge example, *Amer. Math. Monthly* **94:**329–341.

Landau, E. 1908. Über die Approximation einer stetigen Funktion durch eine ganze rationale Funktion, *Rend. Circ. Mat. Palermo.* **25:**337–345.

Pinkus, A. 2000. Weierstrass and approximation theory, *J. Approx. Theory.* **107:**1–66.

Rivlin, T.J. 2003. *An Introduction to the Approximation of Functions*, 2nd Dover reprint, Dover, New York.

Rudin, W. 1976. *Principles of Mathematical Analysis*, 3rd edition, McGraw-Hill.

Runge, C. 1901. Über empirische Funktionen und die Interpolation zwischen äquidistanten Ordinaten, *Z. Math. Phys.* **46**:224–243.

Schoenberg, I.J. 1946a. Contributions to the problem of approximation of equidistant data by analytic functions, Part A: On the problem of smoothing or graduation, *Quart. Appl. Math.* **4**:45–99.

Schoenberg, I.J. 1946b. Contributions to the problem of approximation of equidistant data by analytic functions, Part B: On the problem of osculatory interpolation, *Quart. Appl. Math.* **4**:112–141.

Stone, M.H. 1948. The generalized Weierstrass approximation theorem, *Math. Mag.* **21**:167–184 and 237–254.

Weierstrass, K. 1885. Über die analytische Darstellbarkeit sogenannter willkürlicher Functionen einer reellen Veränderlichen, Sitzungsber. *Akad. Berlin.* 789–805.

Part II
Applicable Mathematics

CHAPTER 7

The Wonderful World of Eigenvalues

Arup Bose[*,**] *and Aloke Dey*[***,†]

[*]Research supported by J.C. Bose Fellowship, DST, Govt. of India
[†]Research supported by the Indian National Science Academy
under its Senior Scientist scheme
[**]Indian Statistical Institute, Kolkata 700 108, India.
[***]Indian Statistical Institute, New Delhi 110 016, India.
e-mail: `bosearu@gmail.com`, `aloke.dey@gmail.com`

1 Introduction

Linear transformations can be represented by matrices which act on vectors. Eigenvalues, eigenvectors and eigenspaces are properties of a matrix[1]. They capture all the essential properties of the matrix or the corresponding transformation. Historically, the importance of eigenvalues and the corresponding eigenvectors arose from studies in physics and in the study of quadratic forms and differential equations. The concepts of eigenvalues and eigenvectors extend to linear transformations in more general spaces. These have applications in many different areas of science; in particular, in economics, engineering, finance, quantum mechanics, mathematics and statistics.

In this article, we first describe some basic facts and mathematical results on eigenvalues and also provide the eigenvalues of k-circulants. We then provide a glimpse of a few applications–the role of eigenvalues and

---

[1]The prefix *eigen* is the German word for innate, distinct, self or proper.

eigenvectors in principal component analysis, some simulations of eigen-value distributions of a few random matrices and, application of the extended idea of eigenvalues for general linear functionals to an important result in asymptotic theory of statistics.

2 Basic concepts and results

2.1 Notation and preliminaries

For simplicity, we initially restrict attention to matrices with real entries. The symbols $\mathbb{R}^n$ and $\mathbb{R}$ will stand respectively for the n-dimensional Euclidean space and the set of reals. All vectors will be written as column vectors and denoted by bold-face lower case letters or numerals. A prime over a matrix or a vector will denote its transpose. For instance, A' is the transpose of A. For positive integers s, t, I_s is the identity matrix of order s and $\mathbf{0}_{st}$ is the $s \times t$ null matrix. The $s \times 1$ vector $\mathbf{0}_{s1}$ will be denoted simply by $\mathbf{0}_s$. The subscripts indicating the order of matrices and vectors may not always be written out when it is obvious what they are in a given context. The length (or Euclidean norm) $\|x\|$, of a vector $x = (x_1, x_2, \ldots, x_n)' \in \mathbb{R}^n$, is the positive square root of $(x_1^2 + x_2^2 + \cdots + x_n^2)$. The diagonal matrix

$$
\begin{bmatrix}
a_1 & 0 & 0 & \cdots & 0 \\
0 & a_2 & 0 & \cdots & 0 \\
\vdots & & & & \\
0 & 0 & 0 & \cdots & a_n
\end{bmatrix}
$$

will be abbreviated as $\mathrm{diag}(a_1, a_2, \ldots, a_n)$. A square matrix A is said to be *symmetric* if $A' = A$ and *idempotent*, if $A^2 = A$. For a square matrix A, $\det(A)$ and $\mathrm{tr}(A)$ will respectively, denote its determinant and trace (the sum of the entries on the principal diagonal of A). A pair of vectors $x, y \in \mathbb{R}^n$ are said to be orthogonal to each other if $x'y = 0$. A square matrix P of order n is said to be *orthogonal* if $PP' = I_n$. As a consequence, for an orthogonal matrix P, we also have $P'P = I_n$. Two $n \times n$ matrices A, B are called *similar* if there exists a non-singular matrix S such that $A = S^{-1}BS$.

2.2 Linear transformation and matrices

A transformation (or map) $T : \mathbb{R}^n \to \mathbb{R}^m$ associates to every vector $x \in \mathbb{R}^n$, a unique vector $y \in \mathbb{R}^m$. T is said to be a *linear map* if $T(\alpha x + y) = \alpha T(x) + T(y)$ for each pair $x, y \in \mathbb{R}^n$ and every $\alpha \in \mathbb{R}$.

Linear maps are intimately connected to matrices. Let $B = (b_{ij})$ be an $m \times n$ matrix and let $y = Bx$ where $x \in \mathbb{R}^n$, and consider the map

$$T_B(x) = Bx. \tag{1}$$

It is obvious that T_B is a linear transformation. In particular, any $n \times n$ matrix A may be viewed as a linear map from $\mathbb{R}^n$ into itself. Conversely, one can show the following.

Lemma 1. *Let $T : \mathbb{R}^n \to \mathbb{R}^m$ be a linear map. Then there exists an $m \times n$ matrix B such that $T(x) = Bx$ for all $x \in \mathbb{R}^n$.*

2.3   Eigenvector, eigenvalue and algebraic multiplicity

Suppose A is an $n \times n$ (real) matrix. Then a non-null $n \times 1$ vector x is called a (right) *eigenvector* (of A) if there is a scalar λ (possibly complex) satisfying

$$Ax = \lambda x. \tag{2}$$

The scalar λ is called an *eigenvalue* of A associated with the eigenvector x. From equation (2) it is clear that λ is an eigenvalue of A if and only if there is a non-null solution to the equation

$$(A - \lambda I_n)x = 0_n. \tag{3}$$

Recall that equation (3) has a non-null solution (i.e., $x \neq 0_n$) if and only if the matrix $A - \lambda I_n$ is *not* invertible, i.e., if and only if

$$\det(A - \lambda I_n) = 0. \tag{4}$$

Note that $\det(A - \lambda I_n)$ is a polynomial of degree n. It is called the *characteristic polynomial* of A and shall be denoted by $\mathrm{ch}_A(\lambda)$. Equation (4) is called the *characteristic equation* of A. Hence the eigenvalues of A are simply the roots of the characteristic polynomial and are also known as the characteristic roots of A. The *algebraic multiplicity* of an eigenvalue is its multiplicity as a root of equation (4).

If λ is an eigenvalue of A, then it turns out that there exists a (left) eigenvector y such that $y'A = \lambda y'$. This implies that A and A' have identical left and right eigenvectors corresponding to an eigenvalue. In this article all eigenvectors are taken to be right eigenvectors.

Equation (2) implies that if $\lambda > 0$, then the vector x has the property that its direction is unchanged by the transformation A and it is only scaled by a factor of λ. Geometrically, this means that eigenvectors experience changes in magnitude and sign and $|\lambda|$ is the amount of "stretch" (if $|\lambda| > 1$) or "shrink" (if $|\lambda| < 1$) to which the eigenvector is subjected to when transformed by A.

Since we are restricting attention to matrices with real entries, all coefficients of $\mathrm{ch}_A(\lambda)$ are real. Its roots of course need not necessarily be real.

Example 1. Let

$$A = \begin{bmatrix} 1 & 0 & 2 \\ 0 & -1 & 1 \\ 0 & -1 & 0 \end{bmatrix}.$$

The characteristic polynomial of A is

$$\mathrm{ch}_A(\lambda) = (1 - \lambda)(\lambda^2 + \lambda + 1).$$

Hence the eigenvalues of A are 1, $(-1 - i\sqrt{3})/2$ and $(-1 + i\sqrt{3})/2$, where $i^2 = -1$.

2.4 Eigenspace and geometric multiplicity

There may be two (or more) different eigenvectors corresponding to the same eigenvalue. If x_1 and x_2 are eigenvectors corresponding to the same eigenvalue λ of A, then for $\alpha, \beta \in \mathbb{R}$, $\alpha x_1 + \beta x_2$ is also an eigenvector with the same eigenvalue λ. In particular if x is an eigenvector then for any α, αx is also an eigenvector with the same eigenvalue.

Consider the set of eigenvectors of A corresponding to the same eigenvalue λ of A, along with the null vector. It follows from the above discussion that, these vectors form a vector subspace, called the *eigenspace* of λ. The dimension of the eigenspace of an eigenvalue is called its *geometric multiplicity*.

Recall that for any $m \times n$ matrix B, the collection of $n \times 1$ vectors $\{x\}$ satisfying $Bx = \mathbf{0}_m$ forms a subspace of $\mathbb{R}^n$, called the null space of B, denoted by $\mathcal{N}(B)$. The dimension of $\mathcal{N}(B)$ is called the nullity of B, denoted by $\nu(B)$. Furthermore, the *rank-nullity theorem* states that

$$\mathrm{Rank}(B) + \mathrm{Nullity}(B) = n, \quad \text{the number of columns in } B.$$

One can verify easily that the geometric multiplicity of an eigenvalue λ of A equals the nullity of $A - \lambda I_n$.

Example 2. Let

$$A = \begin{bmatrix} 1 & 2 & 0 \\ 0 & 1 & 3 \\ 0 & 0 & 1 \end{bmatrix}.$$

The characteristic equation of A is easily seen to be

$$(1 - \lambda)^3 = 0,$$

which shows that 1 is an eigenvalue with algebraic multiplicity 3.

The geometric multiplicity of 1 equals

$$\nu(A - I) = 3 - \text{Rank}(A - I), \text{ by the rank-nullity theorem,}$$
$$= 3 - 2 = 1.$$

Thus, here the algebraic multiplicity of 1 is strictly larger than its geometric multiplicity.

In general, the following holds.

Theorem 1. *For any eigenvalue λ, the geometric multiplicity cannot exceed the algebraic multiplicity.*

2.5   Elementary results on eigenvalues

Lemma 2. *The sum of the eigenvalues of A is $\text{tr}(A)$ and their product is $\det(A)$.*

Lemma 3. *If A is upper or lower triangular, then its entries on the principal diagonal are all its eigenvalues.*

Lemma 4. *Similar matrices have the same characteristic polynomial and hence, the same set of eigenvalues.*

Lemma 5. *Let $f(\cdot)$ be a polynomial and λ be an eigenvalue of A. Then $f(\lambda)$ is an eigenvalue of $f(A)$. Hence, if A is idempotent then each of its eigenvalue is 0 or 1.*

Lemma 6. *Let $A_{m \times n}$ and $B_{n \times m}$ be a pair of matrices, where $m \leq n$. Then, $\text{ch}_{BA}(\lambda) = \lambda^{n-m} \text{ch}_{AB}(\lambda)$.*

It immediately follows from Lemma 6 that the non-zero eigenvalues of *AB* and *BA* are same and the algebraic multiplicity of 0 as an eigenvalue of *BA* is at least $n - m$. In particular, if $m = n$ then the characteristic polynomials of *AB* and *BA* are identical.

2.6   Eigenvalues of a symmetric matrix and the spectral decomposition

As we have observed in Example 1, the eigenvalues of a (real) matrix in general can be complex numbers. However, for symmetric matrices the following holds.

Theorem 2. *If A is symmetric then all its eigenvalues are real and the eigenvectors can be chosen to have real entries.*

Theorem 3. *For a symmetric matrix, the eigenvectors are mutually orthogonal.*

We emphasize that in Theorem 3 the orthogonality holds not only for eigenvectors corresponding to distinct eigenvalues but also for eigenvectors corresponding to the eigenvalues with algebraic multiplicities greater than 1. The following example illustrates this.

Example 3. Let

$$A = \begin{bmatrix} 1 & 2 & 2 \\ 2 & 1 & 2 \\ 2 & 2 & 1 \end{bmatrix}.$$

The characteristic equation is $(\lambda + 1)^2(\lambda - 5) = 0$. Hence, the eigenvalues are $\lambda_1 = 5$ with (algebraic) multiplicity 1 and $\lambda_2 = -1$ with multiplicity 2. It is easy to see that $x_1 = (1, 1, 1)'$ is an eigenvector corresponding to λ_1. Also, it can be seen that $x_{21} = (-2, 1, 1)'$ and $x_{22} = (0, -1, 1)'$ are each an eigenvector corresponding to λ_2. The vectors x_1, x_{21} and x_{22} are clearly, pairwise orthogonal.

Theorem 4. *Let $\lambda_1, \ldots, \lambda_n$ be the eigenvalues of a symmetric matrix A, including multiplicities, and $\xi_1, \ldots, \xi_n$ be the corresponding mutually orthogonal eigenvectors each chosen to be of unit length. Then*

$$A = \lambda_1 \xi_1 \xi_1' + \cdots + \lambda_n \xi_n \xi_n'. \tag{5}$$

Such a representation of a symmetric matrix A is called its *spectral representation*. Equivalently, the above can be written as

$$A = P\Delta P' \tag{6}$$

where the matrix P, with columns $\xi_1, \ldots, \xi_n$, is orthogonal, and

$$\Delta = \text{diag}(\lambda_1, \ldots, \lambda_n).$$

From equation (6), remembering that for an $n \times n$ orthogonal matrix P, $PP' = I_n = P'P$, it is easily seen that $P'AP = \Delta$ and thus, A can be diagonalized by an orthogonal transformation. In other words, A is orthogonally similar to a diagonal matrix, the diagonal entries being the eigenvalues of A.

2.7   Quadratic forms

Quadratic forms arise in the study of conics in geometry and energy in physics. They also have wide applications in several other areas, including statistics. A quadratic form in n variables $x_1, \ldots, x_n$, is an expression of the form

$$Q(x_1, \ldots, x_n) = \sum_{i=1}^{n} \alpha_i x_i^2 + \sum_{1 \le i < j \le n} \beta_{ij} x_i x_j, \tag{7}$$

where the α_i's and β_{ij}'s belong to $\mathbb{R}$. Writing $x = (x_1, \ldots, x_n)'$, the right side of (7) can be written as $x'Ax$, where the $n \times n$ symmetric matrix $A = (a_{ij})$ is given by

$$a_{ij} = \begin{cases} \alpha_i & \text{if } i = j \\ \beta_{ij}/2 & \text{if } i < j \\ \beta_{ji}/2 & \text{if } i > j. \end{cases}$$

Conversely, if $A = (a_{ij})$ is any symmetric matrix of order n, then $x'Ax$ is the right side of (7) with $\alpha_i = a_{ii}$ and for $i < j$, $\beta_{ij} = 2a_{ij}$. Thus, any quadratic form can be uniquely written as $x'Ax$, where A is a symmetric matrix. The symmetric matrix A associated with the quadratic form $x'Ax$ is called the matrix of the quadratic form. Quadratic forms are usually classified as follows:

(i) nonnegative definite (n.n.d.) if $x'Ax \ge 0$ for all $x \in \mathbb{R}^n$;

(ii) positive definite (p.d.) if $x'Ax > 0$ for all $x \ne 0_n$;

(iii) nonpositive definite (n.p.d.) if $x'Ax \leq 0$ for all $x \in \mathbb{R}^n$;

(iv) negative definite (n.d.) if $x'Ax < 0$ for all $x \neq 0_n$.

A quadratic form is called indefinite if it is neither n.n.d nor n.p.d. It is customary to call the matrix of a quadratic form as n.n.d., p.d., etc., if the associated quadratic form is n.n.d., p.d., etc. The following are some important results that characterize a p.d. (or, n.n.d.) matrix.

Theorem 5. *A symmetric matrix A is p.d. (respectively, n.n.d.) if and only if each eigenvalue of A is positive (respectively, nonnegative).*

Theorem 6. *A (real) matrix A is n.n.d. if and only if there exists a real matrix B such that $A = B'B$. A is p.d. if and only if $A = B'B$ for some real non-singular matrix B.*

We now state a result on the bounds of $x'Ax/x'x$, called the *Rayleigh quotient*.

Theorem 7. *Let $\lambda_1 \leq \lambda_2 \leq \cdots \leq \lambda_n$ be the eigenvalues of a real symmetric matrix A. Then*

$$\lambda_1 \leq \frac{x'Ax}{x'x} \leq \lambda_n. \tag{8}$$

Since the extrema of $x'Ax/x'x$ given by (8) are attained by choosing x to be an eigenvector corresponding to λ_1 or λ_n, one can write

$$\lambda_1 = \min_x \frac{x'Ax}{x'x},$$
$$\lambda_n = \max_x \frac{x'Ax}{x'x}.$$

For many other results on the extrema of quadratic forms in terms of the eigenvalues of the matrix of the form, one may refer to Magnus and Neudecker (1988).

2.8 QR algorithm for computing eigenvalues and eigenvectors

The characteristic polynomial is of degree n and the complexity of calculating the eigenvalues increases with increasing n. For large n, one has to resort to appropriate numerical methods to find approximate roots.

The roots of a polynomial equation are extremely sensitive functions of the coefficients and thus computational procedures can often be very inaccurate in the presence of round-off errors.

An efficient and accurate method to compute eigenvalues and eigenvectors is based on the QR algorithm discovered in 1961, independently by (Francis, 1961, 1962) and Kublanovskaya (1963). We briefly describe the essentials of this algorithm as applied to $n \times n$ non-singular matrices. We first state the following important and well-known result.

Theorem 8. *Any non-singular matrix A of order n can be factorized as $A = QR$, where Q is an orthogonal matrix and R is an upper triangular matrix. The factorization is unique if all the diagonal entries of R are assumed to be positive.*

Example 4. Let

$$A = \begin{bmatrix} 1 & 1 & 2 \\ 1 & 0 & -2 \\ -1 & 2 & 3 \end{bmatrix}.$$

Then, with Q and R as given below, one can verify that $A = QR$:

$$Q = \begin{bmatrix} \frac{1}{\sqrt{3}} & \frac{4}{\sqrt{42}} & \frac{2}{\sqrt{14}} \\ \frac{1}{\sqrt{3}} & \frac{1}{\sqrt{42}} & -\frac{3}{\sqrt{14}} \\ -\frac{1}{\sqrt{3}} & \frac{5}{\sqrt{42}} & -\frac{1}{\sqrt{14}} \end{bmatrix}, \quad R = \begin{bmatrix} \sqrt{3} & -\frac{1}{\sqrt{3}} & -\sqrt{3} \\ 0 & \frac{\sqrt{14}}{\sqrt{3}} & \frac{\sqrt{21}}{\sqrt{2}} \\ 0 & 0 & \frac{\sqrt{7}}{\sqrt{2}} \end{bmatrix}.$$

The first step in the QR algorithm is to factorize A as

$$A = A_1 = Q_1 R_1,$$

into a product of an orthogonal matrix Q_1 and an upper triangular matrix R_1 with all diagonal entries positive, by using a numerically stable *Gram-Schmidt orthogonalization* procedure. Next, multiply the two factors in the *reverse order* to obtain

$$A_2 = R_1 Q_1.$$

These two steps are then repeated. Thus, we factor $A_2 = Q_2 R_2$, again using the Gram-Schmidt process and then multiply the factors in the reverse order to get

$$A_3 = R_2 Q_2.$$

The complete algorithm can be written as

$$A = Q_1 R_1, \quad R_k Q_k = A_{k+1} = Q_{k+1} R_{k+1}, \quad k = 1, 2, 3, \ldots,$$

where Q_k, R_k come from the previous step, and the subsequent orthogonal and upper triangular matrices Q_{k+1} and R_{k+1}, respectively, are computed by using a numerically stable form of the Gram-Schmidt algorithm. The surprising fact is that for many matrices, the iterates A_k converge to an upper triangular matrix U whose diagonal entries are the eigenvalues of A. This means that after a sufficient number of iterations, say k', the matrix $A_{k'}$ will have very small entries below the diagonal and the entries on the diagonal are the approximate eigenvalues of A. For each eigenvalue, the computation of the corresponding eigenvectors can be accomplished by solving the appropriate linear homogeneous system of equations.

If the original matrix is (real) symmetric, then the $A_k \to \Lambda$, where Λ is a diagonal matrix containing the eigenvalues of A on its diagonal. Furthermore, if we recursively define

$$S_k = S_{k-1} Q_k = Q_1 Q_2 \ldots Q_{k-1} Q_k, \quad S_1 = Q_1,$$

then the limit of S_k is S, an orthogonal matrix whose columns are the orthonormal eigenvector basis of A.

Software packages like MAPLE or MATLAB can be used to obtain the eigenvalues and corresponding eigenvectors.

2.9   Circulant matrices

There is a class of useful matrices, for which a formula for the eigenvalues is available. This is the class of 1-circulants and its variations.

2.9.1   The 1-Circulant matrix

This is the usual circulant matrix. An $n \times n$ 1-circulant matrix is given by

$$C_n = \begin{bmatrix} a_0 & a_1 & a_2 & \cdots & a_{n-2} & a_{n-1} \\ a_{n-1} & a_0 & a_1 & \cdots & a_{n-3} & a_{n-2} \\ a_{n-2} & a_{n-1} & a_0 & \cdots & a_{n-4} & a_{n-3} \\ & & & \vdots & & \\ a_1 & a_2 & a_3 & \cdots & a_{n-1} & a_0 \end{bmatrix}.$$

Note that the entries in a row are obtained by shifting the elements of the previous row by one position. Its eigenvalues $\{\lambda_i\}$ are (see e.g., Davis 1979 or Brockwell and Davis 2002),

$$\lambda_k = \sum_{l=0}^{n-1} a_l e^{i\omega_k l} = b_k + i c_k \quad \forall k = 1, 2, \dots, n,$$

where

$$\omega_k = \frac{2\pi k}{n}, \quad b_k = \sum_{l=0}^{n-1} a_l \cos(\omega_k l), \quad c_k = \sum_{l=0}^{n-1} a_l \sin(\omega_k l). \tag{9}$$

For $k = 1, 2, \dots, n$, the corresponding eigenvectors are given by

$$\xi_k = \left(x_k^0, x_k, x_k^2, \dots, x_k^{n-1} \right)',$$

where $x_1, \dots, x_n$ are the n roots of unity.

2.9.2 The k-circulant matrix

These are circulant matrices where, instead of shifting the elements in subsequent rows by one position, they are shifted by k positions. Suppose $a = \{a_l\}_{l \geq 0}$ is a sequence of real numbers. For positive integers k and n, a k-circulant matrix of order $n \times n$ with *input* sequence $\{a_l\}$ is defined as

$$A_{k,n}(a) = \begin{bmatrix} a_0 & a_1 & \cdots & a_{n-1} \\ a_{n-k} & a_{n-k+1} & \cdots & a_{n-k-1} \\ a_{n-2k} & a_{n-2k+1} & \cdots & a_{n-2k-1} \\ & & \vdots & \end{bmatrix}.$$

We write $A_{k,n}(a) = A_{k,n}$. The subscripts appearing in the matrix entries above are calculated modulo n and the convention is to start the row and column indices from zero. Thus, the 0-th row of $A_{k,n}(a)$ is $(a_0, a_1, a_2, \dots, a_{n-1})$. For $0 \leq j < n-1$, the $(j+1)$-th row of $A_{k,n}$ is a right-circular shift of the j-th row by k positions (equivalently, $k \bmod n$ positions). Without loss of generality, k may always be reduced modulo n.

Note that $A_{1,n}$ is the 1-circulant matrix C_n. It can be checked that the k-circulant matrix is symmetric for any $\{a_i\}$ if and only if $k = n - 1$.

This $(n-1)$-circulant matrix, also known as the *reverse circulant matrix* has identical elements on each of the *anti-diagonals* and is given by

$$
RC_n = \begin{bmatrix}
a_0 & a_1 & a_2 & \cdots & a_{n-2} & a_{n-1} \\
a_1 & a_2 & a_3 & \cdots & a_{n-1} & a_0 \\
a_2 & a_3 & a_4 & \cdots & a_0 & a_1 \\
& & & \vdots & & \\
a_{n-1} & a_0 & a_1 & \cdots & a_{n-3} & a_{n-2}
\end{bmatrix}_{n\times n}.
$$

2.9.3 Why study k-circulants?

The k-circulant matrix and its block versions arise in many different areas of mathematics and statistics – in multi-level supersaturated design of experiment (Georgiou and Koukouvinos, 2006), spectra of de Bruijn graphs (Strok, 1992) and in $(0,1)$-matrix solutions to $A^m = J_n$ (Wu *et al.*, 2002). See also the book by (Davis, 1979) and the article by Pollock (2002).

A matrix is said to be a *Toeplitz* matrix (see Section 3.2 for a definition) if each diagonal has identical elements. Circulant matrices have deep connection with the Toeplitz matrix. As we have seen, the 1-circulant matrix has an explicit spectral decomposition. However, the spectral analysis of the latter is much harder and challenging in general. If the input sequence $\{a_l\}_{l\geq 0}$ is *square summable*, then the circulant approximates the corresponding Toeplitz in various senses when the dimension $n \to \infty$. See Gray (2009) for a recent and relatively easy account. For an introduction to many interesting results, specially on the eigenvalue structure of Toeplitz matrices, see Grenander and Szegö (1984).

The eigenvalues of the circulant matrices crop up crucially in time series analysis. For example, the *periodogram* of a sequence $\{a_l\}_{l\geq 0}$ is defined as $n^{-1}I_n(\omega_k)$ where

$$
I_n(\omega_k) = \left| \sum_{t=0}^{n-1} a_t e^{-it\omega_k} \right|^2, \quad k = 0, 1, \ldots, n-1, \tag{10}
$$

and $\omega_k = 2\pi k/n$ are the Fourier frequencies. This is a simple function of the eigenvalues of the corresponding circulant matrix. The study of the properties of the periodogram is fundamental in the spectral analysis of time series. See for instance Fan and Yao (2003). The maximum of the periodogram, in particular, has been studied in Davis and Mikosch (1999).

The k-circulant matrices with random input sequence also serve as an important class of "patterned" matrices. Deriving the asymptotic properties of the spectrum of general patterned matrices has drawn significant attention in the recent literature. See Section 3.2 for some simulations on the spectrum of a few patterned random matrices, including the (random) 1-circulant.

2.9.4 Description of eigenvalues of k circulant

One can give a formula for the eigenvalues of the k-circulant for general k. This formula solution was derived by Zhou (1996) and is given below in Theorem 9. A proof may also be found in Bose *et al.* (2010d). Recall the definition of Fourier frequencies $\{\omega_k\}$ defined in equation (10) and the eigenvalues $\lambda_t = \sum_{l=0}^{n-1} a_l e^{i\omega_t l}$, $0 \le t < n$ of the 1-circulant $A_{1,n}$. Let $p_1 < p_2 < \cdots < p_c$ be all the common prime factors of n and k. Then we may write,

$$n = n' \prod_{q=1}^{c} p_q^{\beta_q} \quad \text{and} \quad k = k' \prod_{q=1}^{c} p_q^{\alpha_q}. \tag{11}$$

Here $\alpha_q, \beta_q \ge 1$ and n', k', p_q are pairwise relatively prime. For any positive integer m, let

$$\mathbb{Z}_m = \{0, 1, 2, \ldots, m - 1\}.$$

We introduce the following family of sets

$$S(x) := \left\{ xk^b \bmod n' : b \ge 0 \right\}, \quad x \in \mathbb{Z}_{n'}. \tag{12}$$

Let $g_x = \#S(x)$. Note that $g_0 = 1$. It is easy to see that

$$S(x) = \{xk^b \bmod n' : 0 \le b < g_x\}.$$

Clearly $x \in S(x)$ for every x. Suppose $S(x) \cap S(y) \ne \emptyset$. Then, $xk^{b_1} = yk^{b_2} \bmod n'$, for some integers $b_1, b_2 \ge 1$. Multiplying both sides by $k^{g_x - b_1}$ we see that, $x \in S(y)$ so that, $S(x) \subseteq S(y)$. Hence, reversing the roles, $S(x) = S(y)$. Thus, the distinct sets in $\{S(x)\}_{x \in \mathbb{Z}_{n'}}$ form a partition, called the *eigenvalue partition*, of $\mathbb{Z}_{n'}$. Denote the partitioning sets and their sizes by

$$\mathcal{P}_0 = \{0\}, \mathcal{P}_1, \ldots, \mathcal{P}_{\ell-1} \quad \text{and} \quad k_j = \#\mathcal{P}_j, \quad 0 \le j < \ell. \tag{13}$$

Define

$$\Pi_j := \prod_{t \in \mathcal{P}_j} \lambda_{tn/n'}, \quad j = 0, 1, \dots, \ell - 1. \tag{14}$$

Theorem 9. *(Zhou, 1996). The characteristic polynomial of $A_{k,n}$ is given by*

$$\mathrm{ch}_{A_{k,n}}(\lambda) = \lambda^{n-n'} \prod_{j=0}^{\ell-1} \left(\lambda^{k_j} - \Pi_j \right). \tag{15}$$

In particular, the above formula can be specialised to RC_n. The eigenvalues of RC_n may also be obtained directly (see Bose and Mitra (2002)) and are given by

$$\begin{cases} \lambda_0 = \sum_{t=0}^{n-1} a_t \\ \lambda_{n/2} = \sum_{t=0}^{n-1} (-1)^t a_t, & \text{if } n \text{ is even} \\ \lambda_k = -\lambda_{n-k} = \sqrt{I_n(\omega_k)}, & 1 \leq k \leq [\frac{n-1}{2}], \end{cases}$$

where $\{\omega_k\}$ and $\{I_n(\omega_k)\}$ are as defined in equation (10).

3   Some selected applications

3.1   Principal Component Analysis

Scientific phenomena are often quite complex in nature and because of the complexities involved, investigators often collect observations on many different variables. Multivariate statistical analysis deals with a set of methods to elicit information from such data sets. In this section, we briefly discuss one aspect of multivariate analysis, called *principal component analysis* because of its connection with the eigenvalues and eigenvectors of a matrix.

In multivariate situations, an investigator selects $p > 1$ variables or characters to record, the values of these variables being recorded on each individual, item or experimental unit. Then the multivariate data obtained from n individuals on each of the p variables can be represented by an $n \times p$ matrix, say $X = (x_{ij})$, where x_{ij} is the value of the variable j on the individual i, $1 \leq i \leq n, 1 \leq j \leq p$. The variability among the p variables is conveniently described by a $p \times p$ *variance-covariance* (or *dispersion*) matrix, say Σ, whose ith diagonal element is the variance of the ith variable, Var(X_i) and for $i \neq j$, the (i, j)th element represents the covariance between the ith and jth variables, written as $\text{Cov}(X_i, X_j)$. Since $\text{Cov}(U, V) = \text{Cov}(V, U)$ for any U, V, the dispersion matrix Σ is a real symmetric matrix. It is also n.n.d. We shall henceforth assume Σ to be positive definite.

Now consider a situation where p is large. Although p components are required to reproduce the total variability in the system, often much of the variability can be accounted for by a small number k of the *principal components*, which are linear combinations of the original variables. The k principal components can then replace the original p variables and, in such a case the original $n \times p$ data matrix can be reduced to an $n \times k$ data matrix. Thus, in this sense, the method of principal components can be visualized as a procedure to reduce the dimensionality of the problem.

Algebraically, principal components are specific linear combinations of the p original random variables $X_1, \ldots, X_p$. Geometrically, these linear combinations represent the selection of a new coordinate system obtained by rotating the original system (with $X_1, \ldots, X_p$ as the coordinate axes). The new axes represent the directions with maximum variability and provide a simpler and more transparent description of the variance-covariance structure.

Suppose the random vector $X = (X_1, X_2, \ldots, X_p)'$ has the dispersion matrix Σ and suppose $\lambda_1 \geq \lambda_2 \geq \cdots \geq \lambda_p > 0$ are the eigenvalues of Σ. Let $Y_i = a_i'X$, $1 \leq i \leq p$, be a set of linear combinations of the X_i's. Then, it can be easily seen that

$$\text{Var}(Y_i) = a_i'\Sigma a_i, \quad 1 \leq i \leq p,$$
$$\text{Cov}(Y_i, Y_j) = a_i'\Sigma a_j, \quad 1 \leq i < j \leq p.$$

The first principal component is the linear combination with maximum variance, that is, it maximizes $\text{Var}(Y_i) = a_i'\Sigma a_i$ over i. But, the variance $a_i'\Sigma a_i$ can be made arbitrarily large by multiplying a_i by some constant. In order to eliminate this indeterminacy, it is convenient to restrict attention to vectors of unit length. Thus, the first principal component is defined to be that linear combination $a_1'X$ say, that maximizes $a_i'\Sigma a_i$ subject to $a_1'a_1 = 1$. The second principal component is that linear combination $a_2'X$ that maximizes $a_i'\Sigma a_i$ subject to the conditions $a_2'a_2 = 1$ and $\text{Cov}(a_1'X, a_2'X) = 0$. Proceeding in this way, the mth principal component is $a_m'X$ that maximizes $\text{Var}(a_i'X)$ subject to $a_m'a_m = 1$ and $\text{Cov}(a_m'X, a_k'X) = 0$ for all $k < m$.

We then have the following result.

Theorem 10. *Let Σ denote the dispersion matrix of a $p \times 1$ random vector $X = (X_1, \ldots, X_p)'$. Suppose $\lambda_1 \geq \lambda_2 \geq \cdots \geq \lambda_p > 0$ are the eigenvalues of Σ and let $\boldsymbol{\xi}_1, \boldsymbol{\xi}_2, \ldots, \boldsymbol{\xi}_p$ be the corresponding normalized eigenvectors. Then, the ith principal component is given by $Y_i = \boldsymbol{\xi}_i' X$, $1 \leq i \leq p$. If multiplicities exist among the eigenvalues, then the principal components Y_i are not unique.*

From Theorem 10, it is seen that the principal components are *uncorrelated* random variables and have variances equal to the eigenvalues of Σ. We next have the following result.

Theorem 11. *In the set up of Theorem 10, let $Y_i = \boldsymbol{\xi}_i' X$ be the ith principal component, $1 \leq i \leq p$. Then*

$$\sigma_{11} + \sigma_{22} + \cdots + \sigma_{pp} = \sum_{i=1}^{p} \text{Var}(X_i) = \lambda_1 + \lambda_2 + \cdots + \lambda_p = \sum_{i=1}^{p} \text{Var}(Y_i),$$

where for $1 \leq i \leq p$, σ_{ii} is the ith diagonal element of Σ.

Theorem 11 shows that the total variability, $\sum_{i=1}^{p} \sigma_{ii} = \sum_{i=1}^{p} \lambda_i$. In view of this, *the proportion of total variance explained by the kth principal component* is

$$\frac{\lambda_k}{\lambda_1 + \lambda_2 + \cdots + \lambda_p}, \quad 1 \leq k \leq p.$$

The components of the vectors $\boldsymbol{\xi}_i$ appearing in the computation of principal components also have an interesting implication as given in the next result.

Theorem 12. *If for $1 \leq i \leq p$, $Y_i = \boldsymbol{\xi}_i' X$ are the principal components obtained from a dispersion matrix Σ then the correlation coefficient between Y_i and X_k is given by*

$$\frac{\xi_{ik} \sqrt{\lambda_i}}{\sqrt{\sigma_{kk}}},$$

where $\boldsymbol{\xi}_i = (\xi_{i1}, \xi_{i2}, \ldots, \xi_{ip})'$.

Example 5. Consider three random variables $X = (X_1, X_2, X_3)'$ whose dispersion matrix is given by

$$\Sigma = \begin{bmatrix} 1 & -2 & 0 \\ -2 & 5 & 0 \\ 0 & 0 & 2 \end{bmatrix}.$$

For this Σ, we have the eigenvalues and (normalized) eigenvectors as given below:

$$\lambda_1 = 5.83, \quad \xi_1' = (0.383, -0.924, 0);$$
$$\lambda_2 = 2.00, \quad \xi_2' = (0, 0, 1);$$
$$\lambda_3 = 0.17, \quad \xi_3' = (0.924, 0.383, 0).$$

The principal components are thus

$$Y_1 = \xi_1' X = 0.383X_1 - 0.924X_2,$$
$$Y_2 = \xi_2' X = X_3,$$
$$Y_3 = \xi_3' X = 0.924X_1 + 0.383X_2.$$

The proportion of the total variability accounted for by the first principal component is $\lambda_1/(\lambda_1 + \lambda_2 + \lambda_3) = 5.83/8 = 0.73$. Furthermore, the first two principal components explain $(5.83 + 2)/8 = 0.98 = 98\%$ of the variability. Thus, in this case, the principal components Y_1, Y_2 could replace the original three variables with nearly no loss of information.

3.2   Random matrices

In random matrices the entries are random variables. Such matrices arise in nuclear physics, mathematics, signal processing, statistics and wireless communication. In *random matrix theory*, different properties of these matrices are studied. In particular, properties of eigenvalues of random matrices when the dimension increases, are attempted to be captured through the *empirical distribution*, the *spectral radius*, the *extreme eigenvalues* and the spacings between successive eigenvalues.

In this section we define a few important random matrices and show a few simulations for their empirical spectral distribution. The *empirical spectral distribution* (ESD) of the matrix $A_{n \times n}$ is defined as

$$F_n(x) = n^{-1} \sum_{i=1}^{n} I\{\lambda_i \le x\}.$$

Thus it is a random probability distribution with mass $1/n$ at each λ_i.

Some of the more popular and/or important random matrices are the Wigner matrix, the sample variance covariance matrix, the sample auto-covariance matrix, the Toeplitz and Hankel matrices, matrices of the form XX' used in signal processing and wireless communication models, the so

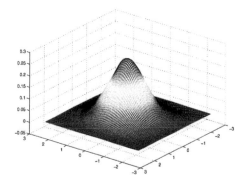

Figure 1: *Distribution of eigenvalues for the scaled 1-circulant matrix of order 400 with Bernoulli entries.*

Color image of this figure appears in the color plate section at the end of the book.

called IID matrix which is probabilistically interesting and, the circulant and k circulant matrices. There are variants of these matrices such as the tridiagonal matrices, band matrices, balanced matrices etc. which are also objects of study for various reasons.

For more information on eigenvalue properties of random matrices and their applications, the reader may consult the books Anderson *et al* (2009), Bai and Silverstein (2006), Mehta (1967) and Tulino and Verdu (2004) and the articles Bai (1999), Bose *et al.* (2010c), Bose and Sen (2008) and Bose *et al.* (2010a).

1-circulant matrix. For some detailed information on the eigenvalue distribution of k-circulant random matrices, see for example Bose *et al.* (2010b) and Bose *et al.* (2010d). Here we provide a simulation result for the 1-circulant matrix.

Wigner matrix. This is a symmetric matrix where the entries on and above the diagonal are i.i.d. with zero mean and variance 1. We scale by dividing by $n^{1/2}$.

$$W_n = n^{-1/2} \begin{bmatrix} x_{11} & x_{12} & x_{13} & \cdots & x_{1(n-1)} & x_{1n} \\ x_{12} & x_{22} & x_{23} & \cdots & x_{2(n-1)} & x_{2n} \\ & & & \vdots & & \\ x_{1n} & x_{2n} & x_{3n} & \cdots & x_{(n-1)n} & x_{nn} \end{bmatrix}. \tag{16}$$

This matrix was introduced by Wigner (1955, 1958) in nuclear physics. His works eventually led to the entire subject of random matrices. See

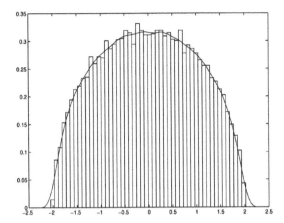

Figure 2: *Histogram and kernel density estimate for the ESD for 15 scaled Wigner matrices of order 400 with iid Bernoulli entries.*

Color image of this figure appears in the color plate section at the end of the book.

Anderson *et al* (2009) for a thorough mathematical treatment of these matrices.

IID matrix. This could be thought of the asymmetric version of the Wigner matrix. Here the entries are *independent and identically distributed* (i.i.d.) with mean zero and variance 1. For the latest results on the spectral distribution of these matrices, see Tao *et al.* (2010).

Sample variance covariance matrix (S matrix). This was originally studied by Wishart (1928). It is an estimate of the population dispersion matrix and as seen earlier, is fundamental in statistical analysis of multivariate data. Suppose $\{x_{jk}, 1 \le j \le p, 1 \le k \le n\}$ is a double array of i.i.d. random variables with mean zero and variance 1. Write $x_j = (x_{j1}, \ldots, x_{jn})$ and define the $p \times n$ matrix $X = [x'_1 \, x'_2 \, \ldots \, x'_p]'$. The *sample variance covariance matrix*, in short the S matrix, is

$$S_n = n^{-1}XX'.$$

If p is held fixed, then standard asymptotic theory applies. The situation becomes interesting if $p \to \infty$ such that p/n converges or, $p/n \to \infty$. For detailed results on this matrix, see (Bai, 1999) and (Bose *et al.*, 2010c).

Toeplitz and Hankel matrix. Let $\{x_0, x_1, \ldots x_n, \ldots\}$ be a sequence of i.i.d. real random variables with mean zero and variance 1. The Toeplitz and

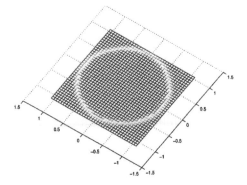

Figure 3: *Distribution of eigenvalues for the scaled iid matrix of order 400 with Bernoulli entries.*

Color image of this figure appears in the color plate section at the end of the book.

Hankel matrices are defined respectively, as

$$
T_n = \begin{bmatrix}
x_0 & x_1 & x_2 & \cdots & x_{n-2} & x_{n-1} \\
x_1 & x_0 & x_1 & \cdots & x_{n-3} & x_{n-2} \\
x_2 & x_1 & x_0 & \cdots & x_{n-4} & x_{n-3} \\
& & & \vdots & & \\
x_{n-1} & x_{n-2} & x_{n-3} & \cdots & x_1 & x_0
\end{bmatrix},
$$

$$
H_n = \begin{bmatrix}
x_0 & x_1 & x_2 & \cdots & x_{n-2} & x_{n-1} \\
x_1 & x_2 & x_3 & \cdots & x_{n-1} & x_n \\
x_2 & x_3 & x_4 & \cdots & x_n & x_{n+1} \\
& & & \vdots & & \\
x_{n-1} & x_n & x_{n+1} & \cdots & x_{2n-3} & x_{2n-2}
\end{bmatrix}.
$$

These matrices and the corresponding operators are important objects in operator theory. For probabilistic properties of these matrices, see Bryc *et al.* (2006), Hammond and Miller (2005) and Bose *et al.* (2010c).

Sample autocovariance matrix. Let $x_0, x_1, x_2, \ldots x_n, \ldots$ be a sequence of random variables. Let the autocovariances be defined as

$$
\hat{\gamma}(k) = n^{-1} \sum_{i=1}^{n-|k|} x_i x_{i+|k|}.
$$

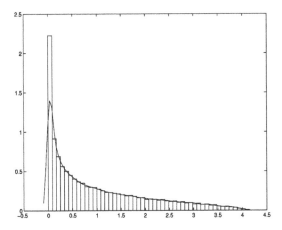

Figure 4: *Histogram and kernel density estimate for the ESD for 15 scaled sample var-covar matrices with n = p = 400 with iid* exp(1) *entries.*

Color image of this figure appears in the color plate section at the end of the book.

The *sample autocovariance matrix* is

$$\hat{\Gamma}_n = ((\hat{\gamma}(i - j)))_{1 \le i, \ j \le n}.$$

Note that this matrix has the Toeplitz structure. The autocovariances and the autocovariance matrix are fundamental object in time series analysis.

4 Eigenvalues/eigenfunctions in function spaces

The concepts of eigenvalue and eigenvector that we have discussed for matrices (i.e., finite dimensional linear transformations) have deep and significant extensions to *linear operators* in abstract spaces. See Dunford and Schwartz (1988) for this development. Here we present an illustrative situation.

Consider the interval $C = [0, 1]$ and the space L_2 of all square integrable (w.r.t. the Lebesgue measure) functions on C. Let $K(x, y)$ be a square integrable (w.r.t. the two dimensional Lebesgue measure) function on $C \times C$. Given K, define T_K

$$T_K(g(x)) = \int_0^1 K(x, y)g(y) \, dy.$$

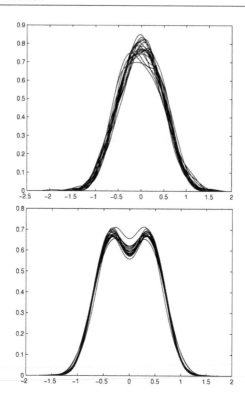

Figure 5: *Individual kernel density estimate for the ESD for 15 simulations from the scaled Toeplitz (top) and Hankel (bottom) matrices of order 400 with Bernoulli entries.*

Color image of this figure appears in the color plate section at the end of the book.

It can be shown that $T : L_2 \to L_2$ is a linear map. Extending the notion of eigenvalue and eigenvector, we may now say that λ is an eigenvalue of T with *eigenfunction $f \in L_2$* if

$$T_K(f(x)) = \lambda f(x) \quad \text{for almost all} \quad x \in [0, 1].$$

It can be shown that corresponding to any K, there is a (possibly infinite) sequence $\{\lambda_i, f_i(\cdot)\}$ which are all the eigenvalues and eigenvectors of T_K with analogous properties that we have described earlier for the finite dimensional case. In particular, the eigenfunctions are *orthogonal*, that is, $\int_0^1 f_k(x) f_{k'}(x)\, dx = 0$ whenever $k \neq k'$. In addition, they can also be chosen to be *orthonormal*, that is, they further satisfy $\int_0^1 f_k^2(x)\, dx = 1$ for all k.

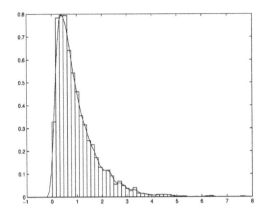

Figure 6: *Histogram with kernel density estimate of the ESD of 15 real-izations of the scaled sample autocovariance of order* 400 *with Bernoulli entries.*

Color image of this figure appears in the color plate section at the end of the book.

The *spectral representation* of the operator takes shape as

$$K(x, y) = \sum_{k=1}^{\infty} \lambda_k f_k(x) f_k(y) \quad \text{(in the } L_2 \text{ sense).}$$

Example 6. As an example, let

$$K(x, y) = \int_0^1 [I(x \le t) - t][I(y \le t) - t] \, dt, \quad x, y \in [0, 1]. \quad (17)$$

Since K is a bounded function it is square integrable. The eigenvalues and eigenfunctions of T_K are given by

$$\lambda_k = \frac{1}{k^2 \pi^2}, \quad f_k(x) = 2 \sin(k\pi x), \quad k = 1, 2, \ldots.$$

To see this, first note that for the constant function c, $T_K(c) = 0$. This implies that if f is an eigenfunction, then so is $g(\cdot) = f(\cdot) - \int_0^1 f(x) \, dx$ (since $f \in L_2$, $\int_0^1 f(x) \, dx$ is well defined). Note that $\int_0^1 g(x) \, dx = 0$. Thus, without loss of generality, we can assume that the eigenfunctions f which we are seeking, satisfy $\int_0^1 f(x) \, dx = 0$. Now start with the eigenvalue equation to obtain

$$T_K(f(x)) = \int_0^1 \int_0^1 [I(x \le t) - t][I(y \le t) - t]f(y)\,dy\,dt, \quad x \in [0, 1] \quad (18)$$

$$= \int_0^1 \int_0^1 [I(\max(x,y) \le t) - tI(x \le t) - tI(y \le t) + t^2]f(y)\,dy\,dt,$$
$$x \in [0, 1] \qquad (19)$$

$$= \int_0^1 \int_0^1 [I(\max(x,y) \le t) - tI(x \le t) - tI(y \le t)]f(y)\,dy\,dt,$$
$$x \in [0, 1] \qquad (20)$$

$$= \int_0^1 \int_x^1 I(y \le t)f(y)\,dy\,dt - \int_0^1 \int_0^x I(x \le t)f(y)\,dy\,dt - 0$$
$$- \int_0^1 \int_0^t tf(y)\,dy\,dt \qquad (21)$$

$$= \int_x^1 (1 - y)f(y)\,dy + (1 - x)\int_0^x f(y)\,dy + C_f \qquad (22)$$
$$(C_f \text{ is a constant})$$

$$= \lambda f(x). \qquad (23)$$

Now for the moment, assume that the eigenfunction we seek are smooth enough so that we can take derivatives. Thus taking derivatives in the above relation,

$$\lambda f'(x) = -(1 - x)f(x) + (1 - x)f(x) - \int_0^x f(y)\,dy \qquad (24)$$

$$= -\int_0^x f(y)\,dy. \qquad (25)$$

Taking another derivative, we see that

$$-f(x) = -\lambda f''(x). \qquad (26)$$

It is easy to see that $f_k(x) = a_k \sin(\pi k x)$, $k = 1, 2, \ldots$ where $\{a_k\}$ are constants, satisfy the above differential equation. Clearly these eigenfunctions are mutually orthogonal, that is $\int_0^1 f_k(x)f_{k'}(x)\,dx = 0$ whenver $k \ne k'$. They are further, orthonormal if we choose $a_k \equiv 2$ for all k. The eigenvalues are easy to obtain from equation (26) and come out as

$$\lambda_k = \frac{1}{k^2 \pi^2}, \quad k = 1, 2, \ldots. \qquad (27)$$

It may be checked that this yields the spectral representation (that is there are no other eigenfunctions).

4.1 Cramér-von Mises statistic

Suppose $X_1, \ldots, X_n \ldots$ are independent uniform and identically distributed over the interval $(0, 1)$.

The Cramer–von Mises statistics defined as (we are considering only the "null" distribution):

$$V_n = n \int_0^1 [F_n(x) - x]^2 \, dx$$

where F_n is defined to be the *empirical* distribution function

$$F_n(x) = \frac{1}{n} \sum_{i=1}^{n} I(X_i \le x).$$

The distribution of V_n is needed for statistical testing purposes but unfortunately cannot be found out in a closed form. After some simple algebra,

$$V_n = \frac{1}{n} \sum_{i=1}^{n} K(X_i, X_i) + \frac{2}{n} \sum_{1 \le i < j \le n} K(X_i, X_j)$$

where K is as defined in equation (17). As a consequence, the *asymptotic* distribution of V_n is closely connected to the linear transformation T_K and the eigenvalues described above. It can be shown that the limiting distribution of V_n is the same as $\sum_{k=1}^{\infty} \lambda_k Y_k$ where $\{\lambda_k\}$ is as in equation (27) and $\{Y_k\}$ are i.i.d. chi-squared variables each with one degree of freedom. The reader may consult Serfling (1980) for further details.

Acknowledgement

We are thankful to Sujatha Ramdorai for inviting us to write this article. We also thank Subhajit Dutta, Debashis Pal and Koushik Saha for the help extended during the preparation of this manuscript.

References

Anderson, G.W., Guionnet, A. and Zeitouni, O. 2009. *An Introduction to Random Matrices*. Cambridge University Press.

Bai, Z.D. 1999. Methodologies in spectral analysis of large dimensional random matrices, a review (with discussion). *Statist. Sinica* **9**, 611–677.

Bai, Z.D. and Silverstein, J. 2006. *Spectral Analysis of Large Dimensional Random Matrices*. Science Press, Beijing.

Bose, A., Hazra, R.S. and Saha, K. 2010a. Patterned random matrices and method of moments. *Proc. Internat. Congr. Mathematicians*. Hyderabad, India. World Scientific, Singapore and Imperial College Press, U.K. pp. 2203–2230 (Invited article).

Bose, A., Hazra, R.S. and Saha, K. 2010b. Product of exponentials and spectral radius of random k circulants. *Ann. Inst. Henri Poincare Probab. Stat.* (to appear).

Bose, A., Gangopadhyay, S. and Sen, A. 2010c. Limiting spectral distribution of XX' matrices. *Ann. Inst. Henri Poincare Probab. Stat.* **46**:677–707.

Bose, A. and Mitra, J. 2002. Limiting spectral distribution of a special circulant. *Stat. Probab. Lett.* **60**, 111–120.

Bose, A., Mitra, J. and Sen, A. 2010d. Large dimensional random k circulants. *J. Theoret. Probab.* DOI:10.1007/s10959-010-0312-9.

Bose, A. and Sen, A. 2008. Another look at the moment method for large dimensional random matrices. *Elec. J. Probab.* **13**:588–628.

Brockwell, P.J. and Davis, R.A. 2002. *Introduction to Time Series and Forecasting*, 2nd. ed. Springer-Verlag, New York.

Bryc, W., Dembo, A. and Jiang, T. 2006. Spectral measure of large random Hankel, Markov and Toeplitz matrices. *Ann. Probab.* **34**:1–38.

Davis, P.J. 1979. *Circulant Matrices*. Wiley, New York.

Davis, R.A. and Mikosch, T. 1999. The maximum of the periodogram of a non-Gaussian sequence. *Ann. Prob.* **27**:522–536.

Dunford, N. and Schwartz, J.T. 1988. *Linear Operators Part I, General Theory*. Wiley-Interscience, New York.

Fan, J. and Yao, Q. 2003. *Nonlinear Time Series: Nonparametric and Parametric Methods*. Springer-Verlag, New York.

Francis, J.G.F. 1961. The QR Transformation, I. *The Computer Journal* **4:** 265–271.

Francis, J.G.F. 1962. The QR Transformation, II. *The Computer Journal* **4:** 332–345.

Georgiou, S. and Koukouvinos, C. 2006. Multi-level k-circulant supersaturated designs. *Metrika* **64**:209–220.

Gray, Robert M. 2009. *Toeplitz and Circulant Matrices: A Review*. Now Publishers, Norwell, Massachusetts.

Grenander, U. and Szegö, G. 1984. *Toeplitz Forms and Their Applications*, 2nd ed. Chelsea, New York.

Hammond, C. and Miller, S.J. 2005. Distribution of eigenvalues for the ensemble of real symmetric Toeplitz matrices. *J. Theoret. Probab.* **18:**537–566.

Kublanovskaya, Vera N. 1963. On some algorithms for the solution of the complete eigenvalue problem. *USSR Computational Mathematics and Mathematical Physics* **1:**637–657. (received Feb 1961).

Magnus, J.R. and Neudecker, H. 1988. *Matrix Differential Calculus with Applications in Statistics and Econometrics*. Wiley, New York.

Mehta, M.L. 1967. *Random Matrices and The Statistical Theory of Energy Levels*. Academic Press, New York.

Pollock, D.S.G. 2002. Circulant matrices and time-series analysis. *Internat. J. Math. Ed. Sci. Tech.* **33:**213–230.

Serfling, R.J. 1980. *Approximation Theorems of Mathematical Statistics*, Wiley.

Strok, V.V. 1992. Circulant matrices and the spectra of de Bruijn graphs. *Ukrainian Math. J.* **44:**1446–1454.

Tao, T., Vu, V. and Krishnapur, M. 2010. Random Matrices: Universality of the ESDs and the circular law. *Ann. Probab.* **38:**2023–2065.

Tulino, A.M. and Verdu, S. 2004. *Random Matrix Theory and Wireless Communications*, Now Publishers, Norwell, Massachusetts.

Wigner, E.P. 1955. Characteristic vectors of bordered matrices with infinite dimensions. *Ann. Math.* (2) **62:**548–564.

Wigner, E.P. 1958. On the distribution of the roots of certain symmetric matrices. *Ann. Math.* (2) **67:**325–327.

Wishart, J. 1928. The generalised product moment distribution in samples from a normal multivariate population. *Biometrika* **20A:**32–52.

Wu, Y.K., Jia, R.Z. and Li, Q. 2002. g-Circulant solutions to the $(0, 1)$ matrix equation $A^m = J_n$. *Lin. Alg. Appln.* **345:**195–224.

Zhou, J.T. 1996. A formula solution for the eigenvalues of g circulant matrices. *Math. Appl. (Wuhan)* **9:**53–57.

CHAPTER 8

On Pták's Nondiscrete Induction

A.S. Vasudeva Murthy

Tata Institute of Fundamental Research,
Centre for Applicable Mathematics,
Bangalore 560 065, India.
e-mail: vasu@math.tifrbng.res.in

1 Introduction

Iterative methods are often used to prove the existence of a solution to linear/nonlinear system of equations, ordinary/partial differential equations and many others. The iterations are usually indexed by integers and the basic tool is the contraction mapping theorem. A continuous version (without the indexing by integers) was proposed by the famous Czech mathematician Vlastimil Pták in 1974 during his quantitative study of the closed graph theorem in functional analysis (Pták, 1974). For a biography on Pták see (Vavřín, 1996). In the following we shall trace a path that will lead to this principle of nondiscrete induction.

2 Linear Systems

Let A be an $n \times n$ matrix with real entries

$$A = ((a_{ij}))_{i,j=1}^{n}.$$

Consider the problem of finding a $x \in \mathbb{R}^n$ such that

$$Ax = b \tag{1}$$

for a given $b \in \mathbb{R}^n$. In our high school we are taught to solve (1) for $n = 2, 3$ using the method of Gaussian elimination which is nothing but decomposing

$$A = LU$$

where L and U are lower and upper triangular matrices, leading to (for a nonsingular A)

$$Ux = L^{-1}b$$

which is easy to solve by back substitution (starting with the last equation which is of the form $ax = b$). For large $n (\gg 10)$ one can perform the above procedure on a computer modulo some pitfalls like round-off error and ill-conditioning.

One could also solve (1) by an iterative procedure by rewriting it as

$$x = (I - A)x + b$$

and set up an iteration (for a given x_0)

$$x_n = (I - A)x_{n-1} + b; \quad n \geq 1 \tag{2}$$

here I is the identity matrix. Clearly if x_n converges then it will converge to the solution of (1). The conditions under which this converges will be given below. Before that we define the norm.

Definition 1. A norm on $\mathbb{R}^n$ is a real valued function $|| \cdot ||$ satisfying

(a) $||x|| \geq 0, ||x|| = 0$ if and only if $x = 0$

(b) $||\alpha x|| = |\alpha| \, ||x|| \quad$ for $\alpha \in \mathbb{R}$

(c) $||x + y|| \leq ||x|| + ||y||$.

Some of the main examples are

$$||x||_2 = \left[\sum_{i=1}^{n} x_i^2 \right]^{\frac{1}{2}} \quad \text{(Euclidean norm)}$$

$$||x||_1 = \sum_{i=1}^{n} |x_i| \quad \text{(Sum norm)}$$

$$||x||_\infty = \max_{1 \leq i \leq n} |x_i|. \quad \text{(Max norm)}.$$

Next we define the induced matrix norm:

Definition 2. Let $\| \cdot \|$ be a norm on $\mathbb{R}^n$. The induced matrix norm of an $n \times n$ matrix A is

$$\|A\| = \max_{\|x\|=1} \|Ax\|.$$

Note that we have used the same notation $\|\cdot\|$ for both vector and matrix norm. The most important property of the induced norm is

$$\|Ax\| \leq \|A\| \, \|x\|. \tag{3}$$

The standard vector norms given above induce the matrix norms

$$\|A\|_1 = \max_j \sum_{i=1}^{n} |a_{ij}|$$

$$\|A\|_\infty = \max_i \sum_{j=1}^{n} |a_{ij}|$$

$$\|A\|_2 = \max_i |\lambda_i|$$

where λ_i is the eigenvalue of $\sqrt{A^*A}$ and A^* is the conjugate transpose of A. For a proof see Strikwerda (2004, p. 404). For the convergence of (2) we have

Lemma 1. *The iteration* (2) *will converge if*

$$\|I - A\| < 1.$$

Proof. Setting $M = I - A$, we will show that $I - M$ is nonsingular and

$$(I - M)^{-1} = \sum_{l=0}^{\infty} M^l.$$

The sequence $\{S_k\}$

$$S_k = \sum_{l=0}^{k} M^l$$

is a Cauchy sequence in $\mathbb{R}^{n \times n}$ for,

$$\|S_k - S_m\| \leq \sum_{l=k+1}^{m} \|M^l\|$$

with $m > k$. From (3) we have

$$\|S_k - S_m\| \le \sum_{l=k+1}^{m} \|M\|^l = \|M\|^{k+1} \left[\frac{1 - \|M\|^{m-k}}{1 - \|M\|} \right].$$

Letting $m, k \to \infty$ and noting the hypothesis $\|M\| < 1$ it follows that $\{S_k\}$ is Cauchy and hence converges to (say) S. Since

$$MS_k + I = S_{k+1}$$

we have

$$MS + I = S$$
$$(I - M)S = I.$$

This shows that $I - M$ is nonsingular and

$$S = (I - M)^{-1}.$$

Further

$$\|A^{-1}\| = \|(I - M)^{-1}\| = \|S\| \le \sum_{l=0}^{\infty} \|M\|^l = \frac{1}{1 - \|M\|}.$$

$\square$

3   Nonlinear systems

Let $f : \mathbb{R}^n \to \mathbb{R}^n$ be a smooth map. We consider the problem of finding an $x \in \mathbb{R}^n$ such that

$$f(x) = 0. \tag{4}$$

There are no direct methods save for $n = 2$ and a few special cases for $n = 3$. So an iterative algorithm is the only alternative. Following (2) we rewrite (4) as

$$x = x - f(x) = g(x)$$

and consider the iteration, for a given x_0,

$$x_n = g(x_{n-1}); \quad n \ge 1. \tag{5}$$

Clearly if x_n converges to an x then $g(x) = x$ and hence x is a solution to (4). To give conditions under which this is possible we need the notion of $\|I - A\| < 1$ for nonlinear maps.

Definition 3. Let $\Omega \subset \mathbb{R}^n$, a mapping $T : \Omega \to \mathbb{R}^n$ is a contraction if there exists a constant α satisfying $0 \le \alpha < 1$ and

$$\|T(x) - T(y)\| \le \alpha \|x - y\|$$

where $\| \cdot \|$ is a norm on $\mathbb{R}^n$.

Some simple examples with obvious ranges are

$$f(x) = e^{-x} \quad \text{on } [0, 1]$$

$$f(x) = \frac{x}{2} - \frac{1}{x} \quad \text{on } [1, 2]$$

$$f(x) = Ax \text{ for } x \in \mathbb{R}^n \text{ and } A \in \mathbb{R}^{n \times n} \text{ such that } \|A\| < 1$$

$$f\begin{pmatrix} x_1 \\ x_2 \end{pmatrix} = \begin{pmatrix} e^{-\frac{x_2^2}{2}} \\ \frac{1}{10} x_1 \end{pmatrix} \quad \text{on } [0, 1] \times (0, 1].$$

Before we state the nonlinear version of Lemma 1 we state a lemma that is intuitively obvious but a proof can be found in any standard text book on analysis.

Lemma 2. *Let $f : \Omega \subset \mathbb{R}^n \to \mathbb{R}^n$ be a continuous function and Ω is a non empty, bounded and closed set. Then f attains its maximum and minimum on Ω.*

The nonlinear version of Lemma 1 is,

Theorem. (Contraction mapping) *Let $\Omega \subset \mathbb{R}^n$ be as in Lemma 2. If $T : \Omega \to \Omega$ is a contraction then T has a unique fixed point.*

Proof. Set

$$f(x) = \|x - T(x)\|.$$

Clearly a zero of f is fixed point of T. Now

$$|f(x) - f(y)| = \Big| \|x - T(x)\| - \|y - T(y)\| \Big|$$
$$\le \|x - T(x) - y + T(y)\|$$

where we have used

$$\|a - b\| \ge \Big| \|a\| - \|b\| \Big|$$

for $a, b \in \mathbb{R}^n$. We then obtain

$$|f(x) - f(y)| \le (1 + \alpha) \|x - y\|.$$

This implies that f is continuous and by the above lemma there exists a $x_0 \in \Omega$ such that $f(x_0)$ is a minimum and

$$f(x_0) \leq f(Tx_0) \leq \alpha f(x_0). \tag{6}$$

The last part of (6) follows from

$$\|Tx_0 - T(Tx_0)\| \leq \alpha\|x_0 - Tx_0\| = \alpha f(x_0).$$

Since $f(x_0) \geq 0$ and $\alpha < 1$ we obtain

$$f(x_0) = 0.$$

$\square$

Remark 1. The above result holds even when Ω is not bounded. One needs to consider
$$\tilde{\Omega} = \{x \in \Omega | f(x) \leq f(q)\}$$
for some $q \in \Omega$. If $x \in \tilde{\Omega}$ then

$$\begin{aligned}\|x - q\| &\leq \|x - Tx\| + \|Tx - Tq\| + \|Tq - q\| \\ &\leq f(x) + \alpha\|x - q\| + f(q) \\ &\leq 2f(q) + \alpha\|x - q\|\end{aligned}$$

and hence
$$\|x - q\| \leq \frac{2f(q)}{(1 - \alpha)}.$$

This shows that $\tilde{\Omega}$ is closed and bounded. From (6)

$$f(Tq) \leq \alpha f(q)$$

it follows that $T(\tilde{\Omega}) \subset \tilde{\Omega}$ and one can proceed as in the proof above.

Remark 2. The fixed point is unique for if x_0 and y_0 are two distinct fixed points then
$$\|x_0 - y_0\| = \|Tx_0 - Ty_0\| \leq \alpha\|x_0 - y_0\|$$

hence
$$\|x_0 - y_0\| = 0$$

which implies $x_0 \equiv y_0$.

Remark 3. If y is a fixed point of T then

$$\|x_n - y\| \le \|x_n - Tx_n\| + \|Tx_n - Ty\|$$
$$\le f(x_n) + \alpha\|x_n - y\|$$
$$\|x_n - y\| \le f(x_n)/(1 - \alpha)$$

From (6)

$$f(Tx_{n-1}) \le \alpha f(x_{n-1})$$
$$f(x_n) \le \alpha f(x_{n-1}) \le \cdots \le \alpha^n f(x_0).$$

Hence we have the estimate

$$\|x_n - y\| \le \frac{\alpha^n}{1 - \alpha}\|x_0 - Tx_0\|$$

and also

$$\|x_{n+1} - x_n\| = \|Tx_n - Tx_{n-1}\| \le \alpha\|x_n - x_{n-1}\|. \tag{7}$$

Pták's first step towards formulating nondiscrete induction was to replace (7) by

$$\|x_{n+1} - x_n\| \le w(\|x_n - x_{n-1}\|)$$

where $w(\cdot)$ is a function. In the case of contraction

$$w(s) = \alpha s$$

for $0 \le \alpha < 1$. One can now estimate

$$\|x_n - x\| \le \|x_{n+1} - x_n\| + \|x_{n+2} - x_{n+1}\| + \cdots$$
$$\le \sum_{j=0}^{\infty} w^j(\|x_{n+1} - x_n\|)$$

where $w^0(t) = t, w^1(t) = w(t), w^j(t) = w(w^{j-1}(t))$ for $j \ge 1$. This leads to the definition

Definition 4. Let $I = (0, t_0)$ for some $t_0 > 0$. A *rate of convergence* or a *small function* is a function w defined on I such that

1. $w(I) \subset I$

2. For each $t \in I, t + w(t) + w^2(t) + \cdots$ is convergent.

The sum of the series is denoted by $\sigma(t)$

$$\sigma(t) = \sum_{j=0}^{\infty} w^j(t) \tag{8}$$

and it satisfies

$$\sigma(t) - t = \sigma(w(t)). \tag{9}$$

A nontrivial example coming from Newton's method is

$$w(t) = \frac{t^2}{2\sqrt{t^2 + d^2}}$$

for $d > 0$ and $I = (0, \infty)$ with

$$\sigma(t) = t + \sqrt{t^2 + d^2} - d.$$

One can directly check that (9) is satisfied but for a more natural way of computing σ from w see (Pták, 1976).

We now describe the abstract model. Let M be a given set (for e.g., it could be the set of all solutions to a nonlinear system of equations or differential equations). Suppose we wish to construct a point $x \in M$. First we replace M by $M(r)$ where r is a small parameter and the inclusion $y \in M(r)$ means that y satisfies the relation $y \in M$ only approximately perhaps by an error r.

To make it more precise, we let (X, d) to be a complete metric space and

$$M(r) \subset X$$

be a family of sets. The point x to be is in some sense a limit point of $M(r)$

$$\underset{r \to 0}{\text{Lt}} \; M(r) = \bigcap_{r>0} \overline{\bigcup_{s \le r} M(s)}.$$

Starting from an initial point $x_r \in M(r)$ one should be able to find a point $x_{r'}$ which is a better approximation, more precisely, we set $r' = w(r)$ where $w(r)$ is the rate of convergence defined above

$$x_{r'} \in M(w(r)).$$

This condition can be stated as if $x_r \in M(r)$ then

$$U(x, r) \cap M(w(r)) \ne \Phi$$

where
$$U(x, r) = \{y \in X | d(y, x) \leq r\}.$$

This can be written in a more compact form
$$M(r) \subset U(M(w(r)), r)$$

and repeatedly applying this we should have
$$M(r) \subset U(Lt_{r \to 0} M(r), \sigma(r))$$

where σ is the sum of the infinite series given in (8).

Theorem. (Nondiscrete induction) *Let (X, d) be a complete metric space and $I = (0, t_0)$ for some $t_0 > 0$ and $w(t)$ be a rate of convergence on I. For $t \in I$ let $M(t) \subset X$ and $M(0)$ be the limit of $M(\cdot)$. Suppose for $t \in I$*

$$M(t) \subset U(M(w(t)), t)$$

then
$$M(t) \subset U(M(0), \sigma(t))$$

for $t \in I$.

Proof. Let $x \in M(t)$, since $M(t) \subset U(M(w(t)), t)$ there exists an $x_1 \in U(x, t) \cap M(w(t))$. Now $x_1 \in M(w(t)) \subset U(M(w^2(t)), w(t))$ so that there exists an $x_2 \in U(x_1, w(t)) \cap M(w^2(t))$. Continuing we obtain a sequence $\{x_n\}$ such that
$$x_n \in U(x_n, w^n(t)) \cap M(w^{n+1}(t)).$$

Since
$$d(x_n, x_{n+1}) < w^n(t)$$

$\{x_n\}$ is a Canchy sequence hence converges to (say) x_∞. Since $x_n \in M(w^n(t))$ and $w^n(t) \to 0$ as $n \to \infty$ we have $x_\infty \in M(0)$. Also
$$d(x, x_\infty) \leq d(x, x_1) + d(x_1, x_2) + \cdots$$
$$\leq t + w(t) + W^2(t) + \cdots$$
$$= \sigma(t)$$

so that
$$x \in U(x_\infty, \sigma(t)) \subset U(M(0), \sigma(t)).$$

$\square$

Remark 4. The reason for calling this result as nondiscrete induction is that it represents a continuous analogue of the method of mathematical induction. The verification of the hypothesis consists of verifying the possibility of passing from a given approximation in $M(r)$ to a better approximation $M(w(r))$. This corresponds to the step n to $n + 1$ in the classical induction process. The conclusion that

$$M(t) \subset U(M(0), \sigma(t))$$

means $M(0)$ is nonempty provided at least one $M(t)$ is nonempty. This corresponds to the first step in the classical induction proof. As an application we prove the following contraction mapping theorem.

Theorem. *Let (X, d) be a complete metric space and $f : X \to X$ satisfy*

$$d(f(x_1), f(x_2)) \le \alpha d(x_1, x_2)$$

for $0 < \alpha < 1$. Then there exists an $x \in X$ such that

$$f(x) = x.$$

Proof. For $t > 0$ let

$$w(t) = \alpha t$$
$$M(t) = \{x | d(x, f(x)) < t\}.$$

Clearly

$$M(0) = \{x | x = f(x)\}$$

and it suffices to show that

$$M(t) \subset U(w(t), t).$$

If $x \in M(t)$ and $x' = f(x)$ then

$$d(x, x') < t,$$

and to show $x' \in M(w(t))$ note that

$$d(x', f(x')) = d(f(x), f(x')) \le \alpha d(x, x') < \alpha t.$$

From the induction theorem it follows that $M(0)$ is nonempty. $\square$

References

Pták, V. 1974. A quantitative refinement of the closed graph theorem, *Czech. Math.* **99**:503–506.

Pták, V. 1976. The rate of convergence of Newton's Process, *Num. Math.* **25**: 279–285.

Strikwerda, J.C. 2004. *Finite difference schemes and partial differential equations*, SIAM Philidelphia.

Vavřín, Z. 1996. Seventy years of Professor Vlastimil Pták., *Math. Bohem.* **121**: 315–327.

CHAPTER 9

Partial Differential Equations

P.N. Srikanth

Tata Institute of Fundamental Research, CAM,
Post Bag 6503, GKVK Post,
Bangalore 560 065, India.
e-mail: srikanth@math.tifrbng.res.in

A Partial Differential Equation (PDE) is an equation involving an unknown function of two or more variables and some of its partial derivatives. Typically any expression of the form:

$$F(D^k u(x), D^{k-1} u(x), \dots, Du(x), u(x), x) = 0$$

is a kth - order PDE. Here **R** denotes the set of real numbers and F is a function

$$F : \mathbf{R}^{n^k} \times \mathbf{R}^{n^{k-1}} \times \dots \times \mathbf{R}^n \times \mathbf{R} \times \mathbf{R}^n \to \mathbf{R}.$$

We will see some typical examples as we go along, also in the above, $\mathbf{R}^n$ could be replaced by an open set in $\mathbf{R}^n$.

Any discussion on Differential Equations, usually starts with a motivation for study of the equation under consideration. The reason for this is not far to seek. Differential Equations are the most important part of Mathematics in understanding physical Sciences and in fact studies on Differential Equations originated that way (Bernoulli and wave equations, Fourier and heat equation). In physical sciences, the aim is to understand the phenomena governing any observation or event. The issues connected with those could be very complex and one would like to understand these through mathematical models. The essence of Mathematical Modelling is simplification and more often that not the model takes the form of a Differential Equation. In many cases, analysis of the model, which may not be entirely rigorous for a mathematician, may satisfy those interested only in understanding the particular phenomena. Given this, the role of a mathematician looking at the same model could become questionable from the point of view of understanding the phenomena. However this is a very narrow view. When a mathematician looks at these equations his aim is not to explain the phenomena which may be behind the formulation of the equation but to understand the solution itself in a global sense. As already mentioned the model is a simplification (approximation) of the real. Since approximations are involved there is need to understand the stability of solutions under perturbations, either of coefficients involved in the equation, or the initial data. If such a stability cannot be proved, there is always doubt about the correctness of the model.

Also a mathematician's desire to understand and study equations similar to the model but in a more general setting can have remarkable influence. An outstanding example is the development of Fourier Analysis and more general orthogonal expansions.

It is not always the case that Differential Equations arise only from physical considerations. Many questions in differential geometry are formulated as questions of existence of solutions of appropriate Differential Equations. An outstanding example is the so called Yamabe problem which has lead to some remarkable studies in the area of Nonlinear elliptic Equations in the non compact case.

The aim of this article is to explain few basic facts and leave it to the reader to explore the ever increasing questions/phenomena modeled through differential equations and the beautiful Mathematical Analysis that arise.

In the next section we will illustrate some of these points.

1   Theorems and Models

Recall that an ordinary differential equation involves an unknown function and its derivatives. In such a case, a reasonable ODE has many solutions (Picard's Theorem). Given this it may seem reasonable to expect PDE's (with no boundary conditions) to have many solutions. In particular, if we look at

$$\sum_{|\alpha|\le k} a_\alpha(x)\partial^\alpha u(x) = f(x)$$

where f and a_α are C^∞, given $x_0 \in \mathbf{R}^n$, can we find a solution in a neighbourhood of "x_0" is the question. If f and a_α are analytic with $a_\alpha(x_0) \ne 0$ for some α with $|\alpha| = k$, the Cauchy - Kowaleski theorem says the answer is yes. However, Lewy came up with the example:

$$Lu = \frac{\partial u}{\partial x} + i\frac{\partial u}{\partial y} - 2i(x + iy)\frac{\partial u}{\partial t} = f$$

for which there exists no solution in a neighbourhood of "0".

This fundamental difference between ODE and PDE has lead to a rich topic of studies in "Local Solvability". The issue involved in these studies is one direction of research on PDE's and an excellent reference is (Hounie, 2002).

Now we turn our attention to the following specific PDE's.

1. $\Delta u = f$

2. $u_t - \Delta u = f$ and

3. $u_{tt} - \Delta u = f$

I am sure most readers of this article have seen these equations before, namely the *Laplace equation, Heat Equation* and the *wave equation*. These three PDEs, in some sense, are considered to be the most fundamental ones and we remark that though they are supposed to describe different physical phenomena, $\Delta = \sum_{i=1}^n \frac{\partial^2}{\partial x_i^2}$ is common to all of them. Hence the question: what is special about Δ? The answer lies in the following two theorems which we now state without proof and refer the reader to (Folland, 1995) for details.

Theorem 1. *If L is a differential operator with constant coefficients on $\mathbf{R}^n$ and $f \in C_0^\infty(\mathbf{R}^n)$ then there exists $u \in C^\infty(\mathbf{R}^n)$ such that $Lu = f$.*

Theorem 2. *Suppose L is a PDE on $\mathbf{R}^n$. Then L commutes with translations and rotations if and only if L is a polynomial in Δ, that is $L = \sum a_J\Delta^J$.*

The importance associated to Theorems 1 and 2 is to be seen in the following context. Consider the equation

$$\begin{cases} -\Delta u = f \text{ in } \Omega \\ u(x) = 0 \text{ on } \partial\Omega \end{cases} \tag{1}$$

where $\Omega \subset \mathbf{R}^n (n \geq 2)$ is a smooth bounded domain with boundary $\partial\Omega$, and f continuous. In general u is not a C^2 function. This is not just a case of a mathematician producing a pathological example but is really in the nature of things. For instance the Scalar conservation law

$$u_t + F(u)_x = 1$$

is a PDE governing different one-dimensional phenomena involving Fluid dynamics and models, the formation and propagation of shock waves (Shock wave is a curve of discontinuity of the solution). Theory of distributions, Sobolev spaces, concept of weak solutions arose out of the need to give meaning to solutions which may not be regular.

Turning to Theorem 2, most readers of the article may have heard the following sometime:

- *Physics is independent of position*, and

- *Physics is independent of Direction*.

A moment's reflection should convince the reader any PDE model representing aspects from physics will involve the Laplacian (Δ) due to Theorem 2.

Some Models:

Consider a fluid, flowing with a velocity V in a thin straight tube whose cross section shall be denoted by C. Suppose the fluid contains a chemical whose concentration at a position y at time t is $u(y, t)$. Then at time t, the amount of chemical in a section of the tube between two points, y_1 and y_2 is given by

$$\int_{y_1}^{y_2} u(y, t)C \, dy.$$

The amount of chemical that flows through a plane located at position y, during the time interval from t_1 to t_2 is

$$\int_{t_1}^{t_2} u(y, t)CV \, dt.$$

Then an equation expressing a chemical balance under ideal conditions is:

$$\int_{y_1}^{y_2} u(y, t_2)C\, dy = \int_{y_1}^{y_2} u(y, t_1)C\, dy + \int_{t_1}^{t_2} u(y, t)CV\, dt - \int_{t_1}^{t_2} u(y_2, t)C\, dt$$

that is, the amount of chemical contained in the section (y_1, y_2) at time t_1 plus the amount of chemical that flowed through the plane at position y during (t_1, t_2) minus the amount that flowed through the plane at position y_2 during the same time interval (t_1, t_2). Recall the fundamental theorem of Calculus: We have

$$\int_{y_1}^{y_2} u(y, t_2)C\, dy - \int_{y_1}^{y_2} u(y, t_1)C\, dy = \int_{y_1}^{y_2}\int_{t_1}^{t_2} \partial_t u(y, t)C\, dt\, dx$$

and

$$\int_{t_1}^{t_2} u(y, t)CV\, dt - \int_{t_1}^{t_2} u(y_2, t)CV\, dt = -\int_{t_1}^{t_2}\int_{x_1}^{x_2} \partial_y u(y, t)CV\, dx\, dt$$

and this leads to

$$\int_{y_1}^{y_2}\int_{t_1}^{t_2} \partial_t[u(y, t)C]\, dt\, dx + \int_{t_1}^{t_2}\int_{y_1}^{y_2} \partial_y[u(y, t)CV]\, dx\, dt = 0.$$

If we assume that the equality holds for every segment (y_1, y_2) and time interval (t_1, t_2) and also the function u and its partial derivatives are continuous then we have that

$$\partial_t[u(y, t)C] + \partial_y[u(y, t)CV] = 0 \quad \forall(y, t).$$

When V and C are constants, this reduces to

$$\partial_t u(y, t) + V\partial_y(y, t) = 0 \quad \forall(y, t).$$

This is the so called *Transport Equation* in one dimension.

Remark 1. If $G : \mathbf{R} \to \mathbf{R}$ is any smooth function then $u(y, t) = G(y - Vt)$. This leads to the fact that a solution to a linear first order PDE contains an arbitrary function in contrast to the general solution for a linear first order ODE which contains an arbitrary constant.

Remark 2. Similar considerations for a region in $\mathbf{R}^n$ leads to the Transport Equation

$$\partial_t u(y, t) + \vec{V}. \text{ grad } u(y, t) = 0.$$

Note that in deriving this equation, a certain boundary integral has to be converted into a volume integral. This is achieved through one of the most profound integral identities, namely, the Gauss divergence theorem.

Physical considerations such as heat flows from regions of high temperature (chemicals flow from regions of high concentration) to regions of low temperature (to regions of low concentration) lead to Heat Equation (Diffusion Equation)

$$\partial_t u(\tilde{y}, t) - \text{div}(\text{grad } u(\tilde{y}, t)) = 0.$$

Now $\text{div}(\text{grad } u(\tilde{y}, t)) = \Delta u$, gives the form stated earlier.

The Equations mentioned in Remark 1 are studied extensively as initial value problems or initial boundary value problems to understand the behaviour of solutions in short time or long time frame work.

Remark 3. In the steady state, the natural differential equations to study are problems which involve only the Laplacian operator and hence there is a vast literature concerning the study of the Laplacian or higher order Laplacian.

Remark 4. Maximum (Minimum) principle involved due to physical reasons have been extended and infact Hopf maximum principle has arisen out of mathematical developments. Also Harnack inequalities, which is a consequence of maximum principle, has been extremely useful in the development of the theory of harmonic functions. Since these provide information on the solutions without explicit knowledge of the solution, they are of importance to numerical analysists in approximating the solutions.

Conclusion: The idea behind this article is to highlight some interactions between mathematics and physical sciences and how these interactions contribute to furtherance of knowledge. To keep the exposition short only very few developments have been highlighted. Also no effort has been made to discuss nonlinear PDEs even though the current research activities are focussed on understanding Nonlinear problems. An excellent reference for general developments in PDE is (Brezis and Browder, 1998).

References

Brezis, H. and Browder, F. 1998. Partial Differential Equations in the 20th Century, *Advances in Mathematics*. **135**:76–144.

Folland, G.B. 1995. *Introduction to Partial Differential Equations*, Second edition. Princeton University Press, Princeton, NJ.

Hounie, J. 2002. Fifty years of Local Solvability, Seventh Workshop on Partial Differential Equations, Part II (Rio de Janeiro, 2001), *Mat. Contemp.* **23**:1–17.

CHAPTER 10

Large Deviations

S.R.S. Varadhan

Courant Institute, New York University
251, Mercer Street
New York, NY, 10012, USA.
e-mail: varadhan@cims.nyu.edu

1 Introduction

If a fair coin is tossed, the probability is $\frac{1}{2}$ that it will land on its head (or tail). If the coin is tossed N times, every specific sequence of heads and tails will have probability 2^{-N}. There are $\binom{N}{r}$ individual sequences of length N that have exactly r heads and $N - r$ tails, each one of them contributing probability 2^{-N}. Therefore probability of obtaining exactly r heads in N tosses is

$$P(N, r) = \binom{N}{r} 2^{-N}. \tag{1}$$

2 What is 'Large Deviations'?

While we can not predict the outcome of a single toss of a coin, if we repeat it a large number times we do expect that heads and tails will occur roughly an equal number of times. Of course we can not really expect that the number of heads and tails to be exactly equal. There will be deviations or fluctuations. If we toss the coin N times the number of heads (or tails) that we expect to have is $\frac{N}{2}$. If X is the actual number obtained $X = \frac{N}{2} + Y$ where Y is the deviation from the expectation. We saw that the probability that $X = r$ is calculated easily and is given by (1)

The magnitude of the fluctuation is $\sqrt{N}$ and we can calculate asymptotically as $N \to \infty$ the probability that $Y = X - \frac{N}{2}$ is in the interval $\left[x_1 \sqrt{N}, x_2 \sqrt{N}\right]$

$$\lim_{N \to \infty} \sum_{x_1 \sqrt{N} \leq r - \frac{N}{2} \leq x_2 \sqrt{N}} P(N, r) = \sqrt{\frac{2}{\pi}} \int_{x_1}^{x_2} e^{-2y^2}\, dy \qquad (2)$$

which is the central limit theorem. Deviations from $\frac{N}{2}$ that are larger than $\sqrt{N}$ in magnitude have probabilities that go to 0 with N. For example if $\alpha > \frac{1}{2}$ and $x > 0$,

$$P\left[X \geq \frac{N}{2} + x N^{\alpha}\right] \to 0$$

as $N \to \infty$. If $\frac{1}{2} < \alpha < 1$ we can show that for $x > 0$,

$$P\left[X \geq \frac{N}{2} + x N^{\alpha}\right] \simeq \sqrt{\frac{2}{\pi}} \int_{xN^{\alpha-\frac{1}{2}}}^{\infty} e^{-2y^2}\, dy = \exp\left[-2x^2 N^{2\alpha-1} + o(N^{2\alpha-1})\right].$$

But if $\alpha = 1$, the answer is different and

$$P[X \geq Nx] = \exp[-Nh(x) + o(N)]$$

where

$$h(x) = x \log(2x) + (1 - x) \log(2(1 - x)). \qquad (3)$$

We note that $h(x) > 0$ if $x > \frac{1}{2}$ and $h(x) \simeq 2(x - \frac{1}{2})^2$ as $x \to \frac{1}{2}$. It is not hard to prove these results. One can approximate $\binom{N}{r} = \frac{N!}{r!(N-r)!}$ by using Stirling's formula

$$n! = \sqrt{2\pi} e^{-n} n^{n+\frac{1}{2}}(1 + o(1)).$$

The central limit theorem requires careful analysis, because the summation over r makes a difference. If $r = \frac{N}{2} + x \sqrt{N}$, terms $P(N, r)$ can be approximated as

$$P(N, r) = \sqrt{\frac{2}{N\pi}} \exp\left[-2x^2 + o(1)\right]$$

The sum over r is identified as a Riemann sum approximation for the integral (2). For the large deviation probabilities, where we are interested only in the rate of exponential decay, the sum involves at most N terms and the exponential rate of decay of the sum is the same as the largest summand. If $r = Nx$, again by Stirling's formula

$$P(N, r) = \exp\left[-Nh(x) + o(N)\right] \tag{4}$$

with $h(x)$ as in (4). The function $h(x)$ is easily seen to be nondecreasing for $x > \frac{1}{2}$. Therefore asymptotically for $x > \frac{1}{2}$

$$P[X \geq Nx] = \sum_{r \geq Nx} P(N, r) = \exp\left[-Nh(x) + o(N)\right].$$

3   Entropy

If $\{p_j\}$, $j = 1, 2, \ldots k$ are probabilities that add up to 1, then the entropy function $H(p_1, p_2, \ldots, p_k)$ is defined as

$$H(p_1, p_2, \ldots, p_k) = -\sum_i p_i \log p_i.$$

Although there are different ways of interpreting this function, one particular point of view is relevant for Large Deviations. Suppose we distribute N balls into k boxes and we want the i-th box to get get n_i balls. Then $(n_1, n_2, \ldots, n_k)$ defines how the balls are to be distributed. There are k^N ways of distributing N balls into k boxes and

$$f(N, n_1, n_2, \ldots, n_k) = \frac{N!}{n_1! n_2! \ldots n_k!}$$

of them will lead to the distribution $(n_1, \ldots, n_k)$ that we want. Using Stirling's approximation once again we can see that if for $i = 1, \ldots, k$, $\frac{n_i}{N} \simeq p_i$,

$$f(N, n_1, n_2, \ldots, n_k) = \exp\left[NH(p_1, \ldots, p_k) + o(N)\right].$$

Let us now distribute N balls randomly into k boxes. The probability that a given ball will be placed in box i is π_i. Different balls are distributed independently of one another. Let us count the number of balls in each box. The probability of having for $i = 1, 2, \ldots, k$, exactly n_i balls in box i is given by a formula analogous to (4),

$$P(N, \pi_1, \ldots, \pi_k; n_1, \ldots, n_k) = \frac{N!}{n_1! n_2! \cdots n_k!} \pi_1^{n_1} \ldots \pi_k^{n_k}$$

$$= \exp\left[N \sum_i p_i \log \pi_i - N \sum_i p_i \log p_i\right]$$

$$= \exp\left[-N \sum_i p_i \log \frac{p_i}{\pi_i}\right]. \tag{5}$$

This leads to a simple version of what is known as Sanov's theorem (Sanov, 1957). If we distribute N balls independently into k boxes with probabilities π_i for any given ball to be placed in box i, then by the law of large numbers the observed proportions $\{\frac{f_i}{N}\}$ will be close to $\{\pi_i\}$ with high probability. However the probability that they are instead close to some other $\{p_i\}$ is not zero, but exponentially small and is given by

$$\exp\left[-N \sum_i p_i \log \frac{p_i}{\pi_i} + o(N)\right].$$

It is not hard to make this precise. Let us take $\delta > 0$ and ask for the probability $P(N, \{p_i\}, \delta)$ that

$$\sup_{1 \le i \le k} \left| \frac{f_i}{N} - p_i \right| \le \delta.$$

Then for N sufficiently large (depending on δ)

$$\exp\left[-N \sum_i p_i \log \frac{p_i}{\pi_i} - c(\delta)N\right] \le P(N, \{p_i\}, \delta)$$

$$\le \exp\left[-N \sum_i p_i \log \frac{p_i}{\pi_i} + c(\delta)N\right]$$

and $c(\delta) \to 0$ as $\delta \to 0$. This can be stated as

$$\limsup_{\delta \to 0} \limsup_{N \to \infty} \left| \frac{1}{N} \log P(N, \{p_i\}, \delta) + \sum_i p_i \log \frac{p_i}{\pi_i} \right| = 0.$$

The quantity

$$H(\{p_i\}, \{\pi_i\}) = \sum_i p_i \log \frac{p_i}{\pi_i}$$

is called relative entropy or Kullback-Leibler information number. $H(\{p_i\}, \{\pi_i\}) \ge 0$ and equality holds if and only if $p_i = \pi_i$ for every i.

4 Cramér's Theorem

If $\{X_i\}$ is a sequence of independent random variables one would expect under suitable assumptions that for $a > E[X]$,

$$\lim_{N\to\infty} P\left[\frac{X_1 + \cdots + X_N}{N} \geq a\right] = 0.$$

Perhaps there is exponential decay and

$$P\left[\frac{X_1 + \cdots + X_N}{N} \geq a\right] = e^{-NI(a)+o(N)}$$

for some function $I(a)$ that can be determined. We can get a bound by estimating for $\lambda > 0$,

$$P\left[\frac{X_1 + \cdots + X_N}{N} \geq a\right] \leq e^{-N\lambda a} E\left[e^{\lambda(X_1 + \cdots + X_N)}\right]$$

$$= e^{-N\lambda a}\left[E\left[e^{\lambda X_1}\right]\right]^N$$

$$= \exp\left[-N\left[\lambda a - \log E\left[e^{\lambda X_1}\right]\right]\right].$$

Thus

$$P\left[\frac{X_1 + \cdots + X_N}{N} \geq a\right] \leq \exp\left[-N\left[I(a)\right]\right]$$

where $I(a) = \sup_{\lambda>0}[\lambda a - \log E[e^{\lambda(X_1)}]]$. By Jensen's inequality for $\lambda \leq 0$, and $a > E[X_1]$

$$\log E\left[e^{\lambda(X_1)}\right] \geq \lambda E[X_1] \geq \lambda a$$

and

$$I(a) = \sup_{\lambda>0}\left[\lambda a - \log E\left[e^{\lambda(X_1)}\right]\right] = \sup_{\lambda}\left[\lambda a - \log E\left[e^{\lambda(X_1)}\right]\right].$$

For example if X_i are independent standard normal random variables, $\log E[e^{\lambda X_1}] = \frac{\lambda^2}{2}$. Then $I(a) = \frac{a^2}{2}$. There is a multi dimensional version of Cramér's theorem.

$$I(a) = \sup_{\theta \in R^d}\left[\langle a, \theta\rangle - \log E[e^{\langle\theta,X\rangle}]\right]$$

It is to be interpreted as

$$P\left[\frac{X_1 + \cdots + X_n}{n} \simeq a\right] = e^{-nI(a)+o(n)}.$$

More precisely if $A^o, \bar{A}$ are the interior and closure of the set A,

$$\exp\left[-n \inf_{a \in A^o} I(a) + o(n)\right] \le P\left[\frac{X_1 + \cdots + X_n}{n} \in A\right]$$

$$\le \exp\left[-n \inf_{a \in \bar{A}} I(a) + o(n)\right].$$

The multinomial distribution can be viewed as the distribution of a random variable X with values in R^k and

$$P[X = e_i] = \pi_i$$

where $\{e_i\}$ are the standard basis vectors, i.e., $e_i = (0, \ldots, 1, \ldots, 0)$ with 1 in the i-th coordinate. Then $\sum_{j=1}^{n} X_j$ will be the frequency count $\{f_i\}$ and will have for its distribution the multinomial distribution. It is easy to make the following calculations.

$$M(\theta) = \sum_i \pi_i e^{\theta_i}$$

$$I(a) = \sup_a \left[\langle a, \theta \rangle - \log \sum_i \pi_i e^{\theta_i}\right]$$

is easily calculated. It is infinite unless $a_i \ge 0$ and $\sum_i a_i = 1$. In that case it is the relative entropy

$$\sum_i a_i \log \frac{a_i}{\pi_i}.$$

In other words Cramér's theorem (Cramér, 1938) is the same as Sanov's theorem for the multinomial distribution.

5 Markov Chains

While Sanov's theorem and Cramér's theorem deal with independent random variables, one would like to understand the corresponding results when we have dependence. The simplest such dependence is that of a Markov chain. For simplicity we will consider a Markov chain with two states. Imagine an experiment where we have two coins, coin one with $P(H) = 1 - P(T) = \frac{3}{4}$ and coin two with $P(H) = 1 - P(T) = \frac{1}{4}$. We start with one of the two coins, and every time after a head appears we choose coin one for the next toss and coin two after a tail. Now the successive tosses are not independent. Head favors a head for the next toss and tail favors a tail. This describes a Markov chain with two states: $\{0, 1\}$. Head corresponds to 1 and tail to 0. Transition probabilities are given by the following 2×2 matrix $\pi = \{\pi_{i,j}\}; i, j = 0, 1$.

$$\begin{pmatrix} 3/4 & 1/4 \\ 1/4 & 3/4 \end{pmatrix}$$

Then $S_N = (X_1 + \cdots + X_N)$ is the number of heads in N tosses. While the nature of the initial coin introduces an initial bias we can remove it by making a random choice of the initial coin with equal probability for either one. The probability of a sequence of heads and tails are now equal and $E[X_i] = \frac{1}{2}$. We expect the law of large numbers to hold with

$$P\left[\left|\frac{S_N}{N} - \frac{1}{2}\right| \geq \delta\right] \to 0$$

as $N \to \infty$. In this Markov model if $X_1, X_2, \ldots, X_N$ is a string, to compute its probability we need to know the number of transitions of each of the four types $(0, 0), (0, 1), (1, 0), (1, 1)$. We will denote them by $n_{00}, n_{01}, n_{10}, n_{11}$ respectively. The probability of the string will be

$$P(X_1, X_2, \ldots, X_N) = \frac{1}{2}\left[\frac{3}{4}\right]^{n_{00}}\left[\frac{1}{4}\right]^{n_{01}}\left[\frac{1}{4}\right]^{n_{10}}\left[\frac{3}{4}\right]^{n_{11}} \tag{6}$$

We will now compute $E[e^{\lambda S_N}] = E\left[e^{\lambda[X_1 + \cdots + X_N]}\right]$. If we define

$$\pi^\lambda = \begin{pmatrix} 3e^\lambda/4 & 1/4 \\ e^\lambda/4 & 3/4 \end{pmatrix}$$

then we can verify either by induction on N or by explicit calculation, that

$$E\left[e^{\lambda S_N}\right] = (1/2 \ 1/2)\begin{pmatrix} 3e^\lambda/4 & 1/4 \\ e^\lambda/4 & 3/4 \end{pmatrix}^N \begin{pmatrix} 1 \\ 1 \end{pmatrix}$$

and the behavior of $\log E[e^{\lambda S_N}]$ is controlled by the largest eigen-value σ_λ of π^λ

$$\sigma_\lambda = \frac{1}{8}\left[3(e^\lambda + 1) + \sqrt{9(e^\lambda - 1)^2 + 4e^\lambda}\right]$$

$$\lim_{N \to \infty} \frac{1}{N} \log E\left[e^{\lambda S_N}\right] = \log \sigma_\lambda$$

If we define

$$J(a) = \sup_\lambda[\lambda a - \log \sigma_\lambda]$$

then

$$P\left[\frac{S_N}{N} \simeq a\right] = \exp\left[-NJ(a) + o(N)\right].$$

If the transition probabilities were given instead by the matrix

$$\pi = \begin{pmatrix} 1 - p_1 & p_1 \\ p_2 & 1 - p_2 \end{pmatrix}$$

then the invariant or stationary probability $(p, 1 - p)$ would be the solution of

$$p(1 - p_1) + (1 - p)p_2 = p$$
$$pp_1 + (1 - p)(1 - p_2) = 1 - p$$

or

$$\frac{p}{1 - p} = \frac{p_2}{p_1}.$$

The ergodic theorem would now say that $\frac{1}{N}\sum_{i=1}^{N} X_i$ converges to p and we want $p = a$. Under the two models the probabilities are as in (6) for what we have and

$$P(X_1, X_2, \ldots, X_N) = \frac{1}{2}(1 - p_1)^{n_{00}} p_1^{n_{01}} p_2^{n_{10}} (1 - p_2)^{n_{11}}$$

for what we would like to have. The relative entropy is calculated as

$$E\left[n_{00} \log \frac{4(1 - p_1)}{3} + n_{01} \log(4p_1) + n_{10} \log(4p_2) + n_{11} \log \frac{4(1 - p_2)}{3}\right]$$

when divided by N this has a limit as $N \to \infty$ which equals

$$a(1 - p_1) \log \frac{4(1 - p_1)}{3} + ap_1 \log(4p_1)$$
$$+ (1 - a)p_2 \log(4p_2) + (1 - a)(1 - p_2) \log \frac{4(1 - p_2)}{3}$$

and we want to minimize this over p_1, p_2 subject to

$$\frac{a}{1-a} = \frac{p_2}{p_1}$$

and this will reproduce $J(a)$.

It is more convenient to consider Markov chains in continuous time, where the jumps occur at random times. The process, if it is in state 0 waits for an exponential random time with parameter λ_0 before jumping to 1 and if it is at 1 waits for an exponential random time with parameter λ_1 before jumping to 0. The generator of the process is the matrix

$$A = \begin{pmatrix} -\lambda_0 & \lambda_0 \\ \lambda_1 & -\lambda_1 \end{pmatrix}$$

and the transition probabilities are given by

$$\begin{pmatrix} p_{0,0}(t) & p_{0,1}(t) \\ p_{1,0}(t) & p_{1,1}(t) \end{pmatrix} = e^{tA}.$$

If v_0, v_1 are two values and $a(t) = \int_0^t v_{x(s)} ds$ then Feynman-Kac formula computes for $i, j = 0, 1$

$$E\left[e^{a(t)} \mathbf{1}_{x(t)=j} | x(0) = i\right] = p_{i,j}^v(t) = e_{i,j}^{t(A+V)}$$

where V is the diagonal matrix

$$\begin{pmatrix} v_0 & 0 \\ 0 & v_1 \end{pmatrix}.$$

If $\lambda_0 = \lambda_1 = 1$, the invariant probability is $p_0 = p_1 = \frac{1}{2}$ and the ergodic theorem asserts as before that for large t, $\frac{a(t)}{t} \simeq \frac{v_0+v_1}{2}$ with high probability. Taking $v_0 = -1$ and $v_1 = 1$, the large deviation rates for the difference between the amounts of time spent at 0 and 1 can be calculated as

$$J(a) = \sup_\lambda [av - \sigma(v)]$$

where

$$\sigma(v) = \lim_{t \to \infty} \log\left[e_{i,0}^{t(A+V)} + e_{i,1}^{t(A+V)}\right]$$

and with

$$V = \begin{pmatrix} -v & 0 \\ 0 & v \end{pmatrix}$$

$\sigma(v)$ is the largest eigen-value of $A + V$ and is the largest root of

$$(-1 - v - \sigma)(-1 + v - \sigma) - 1 = 0$$

or

$$(1 + \sigma^2) - v^2 = 1$$

and

$$\sigma(v) = \frac{-1 + \sqrt{v^2 + 1}}{2}.$$

One can now complete the calculation

$$J(a) = \sup_v[av - \sigma(v)] = 1 - \sqrt{1 - a^2}.$$

Tilting involves changing the rates from $1, 1$ to λ_0, λ_1 so that the invariant probabilities are $(\frac{1-a}{2}, \frac{1+a}{2})$ so that now we expect $\frac{a(t)}{t} \simeq a$. This requires

$$\left(\tfrac{1}{2}(1-a) \quad \tfrac{1}{2}(1+a)\right) \begin{pmatrix} -\lambda_0 & \lambda_0 \\ \lambda_1 & \lambda_1 \end{pmatrix} = 0$$

or

$$\frac{\lambda_0}{\lambda_1} = \frac{1+a}{1-a}.$$

The entropy cost of changing a Poisson process from rate 1 to rate λ is

$$\sum_r e^{-\lambda t} \frac{(\lambda t)^r}{r!} \log \frac{e^{-\lambda t}(\lambda t)^r}{e^{-t} t^r} = t[\lambda \log \lambda - \lambda + 1].$$

We need to minimize the total entropy cost of

$$\frac{1-a}{2}[\lambda_0 \log \lambda_0 - \lambda_0 + 1] + \frac{1+a}{2}[\lambda_1 \log \lambda_1 - \lambda_1 + 1]. \qquad (7)$$

subject to $\frac{\lambda_0}{\lambda_1} = \frac{1+a}{1-a}$. Using Lagrange multipliers, we have

$$\min_{\lambda_0, \lambda_1} \left[\frac{1-a}{2}[\lambda_0 \log \lambda_0 - \lambda_0 + 1] + \frac{1+a}{2}[\lambda_1 \log \lambda_1 - \lambda_1 + 1] \right.$$

$$\left. + \lambda[\lambda_0(1-a) - \lambda_1(1+a)] \right]$$

$$\frac{1-a}{2} \log \lambda_0 + \lambda(1-a) = \frac{1+a}{2} \log \lambda_1 - \lambda(1+a) = 0$$

which simplifies to $\lambda_0 \lambda_1 = 1$. Therefore

$$\lambda_0 = \sqrt{\frac{1+a}{1-a}}; \quad \lambda_1 = \sqrt{\frac{1-a}{1+a}}$$

Subtituting it in (7) we arrive again at $J(a) = 1 - \sqrt{1 - a^2}$. The lecture notes of Kac (1980) and the survey article (Varadhan, 2008) contain additional useful information and references.

References

Cramér, Harald. 1938. Sur un nouveau théorème-limite de la thorie des probabilités, *Actualités Scientifiques et Industrielles*. **736**:5–23.

Kac, Mark. 1980. Integration in function spaces and some of its applications, Accademia Nazionale dei Lincei, Scuola Normale Superiore, *Lezioni Fermiane*.

Sanov, I.N. 1957. On the probability of large deviations of random magnitudes. *Mat. Sb. N. S.* **42(84)**.

Varadhan, S.R.S. 2008. Large deviations. *Ann. Probab.* **36(2)**:397–419

CHAPTER 11

From the Binary Alphabet to Quantum Computation

K.R. Parthasarathy

Indian Statistical Institute, Delhi Centre,
New Delhi 110 016, India.
www.isid.ac.in

1 Some historical remarks on the binary alphabet

The goal of quantum computation is the reduction of numerical computations as well as communication by language and pictures to manipulations of finite length sequences of the letters 0 and 1 by viewing them as labels of 'states' of a number of '2-level quantum systems' and effecting changes of states by the methods of quantum mechanics. Two typical examples of 2-level quantum systems are: (1) the electron with its spin 'up' and 'down' states; (2) the ammonia molecule NH_3 in which the nitrogen atom can be 'above' or 'below' the plane of the three hydrogen atoms. (See Feynman (1965)).

The two point set $\{0, 1\}$ is denoted by $\mathbb{F}_2$. It is called the *binary alphabet* and each element in it is called a *bit*, an abbreviation for binary digit. If 0 and 1 are viewed as real numbers every positive integer j admits a unique binary expansion

$$j = \sum_{r=0}^{n-1} x_r 2^r, \quad x_{n-1} \neq 0, \quad x_r \in \{0, 1\} \quad \forall r \tag{1}$$

for some n. Such a number j is called an *n-bit integer*. On the other hand $\mathbb{F}_2$ is also a field of two elements with the addition and multiplication operations defined by

$$0 \oplus 0 = 1 \oplus 1 = 0, 0 \oplus 1 = 1 \oplus 0 = 1,$$
$$0 \cdot 0 = 0 \cdot 1 = 1 \cdot 0 = 0, 1 \cdot 1 = 1.$$

All arithmetical operations of addition, subtraction, multiplication and division on finite bit integers and fractions can be reduced to finite numbers of operations on bit sequences using operations in $\mathbb{F}_2$ recursively.

According to Sridharan (2005) the discovery of the binary expansion (1) and its algorithmic use in arithmetic can be traced back to 300 B.C. in the work of Pingala on Chandas sastra, constituting the study of the metrical system of syllable sequences in Sanskrit prosody. That all arithmetic can be performed equally well in the binary system (eventhough we are accustomed to the decimal scheme) using $\mathbb{F}_2$-operations can be traced to the essay in French by G.W. Leibnitz (the co-inventor of calculus with Newton) in 1703: *Explication de l'arithmetique binaire, qui se sert des seuls caractèrs 0 et 1: avec des remarques sur son utilité et sur quèlle donne le sens des anciennes figures Chinoises de Fo-hy,* Histoires de l'académie des sciences de Paris, 1713, 58-63. In honour of Leibnitz's discovery of binary arithmetic the Duke of Braunschweig, Rudolf August struck a medal. Leibnitz proposed the problem of computing π in the binary system and an incomplete solution was given by Jacob Bernoulli. In 1879, the binary expansion of $\log \pi$ appeared in Journal de Mathématique élémentaires, Paris. For more details we refer to the notes in pages 139–142 of the Americal Mathematical Monthly, vol. 25, 1918.

In technology, the binary alphabet with letters dot $\cdot$ and dash $-$ (in place of 0 and 1) appeared in 1837 when S.F. Morse exhibited a working model of the electromagnetic telegraph using the 'Morse Code' in New York. It encodes the 26 letters $A, B, \ldots, Z$ of the Latin alphabet and the 10 standard numerals $0, 1, 2, \ldots, 9$ according to

We may denote $\cdot$ and $-$ by 0 and 1 respectively and view the Morse code as a collection of code words from the binary alphabet $\mathbb{F}_2$. The most frequently occuring vowel in the English language is E and is encoded as 0. Similary, the most frequently occuring consonant is T which is encoded as 1. The table in the Morse code indicates the philosophy that a more frequent letter from the alphabet of the language is encoded by a

Table 1:

A	· −	N	− ·	1	· − − − −
B	− · · ·	O	− − −	2	· · − − −
C	− · − ·	P	· − − ·	3	· · · − −
D	− · ·	Q	− − · −	4	· · · · −
E	·	R	· − ·	5	· · · · ·
F	· · − ·	S	· · ·	6	− · · · ·
G	− − ·	T	−	6	− · · · ·
H	· · · ·	U	· · −	7	− − · · ·
I	· ·	V	· · · −	8	− − − · ·
J	· − − −	W	· − −	9	− − − − ·
K	− · −	X	− · · −	0	− − − − −
L	· − · ·	Y	− · − −		
M	− −	Z	− − · ·		

shorter binary word and a less frequent letter is encoded by a longer binary word. This rough idea of choosing a code based on frequencies or statistics was developed into a sophisticated science called 'information theory' by Claude Shannon in 1948. For a leisurely and accessible account of this beautiful subject we suggest (Cover and Thomas, 2006).

Around 1906, the Russian mathematician A.A. Markov made a statistical analysis of the frequencies of vowels and consonants by examining 20,000 letters in a portion of A. Pushkin's poem 'Eugene Onegin' and S.T. Aksarov's novel 'The childhood of Bagrov, the grandson'. Denoting a vowel by 0 and a consonant by 1 his frequency and transition frequency data were as in Table 2.

The top row gives the frequencies of vowels and consonants. If a, $b \in \{0, 1\}$ the ratio of the frequency of the combination ab with respect to the frequency of a in the text is the frequency of transition from a to b. The bottom row gives the matrix of transition frequencies.

Whereas the frequency tables for vowels and consonants in Pushkin's poem and Aksarov's prose are close to each other their transition frequency tables are dramatically different. This motivated Markov to formulate the notion of a Markov Chain based on a transition probability matrix and ini-

Table 2:

	Pushkin			Aksarov	
0	1		0	1	
0.432	0.568		0.449	0.551	

	0	1		0	1
0	0.128	0.872	0	0.552	0.448
1	0.663	0.337	1	0.365	0.635

tiate a theory of dependent random variables in his famous book (Markov, 1913). Today, the theory of Markov Chains and Processes with their transition probabilty operators is a highly developed branch of probability theory and from statistical mechanics through population growth to market fluctuations there is no area of scientific investigation left uninfluenced by this development.

2   Classical computation

Any numerical computation in the real line upto any desired, but fixed, degree of accuracy can be reduced to the realization of a map $f : \mathbb{F}_2^n \rightarrow \mathbb{F}_2^m$ for some positive integers n and m. Such a map is called a *classical gate* or, simply, a *gate* with an n-bit *input* and an m-bit *output*. If $f = (f_1, f_2, \ldots, f_m)$ where $f_j(x)$ is the j-th coordinate of $f(x)$ for any $\underline{x} = (x_1, x_2, \ldots, x_n)$, $1 \leq j \leq m$, $x_i \in \mathbb{F}_2$, we observe that all number crunching ultimately reduces to realizations of gates with an n-bit input and a 1-bit output. The aim of classical computation is to express such gates as compositions of a few elementary gates with atmost 2 input and output bits and keep a control over the number of such compositions in any specific computation.

A gate f with an n-bit input and an m-bit output is expressed in the form of a *circuit diagram* 1:

Example 1. NOT gate

$$x \quad \longrightarrow\!\!\!\!\!\!\rhd\!\!\circ\!\!- \quad 1 \oplus x \qquad f(x) = 1 \oplus x.$$

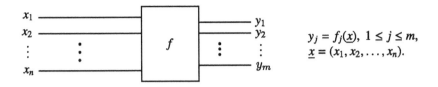

Diagram 1: *gate f*

$$y_j = f_j(\underline{x}), \ 1 \le j \le m,$$
$$\underline{x} = (x_1, x_2, \dots, x_n).$$

This is a gate with a 1-bit input and a 1-bit output. If the input is 0, the output is 1 and vice versa and hence the name NOT.

Example 2. AND gate

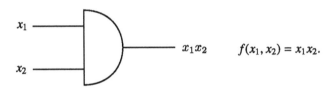

$$x_1 x_2 \qquad f(x_1, x_2) = x_1 x_2.$$

This is a gate with a 2-bit input and a 1-bit output with the property that the output is 1 if and only if both the input bits are 1. Since f is not invertible this is an example of an *irreversible* gate.

Example 3. OR gate

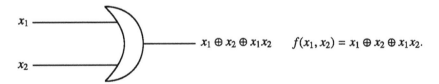

$$x_1 \oplus x_2 \oplus x_1 x_2 \qquad f(x_1, x_2) = x_1 \oplus x_2 \oplus x_1 x_2.$$

This is again a gate with a 2-bit input and a 1-bit output. The output bit is 1 if and only if at least one of the input bits is 1. It is also an irreversible gate.

Example 4. XOR gate

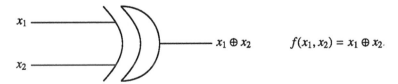

$$x_1 \oplus x_2 \qquad f(x_1, x_2) = x_1 \oplus x_2.$$

Here the output bit is 1 if and only if exactly one of the input bits is 1.

Example 5. NAND gate

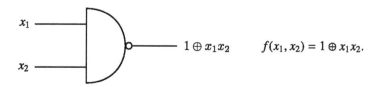

$1 \oplus x_1 x_2 \qquad f(x_1, x_2) = 1 \oplus x_1 x_2.$

This can also be expressed as a composition of Example 2 and 1:

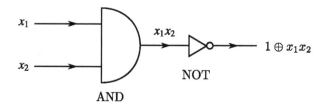

Example 6. FAN OUT

$f(x) = (x, x)$

This is a gate with a 1-bit input and a 2-bit output. It makes two copies of the input bit.

Since

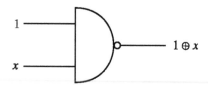

$1 \oplus x$

it follows that the output of NOT can be achieved by NAND with the input bits 1 and x. Applying NAND to the input (x_1, x_2) and following by NOT one can achieve AND. Thus AND can be achieved by repetitions of NAND. It is an interesting elementary exercise to show that Example 1–4 can be achieved through NANDs alone. What is remarkable is that any gate $f : \mathbb{F}_2^n \to \mathbb{F}_2^m$ can be achieved by repetitions of NAND and FAN OUT alone. We say that NAND and FAN OUT are *universal* for classical computation. We present a proof of this statement inductively in the form of a circuit diagram 2. Consider a gate $f : \mathbb{F}_2^n \to \mathbb{F}_2$. As mentioned at the beginning of this section it suffices to prove the result in this case. The result is trivial when $n = 1$. When $n > 1$, define

$$f_0(x_1, x_2, \ldots, x_{n-1}) = f(0, x_1, x_2, \ldots, x_{n-1}),$$
$$f_1(x_1, x_2, \ldots, x_{n-1}) = f(1, x_1, x_2, \ldots, x_{n-1}).$$

Assume that f_0 and f_1 gates have been realized. Now diagram 2 gives the realization of f.

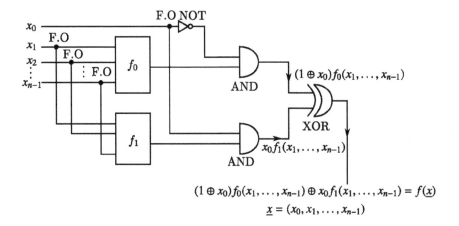

$$(1 \oplus x_0)f_0(x_1, \ldots, x_{n-1}) \oplus x_0 f_1(x_1, \ldots, x_{n-1}) = f(\underline{x})$$
$$\underline{x} = (x_0, x_1, \ldots, x_{n-1})$$

Diagram 2:

F.O. stands for FAN OUT.

In the inductive pictorial argument given above it is clear that the number of simple gates used in achieving f grows exponentially with the number of bits n. The number of simple gates used is also called the *computing time* of the circuit or the algorithm. For some maps it may be possible to restrain this exponential growth of computing time.

As an example we cite a famous and recent example. For any integer $N > 0$, let $f(N) = 1$ if N is a prime and $= 0$ otherwise. If N has n bits in the binary expansion one would like to have a circuit to compute $f(N)$ in which the number of elementary gates grows as a power of n. In 2002, M. Agrawal, N. Kayal and N. Saxena announced in an electronic preprint of IIT Kanpur a proof of the result that the computing time for $f(N)$ is $0((\log N)^{12})$. In their 2004 paper (Agrawal *et al.*, 2004) they incorporated later improvements and showed that the computing time for $f(N)$ is $0((\log N)^6)$. This paper got them the 2006 Gödel Prize.

3   Some aspects of finite level quantum systems and the Dirac notation

A *d-level quantum system A* is described by a *d*-dimensional complex vector space $\mathcal{H}^A$ equipped with a scalar product $\langle u|v\rangle$ between any two elements u, v in $\mathcal{H}^A$. It has the following properties: (1) $\langle u|v\rangle \epsilon \mathbb{C}$, is linear in v, conjugate linear in u and $\langle u|v\rangle = \overline{\langle v|u\rangle}$; (2) $\langle u|u\rangle \geq 0$ and $= 0$ if and only if $u = 0$. The quantity $\langle u|u\rangle^{1/2}$ is denoted by $\|u\|$ and called the *norm* of u. $\|cu\| = |c| \|u\|$ for any scalar c and $\|u + v\| \leq \|u\| + \|v\|$. $\mathcal{H}^A$ is called the *Hilbert space* of the quantum system A. Any element u in $\mathcal{H}^A$ is called a *ket vector* and denoted also as $|u\rangle$. Any linear functional λ on $\mathcal{H}^A$ is of the form $\lambda(v) = \langle u_\lambda|v\rangle \forall v$ for a unique u_λ in $\mathcal{H}^A$. Thus the dual of $\mathcal{H}^A$ can also be identified with $\mathcal{H}^A$. Any u in $\mathcal{H}^A$ considered as an element of this dual space is denoted by $\langle u|$, read as bra u and called a *bra vector*. The bra vector u evaluated at the ket vector v is the bracket $\langle u|v\rangle$, the scalar product between u and v.

Any linear transformation T of $\mathcal{H}^A$ into itself is called an *operator* on $\mathcal{H}^A$ and its adjoint is denoted by $T^\dagger$. We have for all u, v in $\mathcal{H}^A$

$$\langle u|Tv\rangle = \langle T^\dagger u|v\rangle.$$

This common value is denoted by $\langle u|T|v\rangle$. We also write

$$T|u\rangle = |Tu\rangle, \quad \langle v|T^\dagger = \langle Tv|.$$

The product $|u\rangle\langle v|$ is defined to be the operator satisfying

$$|u\rangle\langle v||w\rangle = \langle v|w\rangle|u\rangle \quad \forall w.$$

One has $(|u\rangle\langle v|)^\dagger = |v\rangle\langle u|$ and for any u_i, v_i, $1 \leq i \leq n$,

$$|u_1\rangle\langle v_1||u_2\rangle\langle v_2|\cdots|u_n\rangle\langle v_n| = c|u_1\rangle\langle v_n|$$

where $c = \prod_{j=1}^{n-1}\langle v_j|u_{j+1}\rangle$. If $u \neq 0$, $v \neq 0$, $|u\rangle\langle v|$ is a rank 1 operator with range $\mathbb{C}u$.

Denote by $\mathcal{B}(\mathcal{H}^A)$ the algebra of all operators on $\mathcal{H}^A$ equipped with the involution †. An operator T is said to be *normal* if $TT^\dagger = T^\dagger T$, *hermitian* or *selfadjoint* if $T = T^\dagger$, a *projection* if $T = T^\dagger = T^2$, *unitary* if $T^\dagger T = I$, the identity operator, and *positive* if $\langle u|T|u \rangle \geq 0 \forall u$.

According to the spectral theorem every normal operator T on $\mathcal{H}^A$ can be uniquely expressed as

$$T = \sum_{\lambda \in \sigma(T)} \lambda E_\lambda \qquad (2)$$

where $\sigma(T)$ denotes the *spectrum* or the set of all eigenvalues of T and E_λ is the (orthogonal) projection on the subspace $\{u|Tu = \lambda u\}$. If T is hermitian $\sigma(T) \subset \mathbb{R}$ and if T is unitary $\sigma(T)$ is contained in the unit circle of the complex plane. A hermitian operator T is called a real-valued *observable* of the quantum system assuming the values in $\sigma(T)$ and the eigen projection E_λ in (2) is called the 'event that the observable T assumes the value λ'. Any projection E is called an *event*. Thus an event is an observable assuming only the values 0 or 1.

Any set $\{u_j\} \subset \mathcal{H}^A$ is called an *orthonormal basis* for $\mathcal{H}^A$ if the $u'_j s$ span $\mathcal{H}^A$, $\langle u_i|u_j \rangle = \delta_{ij} \forall i, j$. The cardinality of any such orthonormal basis is the dimension d of $\mathcal{H}^A$. For any operator T we write

$$\mathrm{Tr}\, T = \sum_{i=1}^{d} \langle u_i|T|u_i \rangle$$

where the summation is over an orthormal basis. We observe that Tr T is independent of the choice of the basis and call it the trace of T. Any positive operator ρ of unit trace is called a *state* of the quantum system A. If E is an event, i.e., E is a projection and ρ is a state then $1 \geq \mathrm{Tr}\rho E \geq 0$ and $\mathrm{Tr}\rho E$ is called the *probability of E in the state ρ*. If T is an observable and (2) is its spectral resolution then, in the state ρ, T assumes the value λ with probability $\mathrm{Tr}\rho E_\lambda$ and the expectation of T is equal to $\sum_{\lambda \in \sigma(T)} \lambda \mathrm{Tr}\rho E_\lambda = \mathrm{Tr}\rho T$. The nth moment of T in the state ρ is $\mathrm{Tr}\rho T^n$ and its variance is $\mathrm{Tr}\rho T^2 - (\mathrm{Tr}\rho T)^2$.

By choosing an orthonormal basis in the range of E_λ for each $\lambda \in \sigma(T)$ in (2) we can express any normal operator T as $\sum_{i=1}^{d} \lambda_i |u_i\rangle\langle u_i|$ where $\{|u_i\rangle, 1 \leq i \leq d\}$ is an eigenbasis for T with u_i being unit vectors so that each $|u_i\rangle\langle u_i|$ is a one dimensional projection. Here the eigenvalues λ_i are inclusive of multiplicities and therefore a representation of T in this form need not be unique. In particular, any state ρ can be expressed as

$$\rho = \sum_{j=1}^{d} p_j |u_j\rangle\langle u_j\rangle$$

where $p_j \geq 0$, $\sum p_j = 1$ and $\{|u_j\rangle\}$ is an orthonormal eigenbasis for ρ. This shows that the set of all states of a system A is compact and convex and its extreme points are one-dimensional projections. One dimensional projections are called *pure states*. Thus every pure state is of the form $|u\rangle\langle u|$ where u is a unit vector. If u is changed to λu where λ is a scalar of modulus unity then $|\lambda u\rangle\langle \lambda u| = |u\rangle\langle u|$. Thus every pure state can be identified with a unit ray in $\mathcal{H}^A$ and vice versa. In spite of this ambiguity a unit vector $|u\rangle$ is called a *pure state* but what one really means is the one-dimensional projection $|u\rangle\langle u|$.

The expectation of an observable T in a pure state $|u\rangle$ is equal to $\mathrm{Tr}\,|u\rangle\langle u|T = \langle u|T|u\rangle$. If λ is an eigenvalue of T and $|u\rangle$ is a unit eigenvector of T for this eigenvalue then, in the pure state $|u\rangle$, T assumes the value λ with probability $\mathrm{Tr}\,|u\rangle\langle u|E_\lambda = \langle u|E_\lambda|u\rangle = \langle u|u\rangle = 1$.

We now come to the notion of measurement. If S is a finite set of cardinality not exceeding d and $\{E_s|s\epsilon S\}$ is a set of projections satisfying $E_s E_{s'} = \delta_{ss'} E_s$, $\sum_s E_s = I$ then we say that $M = \{E_s|s \in S\}$ is a *measurement* on the quantum system A with *values* in S. If A is in the state ρ and the measurement M is made on A then the *measurement value is s* with probability $\mathrm{Tr}\,\rho E_s$. In other words, the measurement M produces a classical random variable with values in S such that the probability for s is $\mathrm{Tr}\,\rho E_s$.

If A, B are two finite level quantum systems with respective Hilbert spaces $\mathcal{H}^A$, $\mathcal{H}^B$ then the composite system AB, i.e., the two systems considered together as a single system has the Hilbert space

$$\mathcal{H}^{AB} = \mathcal{H}^A \otimes \mathcal{H}^B$$

where the right hand side is the Hilbert space tensor product of $\mathcal{H}^A$ and $\mathcal{H}^B$. If $\mathcal{H}^A$ and $\mathcal{H}^B$ have dimension d, d' respectively then $\mathcal{H}^{AB}$ has dimension dd'. If $\{u_i, 1 \leq i \leq d\}$ and $\{u'_j, 1 \leq j \leq d'\}$ are orthonormal bases for $\mathcal{H}^A$ and $\mathcal{H}^B$ respectively then the product vectors $\{u_i \otimes u'_j, 1 \leq i \leq d, 1 \leq j \leq d'\}$ constitute an orthonormal basis for $\mathcal{H}^{AB}$. For any

$u, v \in \mathcal{H}^A$, $u', v' \in \mathcal{H}^B$ we have

$$\langle u \otimes u' | v \otimes v' \rangle = \langle u|v \rangle \langle u'|v' \rangle$$

We write

$$u \otimes v = |u \otimes v\rangle = |u\rangle|v\rangle,$$
$$\langle u \otimes v| = \langle u|\langle v|$$

for the product ket and bra vectors in $\mathcal{H}^{AB}$. If T is an operator in $\mathcal{H}^A \otimes \mathcal{H}^B$ and $\{|u'_j\rangle, 1 \le j \le d'\}$ is an orthonormal basis in $\mathcal{H}^B$ consider the sesquilinear form

$$\Lambda(u, v) = \sum_{j=1}^{d'} \langle u \otimes u'_j | T | v \otimes v'_j \rangle .$$

Then there exists a unique linear operator T^A in $\mathcal{H}^A$ such that

$$\langle u | T^A | v \rangle = \Lambda(u, v) \quad \forall u, v \in \mathcal{H}^A.$$

The operator T^A is independent of the choice of the orthonormal basis and called the *relative trace* of T over $\mathcal{H}^B$. Similarly, one can define the relative trace T^B of T in $\mathcal{H}^B$ over $\mathcal{H}^A$. Then

$$\text{Tr } T = \text{Tr } T^A = \text{Tr } T^B.$$

This is a 'Fubini's Theorem' in which tracing is looked upon as integration!

If S, S' are abstract sets of cardinality d, d', respectively,

$$\{|s\rangle | s \in S\}, \quad \{|s'\rangle | s' \in S'\}$$

are labeled orthonormal bases in $\mathcal{H}^A$, $\mathcal{H}^B$ respectively and

$$E_s = |s\rangle\langle s| \otimes I_B$$

is the tensor product of the 1-dimensional projection $|s\rangle\langle s|$ in $\mathcal{H}^A$ and the identity operator I_B in $\mathcal{H}^B$ then $\{E_s | s \in S\} = M^A$ is a measurement with values in S. Similarly, one can do a measurement M^B with values in S'.

If ρ is a state in $\mathcal{H}^{AB}$ then its relative traces ρ^A and ρ^B are states in $\mathcal{H}^A$ and $\mathcal{H}^B$ respectively, called the *marginal states*. If $\rho = \rho^A \otimes \rho^B$ then ρ is called a *product state*.

If the composite system AB is in a product state of the form $|s_0\rangle\langle s_0|\otimes\rho^B$ and the measurement M^A is the one described above through the labeled basis $\{|s\rangle, s \in S\}$ in $\mathcal{H}^A$ then the probability of getting the value s is

$$\mathrm{Tr}(|s_0\rangle\langle s_0| \otimes \rho^B)(|s\rangle\langle s| \otimes I_B) = \delta_{s_0 s}.$$

In other words, the measured value is s_0 with probability one. This interpretation of measurement turns out to be important in 'reading the output register of a quantum computer.' (See Section 4)

What we have done with a pair of quantum systems can be extended to any composite system consisting of several quantum systems A_j with Hilbert spaces $\mathcal{H}^{A_j}$ of dimension d_j respectively for $1 \leq j \leq m$. The composite system $A_1 A_2 \ldots A_m$ is described by the tensor product Hilbert space

$$\mathcal{H}^{A_1 A_2 \ldots A_m} = \mathcal{H}^{A_1} \otimes \mathcal{H}^{A_s} \otimes \cdots \otimes \mathcal{H}^{A_m}$$

spanned by product vectors $u_1 \otimes u_2 \otimes \cdots \otimes u_m$, $u_j \in A_j \forall j$ where

$$u_1 \otimes u_2 \otimes \cdots \otimes u_m = |u_1\rangle|u_2\rangle \cdots |u_m\rangle = |u_1 \otimes u_2 \otimes \cdots \otimes u_m\rangle$$
$$\langle u_1|\langle u_2| \cdots \langle u_m| = \langle u_1 \otimes u_2 \otimes \cdots \otimes u_m|$$

satisfying

$$\langle u_1 \otimes \cdots \otimes u_m|v_1 \otimes \cdots \otimes v_m\rangle = \prod_{j=1}^{m} \langle u_j|v_j\rangle.$$

If $\mathcal{H}^{A_i}$ has a labeled orthonormal basis $\{|s_i\rangle, s_i \in S_i\}$, $\#S_i = d_i$ then we write

$$|s_1\rangle|s_2\rangle \cdots |s_m\rangle = |s_1 s_2 \ldots s_m\rangle$$

so that $\mathcal{H}^{A_1 A_2 \ldots A_m}$ has the labeled orthonormal basis

$$\{|s_1 s_2 \ldots s_m\rangle, (s_1, s_2, \ldots, s_m) \in S_1 \times S_2 \times \cdots \times S_m\}.$$

We conclude this section with the dynamics of changes in quantum states. If U is a unitary operator in the Hilbert space $\mathcal{H}^A$ of a quantum system we call U a *quantum gate*. If a state ρ passes through this gate it changes to a new state $U\rho U^\dagger$. Note that $U\rho U^\dagger$ is also a state. If ρ is a pure state of the form $|u\rangle\langle u|$ where $|u\rangle$ is a unit vector then the changed state $U\rho U^\dagger$ is of the form $|v\rangle\langle v|$ where $|v\rangle = U|u\rangle$. Thus the gate U changes the pure state $|u\rangle$ to the pure state $U|u\rangle$. We describe this in the form of a circuit diagram:

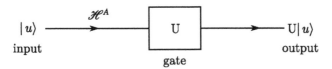

Denote the unitary group of $\mathcal{H}^A$, i.e., the group of all unitary operators on $\mathcal{H}^A$ by $\mathcal{U}^A$.

Consider a composite system $A_1 A_2 \ldots A_m$ with Hilbert space

$$\mathcal{H}^{A_1 A_2 \ldots A_m} = \mathcal{H}^{A_1} \otimes \mathcal{H}^{A_2} \otimes \cdots \otimes \mathcal{H}^{A_m}.$$

If U is in $\mathcal{U}^{A_j}$ denote by the same symbol U the product operator $I_{A_1} \otimes \cdots \otimes I_{A_{j-1}} \otimes U \otimes I_{A_{j+1}} \otimes \cdots \otimes I_m$. Thus $\mathcal{U}^{A_j}$ can be viewed as a subgroup of the unitary group $\mathcal{U}^{A_1 A_2 \ldots A_m}$ in $\mathcal{H}^{A_1 A_2 \ldots A_m}$. In a similar manner using permutation group it is possible to define the subgroup $\mathcal{U}^{A_i A_j}$ of $\mathcal{U}^{A_1 A_2 \ldots A_m}$. Any $U \in \mathcal{U}^{A_i A_j}$ exerts its influence only on the factor $\mathcal{H}^{A_i} \otimes \mathcal{H}^{A_j} = \mathcal{H}^{A_i A_j}$. In $\mathcal{H}^{A_1 A_2 \ldots A_m}$ with this convention we have the following theorem.

Theorem 1. *Any element U in $\mathcal{U}^{A_1 A_1 \ldots A_m}$ can be expressed as a finite product $U_1 U_2 \ldots U_n$ where n depends on U and each U_j is an element of $\mathcal{U}^{A_r A_s}$ for some $r, s \in \{1, 2, \ldots, m\}$.*

Remark. For a proof we refer to (Parthasarathy, 2003). A change in the state of a composite system $A_1 A_2 \ldots A_m$ is produced by changes in the individual systems $A_1, A_2, \ldots, A_m$ and interactions among the component systems. According to the theorem any such general change can be achieved through a finite sequence of changes each one of which is in a single system of the form A_j or an interactive change in a composite pair of the form $A_i A_j$ with $i \neq j$.

4 Quantum computing

In Section 2 we have already noted that all number crunching ultimately reduces to realizations of maps from $\mathbb{F}_2^n$ into $\mathbb{F}_2$. We shall now describe how a map $f : \mathbb{F}_2^n \to \mathbb{F}_2$ can be realized through a quantum gate, i.e., a unitary operator on the tensor product of $n + 1$ copies of a 2-dimensional Hilbert space and an appropriate measurement (Feynman, 2000). The 2-dimensional Hilbert space $\mathbb{C}^2$ can be looked upon as the Hilbert space of a

2-level quantum system. It is called a 1-*qubit* Hilbert space, qubit being an abbreviation for a quantum binary digit. The n-fold tensor product $(\mathbb{C}^2)^{\otimes^n}$ is called a n-*qubit* Hilbert space. $\mathbb{C}^2$ has the canonical labeled orthonormal basis

$$|0\rangle = \begin{bmatrix} 1 \\ 0 \end{bmatrix}, \quad |1\rangle = \begin{bmatrix} 0 \\ 1 \end{bmatrix}$$

with the label set $\{0, 1\}$ identified with $\mathbb{F}_2$. Thus $(\mathbb{C}^2)^{\otimes^n}$ has the product orthonormal basis

$$\{|x_1 x_2 \dots x_n\rangle, x_j \in \mathbb{F}_2, 1 \leq j \leq n\}. \tag{3}$$

We write

$$|\underline{x}\rangle = |x, x_2 \dots x_n\rangle = |x_1\rangle |x_2\rangle \cdots |x_n\rangle$$
$$= |x_1\rangle \otimes |x_2\rangle \otimes \cdots \otimes |x_n\rangle$$

for any $\underline{x} = (x_1, x_2, \dots, x_n) \in \mathbb{F}_2^n$ and call $|x_j\rangle$ the j-th qubit in $|\underline{x}\rangle$.

Now, for any map $f : \mathbb{F}_2^n \to \mathbb{F}_2$ observe that the correspondence

$$|x_0 x_1 x_2 \dots x_n\rangle \to |x_0 \oplus f(x_1, x_2, \dots, x_n)\rangle |x_1 x_2 \dots x_n\rangle,$$

as the x_j's vary in $\mathbb{F}_2$, is just a permutation of the labeled product orthonormal basis $\{|x_0 x_1 x_2 \dots x_n\rangle, x_j \in \mathbb{F}_2, 0 \leq j \leq n\}$ in the $(n + 1)$-qubit Hilbert space. Hence there exists a unique unitary operator U_f in $(\mathbb{C}^2)^{\otimes^{(n+1)}}$ satisfying

$$U_f |x_0 x_1 x_2 \dots x_n\rangle = |x_0 \oplus f(x_1, x_2, \dots, x_m)\rangle |x_1 x_2 \dots x_n\rangle$$

for all $x_j \in \mathbb{F}_2, 0 \leq j \leq n$. In particular,

$$U_f |0 x_1 x_2 \dots x_n\rangle = |f(\underline{x})\rangle |x_1 x_2 \dots x_n\rangle \quad \forall \underline{x} \in \mathbb{F}_2^n.$$

In other words $f(\underline{x})$ appears as the label in the 0-th qubit of the output state if the input state $|0\rangle |\underline{x}>$ passes through the quantum gate U_f. In this output state if we make the measurement $M = \{|0\rangle\langle 0| \otimes I, |1\rangle\langle 1| \otimes I\}$, I being the identity on the last n qubits the outcome will be the label $f(\underline{x})$. According to Theorem 1, U_f can be expressed as a composition of a finite sequence of quantum gates in 2-qubit Hilbert spaces $\mathbb{C}^2 \otimes \mathbb{C}^2$. Thus, in principle, all number crunching can be reduced to applications of one and two qubit quantum gates.

The execution of the programme outlined above is done in the language of elementary quantum gates and circuits. Each 1-qubit Hilbert space is indicated by a straight line or a 'wire'. Any n-qubit Hilbert space is indicated by n parallel wires. Consider the diagram:

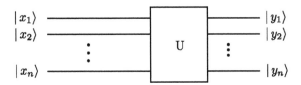

This means that the input state $|x_1 x_2 \ldots x_n\rangle$ passes through the gate U and the output state is $|y_1 y_2 \ldots y_n\rangle$, i.e.,

$$U|x_1 x_2 \ldots x_n\rangle = |y_1 y_2 \ldots y_n\rangle.$$

If U is a unitary operator on the first m qubits then $U \otimes I$ with identity on the remaining $n - m$ qubits is expressed as:

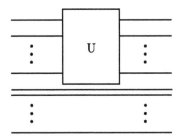

We shall now introduce some examples of elementary gates on qubits of order not exceeding 3.

(1) **Pauli gates** (X, Y, Z)

$$X = \begin{bmatrix} 0 & 1 \\ 1 & 0 \end{bmatrix}, \quad Y = \begin{bmatrix} 0 & -i \\ i & 0 \end{bmatrix}, \quad Z = \begin{bmatrix} 1 & 0 \\ 0 & -1 \end{bmatrix}$$

are hermitian and unitary matrices of determinant -1. Since $X|0\rangle = |1\rangle$, $X|1\rangle = |0\rangle$, X changes the labels of the basis gates and is therefore called also a NOT gate.

(2) **Hadamard gate** (*H*)

$$H = \frac{1}{\sqrt{2}} \begin{bmatrix} 1 & 1 \\ 1 & -1 \end{bmatrix}$$

H is again hermitian, unitary and has determinant -1. $H|0\rangle = \frac{|0\rangle + |1\rangle}{\sqrt{2}}$, $H|1\rangle = \frac{|0\rangle - |1\rangle}{\sqrt{2}}$. Thus *H* changes the basis states $|0\rangle$ and $|1\rangle$ into a new pair of basis states which are superpositions of $|0\rangle$ and $|1\rangle$.

(3) $\pi/8$ gate (*T*)

$$T = \begin{bmatrix} 1 & 0 \\ 0 & e^{i\pi/4} \end{bmatrix}$$

Modulo multiplication by a phase factor $e^{i\pi/8}$, *T* is $\exp(-i\pi Z/8)$ and hence called $\pi/8$ gate.

It is a somewhat delicate result that the subgroup generated by *iH* and $\exp(-i\pi Z/8)$ is dense in the group $SU(2)$ of all 2×2 unitary matrices of unit determinant.

(4) **Controlled not** (CNOT) **gate**

This is a 2-qubit gate expressed diagrammatically as

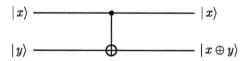

If the imput state is $|xy\rangle$ the output state is $|x\rangle|x \oplus y\rangle$. Thus measurement of the second qubit yields the label $x \oplus y$.

(5) **Controlled** *U*.

In diagrams it is

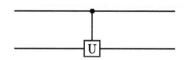

If the input state is $|0y\rangle$ the output is the same. If the input state is $|1y\rangle$ the output state is $|1\rangle U|y\rangle$.

(6) **Toffoli gate**

This is a 3-qubit gate with the circuit diagram

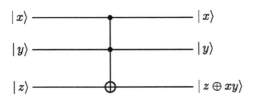

This is a 'doubly controlled' gate in the sense that the output is different from the input if and only if the first two qubits have label 1. Such controls may be put on the 2nd and 3rd or 1st and 3rd qubits too. If $z = 0$ and measurement is made on the 3rd qubit one obtains the product. Thus controlled not and Toffoli gates yield the addition and multiplication operations in the field $\mathbb{F}_2$.

This shows that by compositions of CNOT and Toffoli gates and making measurements one can achieve addition of arbitrary n-bit integers.

We have the following sharpened version of Theorem 2.

Theorem 2. *In a n-qubit Hilbert space every gate can be expressed as a finite composition of* 1*-qubit gates and CNOT gates.*

For a proof of the theorem we refer to (Nielsen and Chuang, 2000), (Parthasarathy, 2006). Combining this result with the discussion at the beginning of this section it follows that any number crunching can be achieved on a 'quantum computer' through a sequence of gates, each of which is either a 1-qubit gate or a CNOT gate. Thus it is important to seek practically feasible means of implementing the 2-qubit controlled not and all 1-qubit gates.

In order to compute a function f on the set of natural numbers it is desirable to have a quantum circuit in which the number of 1-qubit and CNOT gates required for evaluating $f(n)$ grows as a fixed power of $\log n$. Quantum computation attracted the attention of computer scientists and mathematicians when, in 1995, P. Shor announced the existence of such a circuit for the problem of factorizing an integer n into its prime factors. For details see (Nielsen and Chuang, 2000), (Parthasarathy, 2006).

References

Agrawal, M., Kayal, N. and Saxena, N. 2004. Primes is in *P*, *Ann. Math.* **160**:781–793.

Cover, T.M. and Thomas, J.A. 2006. *Elements of Information Theory*, 2nd edn., Wiley, New York.

Feynman, R.P. 1965. *Lectures on Physics*, Vol. III, World Students' edn., Addison Wesley, New York.

Feynman, R.P. 2000. *Lectures on Quantum Computation* (eds., T. Hey, R. Allen) Westview, USA.

Markov, A.A. 1913. *Izchisleniye Veroyatnostei* (in Russian), 3rd edn.

Nielsen, M.A. and Chuang, I.L. 2000. *Quantum Computation and Quantum Information*, Cambridge University Press, Cambridge.

Parthasarathy, K.R. 2003. A remark on the unitary group of a tensor product of n finite dimensional Hilbert spaces, *Proc. Ind. Acad. Sci.* (Math. Sci.) **113**:3–13.

Parthasarathy, K.R. 2006. *Lectures on Quantum Computation, Quantum Error Correcting Codes and Information Theory*, Tata Institute of Fundamental Research, Mumbai, Narosa Publishing House, New Delhi.

Sridharan, R. 2005. *Sanskrit prosody, Pingala sutras and binary arithmetic, in Contributions to the History of Indian Mathematics* (eds: G.G. Emch, R. Sridharan and M.D. Srinivas), Hindustan Book Agency, New Delhi. 33–62.

CHAPTER 12

Is There a Science Behind Opinion Polls?

Rajeeva L. Karandikar

Director,
Chennai Mathematical Institute
Chennai 603103, India.
e-mail: rlk@cmi.ac.in

1 Introduction

Is there a science behind opinion polls or is it just guess work? We will discuss this question and other issues involved in psephology with focus on the Indian context. We will see that simple mathematics and statistics, lots of common sense and a good understanding of the ground reality or domain knowledge together can yield very good forecast or predictions of the outcome of elections based on opinion polls and exit polls.

We will address the following questions:

- How can obtaining opinion of a small fraction of voters be sufficient to predict the outcome?

- Is a sample size of, say 20,000 sufficient to predict the outcome in a country with over 71.41 crore voters?

- Do opinion polls serve any purpose other than giving an idea of the election result?

- Do opinion polls conducted well ahead (say a month) of the polling date have predictive powers as far as final results are concerned?

We begin with a brief introduction to Sampling theory with relevant probability theory and statistical inference.

2   Probability theory and Inference

Consider an urn that contains M balls identical in all aspects except for colour, with K of them being orange and rest being green. If the balls are mixed and without looking, one of them is drawn, then the chance of it being orange is $\frac{K}{M}$.

This is based on the premise that each ball has an equal probability of being drawn and since there are K orange balls, the required probability is K/M.

Suppose $M = 10000$ and K is either 9900 or 100, so either 9900 are orange or 9900 are green. One ball is drawn (after mixing, without looking) and its colour is found to be green. We are to make a decision about K-choose out of the two possibilities: $K = 100$ or $K = 9900$.

If $K = 100$, then probability of drawing a green ball is 0.99 whereas if $K = 9900$, then probability of drawing a green ball is 0.01. So if the colour of the ball drawn is green, it is 99 times more likely that most balls are green as compared to the possibility that most balls are orange and hence if we have to choose between the two possibilities - namely *most balls are green* and *most balls are orange*, we should choose *most balls are green*.

This is the only idea from probability theory or statistics that is needed to answer most of the questions of skeptics as we will see.

Now consider a constituency, say Chennai South and to make matters simple, suppose there are two candidates, A and B. Suppose there are M voters, $V_1, V_2, \ldots, V_M$.

Suppose $a_i = 1$ if V_i prefers A (to B) and $a_i = 0$ if V_i prefers B (to A). Suppose p is the proportion of voters who prefer A to B *i.e.*

$$K = a_1 + a_2 + \cdots + a_M, \quad p = \frac{K}{M}.$$

The interest is to know *Who will win the election A or B, i.e.* in deciding if $K > (M/2)$ or $K < (M/2)$ (for simplicity, let us assume M is an odd integer).

Of course if we observe each a_i, we will know the answer. *Can we have a decision rule even if we observe only a few a_i's?*

Let S denote the collection all n-tuples $(i_1, i_2, \ldots, i_n)$ of elements of $\{1, 2, \ldots, M\}$. For $(i_1, i_2, \ldots, i_n) \in S$ let

$$f(i_1, i_2, \ldots, i_n) = a_{i_1} + a_{i_2} + \cdots + a_{i_n},$$

$$g(i_1, i_2, \ldots, i_n) = \frac{1}{n} f(i_1, i_2, \ldots, i_n).$$

Note that $g(i_1, i_2, \ldots, i_n)$ is the proportion of 1's in $\{i_1, i_2, \ldots, i_n\}$.

We are going to estimate

$$\frac{\#\{(i_1, i_2, \ldots, i_n) \in S : | g(i_1, i_2, \ldots, i_n) - p | > \epsilon\}}{M^n}$$

It can be easily seen (high school algebra) that

$$\sum_S f(i_1, i_2, \ldots, i_n) = nKM^{n-1}$$

and

$$\sum_S (f(i_1, i_2, \ldots, i_n))^2 \Psi = nKM^{n-1} + n(n - 1)K^2 M^{n-2}.$$

Thus

$$\frac{1}{M^n} \sum_S g(i_1, i_2, \ldots, i_n) = \frac{K}{M} = p,$$

$$\frac{1}{M^n} \sum_S (g(i_1, i_2, \ldots, i_n))^2 = \frac{p}{n} + \frac{(n - 1)p^2}{n}.$$

As a consequence,

$$\frac{1}{M^n} \sum_S \left(g(i_1, i_2, \ldots, i_n) - p\right)^2 = \frac{p(p - 1)}{n}. \tag{1}$$

Thus

$$\frac{\#\{(i_1, i_2, \ldots, i_n) \in S : | g(i_1, i_2, \ldots, i_n) - p | > \epsilon\}}{M^n} \leq \frac{1}{\epsilon^2} \frac{p(1 - p)}{n}. \tag{2}$$

Indeed, if (2) is not true then the contribution of these terms in the LHS of (1) would exceed the RHS.

Now let us imagine that for each $(i_1, i_2, \ldots, i_n) \in S$, we have a slip of paper on which $(i_1, i_2, \ldots, i_n)$ is written and on the slip $g(i_1, i_2, \ldots, i_n)$ is recorded. All the slips are put in a box, mixed and one slip is drawn and then $g(i_1, i_2, \ldots, i_n)$ is noted. This is like drawing one ball and noting its colour. From equation (2), it follows that the observed $g(i_1, i_2, \ldots, i_n)$ and the true (unobserved) p are close to each other with large probability.

In practice, $(i_1, i_2, \ldots, i_n)$ is chosen (known as sample) and then $a_{i_1}, \ldots, a_{i_n}$ are observed and $g(i_1, i_2, \ldots, i_n)$ is computed. It is called *sample pro-portion* and let us denote it by $\hat{p}_n$. The equation (2) can be recast as

$$\text{Prob}(|\hat{p}_n - p| > \epsilon) \le \frac{1}{\epsilon^2} \frac{p(1-p)}{n}. \tag{3}$$

This estimate (3) for $p = 0.5, \epsilon = 0.05$ yields

$$\text{Prob}(|\hat{p}_n - p| > 0.05) \le \frac{100}{n}.$$

Note that the estimate does not depend on M. The probability estimate given above can be improved significantly (using central limit theorem). It can be shown that, to ensure that

$$\text{Prob}(|\hat{p}_n - p| > 0.05) \le 0.01,$$

we need to take $n = 640$ while to ensure that

$$\text{Prob}(|\hat{p}_n - p| > 0.02) \le 0.01,$$

we need to take $n = 4000$, to ensure that

$$\text{Prob}(|\hat{p}_n - p| > 0.01) \le 0.01,$$

we need to take $n = 16000$.

Note that the n needed to achieve a given level of precision does not depend upon the population size M. Thus sampling fraction $(\frac{n}{M})$ is not important but n is important. This may be somewhat counter intuitive but true. Indeed, if we wish to estimate the percentage votes for the political parties across the nation, a sample of size 16000 will give an estimate correct up to 1% with probability 99% even though the size of the electorate across the country is over 70 crores (700 million).

What the probability estimates given above mean is that if in a constituency the proportion of voters supporting the winning candidate is say 0.525 or more and sample size is 4000 then whoever gets more votes in the sample is likely to be the winner with high probability. In other words, suppose $p > 0.525$ or $p < 0.475$ and the sample size is 4000. For the chosen sample if we see that $\hat{p}_n > 0.5$ then we can conclude that A will win and if we see that $\hat{p}_n < 0.5$ then we can conclude that B will win and we will be correct with 0.99 probability.

The arguments given above can be summarized as: *Most samples with size 4000 are representative of the population and hence if we select one randomly, we are likely to end up with a representative sample.*

Here *select one sample randomly* means all possible samples of the given size have the same probability of being selected under the scheme. One way of achieving this is to first choose one voter out of M randomly and repeating the same procedure n times. This sampling procedure is called *simple random sampling with replacement (SRSWR)*. In practice, one uses procedure called *random sampling without replacement (SRSWOR)* where one chooses randomly a subset of $\{1, 2, \ldots, M\}$ of size n. This is implemented as follows: first choose one voter out of M randomly (i.e. probability of each voter being selected is the same), then choose one voter out of the remaining ones randomly (i.e. probability of each voter being selected is the same) and so on till we have n voters in the sample. When M is much larger than n, the probability estimates given above are valid for the SRSWOR scheme as well.

3   Opinion polls and Random sampling

In colloquial English, the word *random* is also used in the sense of *arbitrary* (as in Random Access Memory-RAM). So some people think of a random sample as any arbitrary subset. It should be noted that in this context *random* is not the property of a given subset but *random* in random sample refers to the procedure used in selecting the subset (sample).

Failure to select a random sample can lead to wrong conclusions. In 1948, all opinion polls in US predicted that Thomas Dewey would defeat Harry Truman in the presidential election. The problem was traced to choice of sample being made on the basis of randomly generated telephone numbers and calling the numbers. In 1948, the poorer sections of the society went unrepresented in the survey.

Today, the penetration of telephones in US is almost universal and so the method now generally works in US. It would not work in India even after the unprecedented growth in telecom sector in the last few years, as poorer section are highly under represented among people with telephone and thus a telephone survey will not yield a representative sample.

Another method used by market research agencies is called quota sampling, where they select a group of respondents with a given profile - a profile that matches the population on several counts, such as male/female, rural/urban, education, caste, religion etc. Other than matching the sample profile, no other restriction on choice of respondents is imposed and is left to the enumerator. However, in my view, *the statistical guarantee* that the sample proportion and population proportion do not differ significantly doesn't apply unless the sample is chosen via randomization.

In the Indian context, we know that there is a lot of variation in support for political parties across states. For example, for CPM the support is concentrated in West Bengal and Kerala while for Congress it is spread across the country, though there is variation. For BJP, the support is negligible in several states. So we should ensure that each state is represented proportionately in our sample. This is called stratified sampling (where each state is a strata) and is an improvement on the simple random sampling method. The sample should be chosen by suitable randomization, perhaps after suitable stratification. This costs a lot more than the quota sampling! But in my view randomization is a must. We also know that urban and rural voters do vote differently (based on past data) and this should be reflected in choice of strata. We also know that other socio-economic attributes also have influence on voting behavior but factoring them is not easy as the voters list does not capture this information.

The method adopted by us in sampling addresses the issue to some extent. We first choose some constituencies, then we choose some polling booths and then choose voters from the voters list. This is done as follows. The list of constituencies prepared by election commission is such that all the constituencies in a state occur together. Indeed, adjacent constituencies come together in most cases. Thus choosing constituencies by circular random sampling or systematic sampling generates a representative sample. For example if we decide to first choose 20% of constituencies (from where we will draw our sample) then we choose a number between 1 and 5 at random (say we have chosen 2) and then we choose every 5th constituency $(2, 7, 12, 17, \ldots)$. At the next stage we get the list of polling booths in each of chosen constituencies and again using the fact that adjacent areas fall in the same polling booth or polling booths that are adjacent to each other in the list. Thus once again choosing required number of booths by circular random sampling method gives a representative sample of booths. From each chosen booth, the required number of voters are chosen again by circular random sampling.

We have seen on several occasions that sample chosen via such a multistage circular random sampling yields a sample whose profile on various socio-economic variables such as caste, religion, economic status, urban/rural, male/female is close to the population profile available via census.

4 Predicting the number of seats

Following statistical methodology, one can get a fairly good estimate of percentage of votes of the major parties at the time the survey is conducted. However, the public interest is in prediction of number of seats and not percentage votes for parties.

It is possible (though extremely unlikely) even in a two party system for a party A with say 26% votes to win 272 (out of 543) seats (majority) while the other party B with 74% votes winning only 271 seats: A gets just over 50% votes in 272 seats winning them, while B gets 100% votes in the remaining 271 seats. In other words, in the first past the post parliamentary system as practiced in India, the number of seats for a party is a function of the distribution of votes for major parties across the country. Thus good estimate of vote percentages for major parties across the country does not automatically translate to a good estimate of number of seats for major parties. So in order to predict the number of seats for parties, we need to estimate not only the percentage votes for each party, but also the distribution of votes of each of the parties across constituencies. And here, independents and smaller parties that have influence across few seats make the vote-to-seat translation that much more difficult.

Let us first examine prediction of a specified seat. Consider the task of predicting the winner in a Lok Sabha election in Chennai south constituency (which has over 20 lakh voters). Suppose that the difference between true supports for the two leading candidates is over 4 percent votes. By generating a random sample of size 4000 and getting their opinion, we can be reasonably sure (with 99% probability) that we can pick the winner. Indeed the same is true even if the constituency had larger number of voters.

So if we can get a random sample of size 4000 in each of the 543 constituencies, then we can predict winner in each of them and we will be mostly correct (in constituencies where the contest is not a very close one). But conducting a survey with more than 21 lakh respondents is very difficult: money, time, reliable trained manpower . . . each resource is limited.

One way out is to construct a model of voter behavior based on socio-economic attributes. While such a model can be built, estimating various parameters of such a model would itself require a very large sample size. Another approach is to use past data in conjunction with the opinion poll data. In order to do this, we need to build a suitable *model* of voting behavior- not of individual voters but for percentage votes for a party in a constituency.

To make a model, let us observe some features of the Indian democracy.

Voting intentions are volatile- in a matter of months they can undergo big change. (Example: Delhi in March 98, November 98, October 99). This is very different from the situation in UK where voting intentions are very stable, and thus methods used in UK can not be used in India, though superficially, the Indian political system resembles the one in UK. This is where domain knowledge plays an important role. A model which works in the west may not work in Indian context if it involves human behavior. And having all the data relating to elections in India (since 1952) will not help. The point is that large amount of data cannot substitute understanding of ground realities. Moreover, the dats from 50's and 60's is irrelevant today- in that era, a single party namely Congress dominated the elections in most of the states. It can be argued that perhaps data from last 3 or 4 elections alone is relevant in Indian context.

While the behavior of voters in a constituency may be correlated with that in adjacent constituencies in the same state, the voting behavior in one state has no (or negligible) correlation with that in another state. (The behavior is influenced by many local factors.)

The socio-economic factors do influence the voting pattern significantly. However, incorporating it directly in a model will require too many parameters.

It seems reasonable to assume that the socio-economic profile of a constituency does not change significantly from one election to the next. So while the differences in socio-economic profiles between two constituencies are reflected in the differences in voting pattern in a given election, by and large the change from one election to the next in a given constituency does not depend on the socio-economic profile.

So we make an assumption that *the change in the percentage of votes for a given party from the previous election to the present is constant across a given state.*

The resulting model is not very accurate if we look at historical data, but is a reasonably good approximation- good enough for the purpose- namely to predict the seats for major parties at national level.

The change in the percentage of votes is called *swing.* Under this model, all we need to do via sampling is to estimate the swing for each party in each state and then using the past data we will have an estimate of percentage votes for each party in each state. We can refine this a little- we can divide the big states in regions and postulate that the swing in a seat is a convex combination of swing across the state and swing across the region.

Here enters one more element. We need to predict the winner in each constituency and then give number of seats for major parties. Suppose in one constituency with only two candidates, we predict A gets 50.5%, B gets 49.5%, in another constituency we predict that C gets 58% votes, D gets 42% votes, in both cases, the sample size is say 625. It is clear that while winner between A and B is difficult to call, we can be lot more sure that C will win the second seat. Thus we need to factor the difference in predicted votes for leading candidates into account while predicting seats.

What is the best case scenario for B-that indeed A and B have nearly equal support with B having a very thin lead, and yet a random sample of size 625 gives a 1% lead to A. This translates to: in 625 tosses of a fair coin, we observe 316 or more heads. The probability of such an event is 0.405 (using normal approximation). So we assign B a winning probability of 0.402 and A a winning probability of $1 - 0.405 = 0.595$.

This can be extended to cover the case when there are three candidates A,B and C getting significant votes, say 36%, 33%, 31% respectively. Now we will asiign probabilities to the three candidates, adding upto one. First the best case scenario for C, then the best case scenario for B.

Summing over the probabilities over all the 543 seats we get the expected number of seats for each party. This method gives reasonable predictions at state level and good predictions at the national level. The methodology described above was developed in 1997–98 for the 1998 Lok Sabha elections and our findings and observations are published in

Karandikar *et al.* (2002). Also see (Karandikar, 1999).

The same methodology can be adapted to an election for state legislature. Each time it has to be tweaked to take into account some ground realities - such as the alliances may have changed since last elections or as happened in 2004, the constituency boundaries were redrawn etc. Over last 12 years we have had good success in predicting winning party/alliance in national or state elections - getting it right in every instance. The extent of victory or exact number of seats for the winning party or alliance is more difficult to predict and we have had mixed results. In some cases we got it very close to the final outcome while in some cases we predicted the party getting highest number of seats correctly but failed to predict that the party will actually get majority.

5   Limitations of opinion polls

Opinion polls conducted well ahead of actual polls can only measure the mood of the electorate at the time the poll is conducted and cannot yield any insight into how voters are going to vote on the election day. It is believed by political scientists that in India, lot of churn takes place close to the election day. In 1998, we had conducted opinion poll before start of the Lok Sabha election and the same respondents were also asked a day after they voted as to whom they had voted for. The gap between the two polls was 14 days on the average and we found that about 30% of the respondents had changed their mind. This puts a big question mark on predictive power of any opinion poll conducted in India ahead of the voting day.

Another point is that while the opinion polls measure the opinion of the entire electorate, only about 55%–65% voters actually turn up for voting introducing bias in the opinion poll predictions. One could ask respondents if they intend to vote but this does not help as in such cases 80%–90% respondents say they are going to vote. Indeed, we also ask people if they voted in the last election. We have found that lot more people seem to recall that they voted in the last election as compared to actual percentage.

This limitation is addressed by an Exit poll where voters are interviewed as they exit from the booth. This way we get a sample of people who have voted. However, an Exit poll makes it difficult to randomly choose the respondents and one has to leave it to the interviewer to select the respondents. This may create bias.

Over the years we have found that *day after poll* where respondents are chosen via a proper statistical methodology and interviewed the day after polling day is most satisfactory. Here we can ignore the opinion of voters who did not vote and take into account the opinion of only those who voted. Of course here the media (who funds such polls) has to wait before the projections would be available while in case of exit poll, the predictions are available on the same night.

6   Some remarks

Opinion polls followed by election is a rare occasion where statistics is put to a test: namely a sample survey is conducted and results made public and is soon followed by a complete enumeration. This shows the power of statistical tools if used correctly and at the same time it also shows its limitations. Data collected via sample survey is the basis of lot of crucial decisions and decision makers should be aware of power and limitations of sample survey methodology.

Opinion polls also give an insight into why people voted the way they did. Also only opinion polls can give the opinion of a specific socio-economic class such as SC or ST or OBC voters or of poor voters. It also gives insight into reasons for defeat for ruling party- which socio-economic group of voters got turned away by its policies. A well conducted opinion poll can also serve as a mirror for ruling party to gauge public mood on important policy matters.

References

Karandikar, R. 1999. Opinion Polls and Statistics, *Calcutta Statistical Association Bulletin* **49:**193–194.

Karandikar, Payne, R.C. and Yadav, Y. 2002. Predicting the 1998 Indian parliamentary election, *Electoral studies* **21:**66–89.

CHAPTER 13

An Introduction to Financial Mathematics

Sandeep Juneja

Tata Institute of Fundamental Research,
Mumbai 400 005, India.
e-mail: juneja@tifr.res.in

1 Introduction

A wealthy acquaintance when recently asked about his profession reluctantly answered that he is a middleman in drug trade and has made a fortune helping drugs reach European markets from Latin America. When pressed further, he confessed that he was actually a 'quant' in a huge wall street bank and used mathematics to price complex derivative securities. He lied simply to appear respectable! There you have it. Its not fashionable to be a financial mathematician these days. On the plus side, these quants or financial mathematicians on the wall street are sufficiently rich that they can literally afford to ignore fashions.

On a serious note, it is important to acknowledge that financial markets serve a fundamental role in economic growth of nations by helping efficient allocation of investment of individuals to the most productive sectors of the economy. They also provide an avenue for corporates to raise capital for productive ventures. Financial sector has seen enormous growth over the past thirty years in the developed world. This growth has been led by the innovations in products referred to as financial derivatives that require great deal of mathematical sophistication and ingenuity in pricing and in creating an insurance or a hedge against associated risks. This chapter briefly discusses some such popular derivatives including those that played a substantial role in the economic crisis of 2008. Our primary focus are the key underlying mathematical ideas that are used to price such derivatives. We present these in a somewhat simple setting.

Brief history: During the industrial revolution in Europe there existed great demand for setting up huge industrial units. To raise capital, entrepreneurs came together to form joint partnerships where they owned 'shares' of the newly formed company. Soon there were many such companies each with many shares held by public at large. This was facilitated by setting up of stock exchanges where these shares or stocks could be bought or sold. London stock exchange was first such institution, set up in 1773. Basic financial derivatives such as futures have been around for some time (we do not discuss futures in this chapter; they are very similar to the forward contracts discussed below). The oldest and the largest futures and options exchange, The Chicago Board of Trade (CBOT), was established in 1848. Although, as we discuss later, activity in financial derivatives took off in a major way beginning the early 1970's.

Brief introduction to derivatives (see, e.g., (Hull, 2008) for a comprehensive overview): A derivative is a financial instrument that derives its value from an 'underlying' more basic asset. For instance, consider a forward contract, a popular derivative, between two parties: One party agrees to purchase from the other a specified asset at a particular time in future for a specified price. For instance, Infosys, expecting income in dollars in future (with its expenses in rupees) may enter into a forward contract with ICICI bank that requires it to purchase a specified amount of rupees, say Rs. 430 crores, using specified amount of dollars, say, $10 crore, six months from now. Here, the fluctuations in more basic underlying exchange rate gives value to the forward contract.

Options are popular derivatives that give buyer of this instrument an option but not an obligation to engage in specific transactions related to the underlying assets. For instance, a call option allows the buyer of this instrument an option but not an obligation to purchase an underlying asset at a specified *strike* price at a particular time in future, referred to as *time to maturity*. Seller of the option on the other hand is obligated to sell the underlying asset to the buyer at the specified price if the buyer exercises the option. Seller of course receives the option price upfront for selling this derivative. For instance, one may purchase a call option on the Reliance stock, whose current value is, say, Rs. 1055, that gives the owner the option to purchase certain number of Reliance stocks, each at price Rs. 1100, three months from now. This option is valuable to the buyer at its time of maturity if the stock then is worth more than Rs. 1100. Otherwise this option is not worth exercising and has value zero. In the earlier example, Infosys may instead prefer to purchase a call option that allows it the option to pay $10 crore to receive Rs. 430 crore six months from now. Infosys would then exercise this option if each dollar gets less than Rs. 43 in the market at the option's time to maturity.

Similarly, a put option gives the buyer of the instrument the option but not an obligation to sell an asset at a specified price at the time to maturity. These options are referred to as European options if they can be exercised only at the time to maturity. American options allow an *early exercise feature*, that is, they can be exercised at any time up to the time to maturity. There exist variants such as Bermudan options that can be exercised at a finite number of specified dates. Other popular options such as interest rate swaps, credit debt swaps (CDS's) and collateralized debt obligations (CDOs) are discussed later in the chapter. Many more exotic options are not discussed in this chapter (see, e.g, (Hull, 2008), (Shreve, 2004)).

1.1   The no-arbitrage principle

Coming up with a fair price for such derivatives securities vexed the financial community right up till early seventies when Black and Scholes (1973) came up with their famous formula for pricing European options. Since then, the the literature on pricing financial derivatives has seen a huge explosion and has played a major role in expansion of financial derivatives market. To put things in perspective, from a tiny market in the seventies, the market of financial derivatives has grown in notional amount to about $600 trillion in 2007. This compared to the world GDP of the order $45 trillion. Amongst financial derivatives, as of 2007, interest rate based derivatives constitute about 72% of the market, currencies about 12%, and equities and commodities the remaining 16% (See, e.g., (Baaquie, 2010)). Wall street employs thousands of PhDs that use quantitative methods or 'rocket science' in derivatives pricing and related activities.

Figure 1: *No Arbitrage Principle: Price of a two liter ketchup bottle equals twice the price of a one liter ketchup bottle, else ARBITRAGE, that is, profits can be made without any risk.*

Color image of this figure appears in the color plate section at the end of the book.

'No-arbitrage pricing principle' is the key idea used by Black and Scholes to arrive at their formula. It continues to be foundational for financial mathematics. Simply told, and as illustrated in Figure 1, this means that price of a two liter ketchup bottle should be twice the price of a one liter ketchup bottle, otherwise by following the sacred mantra of *buy low and sell high* one can create an arbitrage, that is, instantaneously produce profits while taking zero risk. The no arbitrage principle precludes such free lunches and provides a surprisingly sophisticated methodology to price complex derivatives securities. This methodology relies on replicating pay-off from a derivative in every possible scenario by continuously and appropriately trading in the underlying more basic securities (transaction costs are assumed to be zero). Then, since the derivative and this trading strategy have identical payoffs, by the no-arbitrage principle, they must have the same price. Hence, the cost of creating this trading strategy provides the price of the derivative security.

In spite of the fact that continuous trading is an idealization and there always are small transaction costs, this pricing methodology approximates the practice well. Traders often sell complex risky derivatives and then dynamically trade in underlying securities in a manner that more or less cancels the risk arising from the derivative while incurring little transactional cost. Thus, from their viewpoint the price of the derivative must be at least the amount they need to cancel the associated risk. Competition ensures that they do not charge much higher than this price.

In practice one also expects the no-arbitrage principle to hold as large banks typically have strong groups of arbitragers that identify and quickly take advantage of such arbitrage opportunities (again, by buying low and selling high) so that due to demand and supply, the prices adjust and these opportunities become unavailable to common investors.

1.2   Popular derivatives

Interest rate swaps and swaptions, options on these swaps, are by far the most popular derivatives in the financial markets. The market size of these instruments was about $310 trillion in 2007. Figure 2 shows an example of cash flows involved in an interest rate swap. Typically, for a specified duration of the swap (e.g., five years) one party pays a fixed rate (fraction) of a pre-specified notional amount at regular intervals (say, every quarter or half yearly) to the other party, while the other party pays variable floating rate at the same frequency to the first party. This variable rate may be a function of prevailing rates such as the LIBOR rates (London Interbank Offered Rates; inter-bank borrowing rate amongst banks in London). This is used by many companies to match their revenue streams to liability streams. For instance, a pension fund may have fixed liabilities. However, the income they earn may be a function of prevailing interest rates. By entering into a swap that pays at a fixed rate they can reduce the variability of cash-flows and hence improve financial planning.

Swaptions give its buyer an option to enter into a swap at a particular date at a specified interest rate structure. Due to their importance in the financial world, intricate mathematical models have been developed to accurately price such interest rate instruments. Refer to, e.g., (Brigo and Mercurio, 2006), (Cairns, 2004) for further details.

Credit Default Swap is a financial instrument whereby one party (A) buys protection (or insurance) from another party (B) to protect against default by a third party (C). Default occurs when a debtor C cannot meet its legal debt obligations. A pays a premium payment at regular intervals (say, quarterly) to B up to the duration of the swap or until C defaults. During

Fixed payment leg: Example, 6% of notional amount

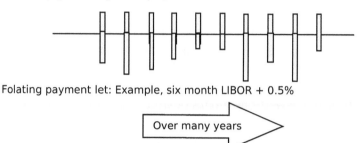

Folating payment let: Example, six month LIBOR + 0.5%

Over many years

Figure 2: *Example illustrating interest rate swap cash-flows.*

Color image of this figure appears in the color plate section at the end of the book.

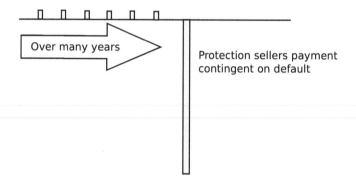

Over many years

Protection sellers payment
contingent on default

Figure 3: *CDS cash flow.*

Color image of this figure appears in the color plate section at the end of the book.

the swap duration, if C defaults, B pays A a certain amount and the swap terminates. These cash flows are depicted in Figure 3. Typically, A may hold a bond of C that has certain nominal value. If C defaults, then B provides protection against this default by purchasing this *much devalued* bond from A at its higher nominal price. CDS's were initiated in early 90's but the market took-of in 2003. By the year 2007, the amount protected by CDS's was of order $60 trillion. Refer to (Duffie and Singleton, 2003), (Lando, 2005) and (Schonbucher, 2003) for a general overview of credit derivatives and the associated pricing methodologies for CDSs as well as for CDOs discussed below.

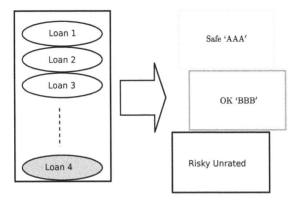

Figure 4: *Underlying loans are tranched into safe and less safe securities that are sold to investors.*

Color image of this figure appears in the color plate section at the end of the book.

Collateralized Debt Obligation is a structured financial product that became extremely popular over the last ten years. Typically CDO's are structures created by banks to offload many loans or bonds from their lending portfolio. These loans are packaged together as a CDO and then are sold off to investors as CDO tranche securities with varying levels of risk. For instance, investors looking for safe investment (these are typically the most sought after by investors) may purchase the super senior tranche securities (which is deemed very safe and maybe rated as AAA by the rating agencies), senior tranche (which is comparatively less safe and may have a lower rating) securities may be purchased by investors with a higher appetite for risk (the amount they pay is less to compensate for the additional risk) and so on. Typically, the most risky equity tranche is retained by the bank or the financial institution that creates this CDO. The cash flows generated when these loans are repaid are allocated to the security holders based on seniority of the associated tranches. Initial cash flows are used to payoff the super senior tranche securities. Further generated cash-flows are used to payoff the senior tranche securities and so on. If some of the underlying loans default on their payments, then the risky tranches are the first to absorb these losses. See Figures 4 and 5 for a graphical illustration.

Pricing CDOs is a challenge as one needs to accurately model the dependencies between somewhat rare but catastrophic events associated with many loans defaulting together. Also note that more sophisticated CDOs,

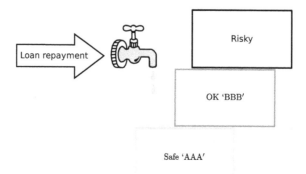

Figure 5: *Loan repayments are first channeled to pay the safest tranche, then the less safe one and so-on. If their are defaults on the loans then the risky tranches are the first to suffer losses.*

Color image of this figure appears in the color plate section at the end of the book.

along with loans and bonds, may include other debt instruments such as CDS's and tranches from other CDO's in their portfolio.

As the above examples indicate, the key benefit of financial derivatives is that they help companies reduce risk in their cash-flows and through this improved risk management, aid in financial planning, reduce the capital requirements and therefore enhance profitability. Currently, 92% of the top Fortune 500 companies engage in derivatives to better manage their risk.

However, derivatives also make speculation easy. For instance, if one believes that a particular stock price will rise, it is much cheaper and logistically efficient to place a big bet by purchasing call options on that stock, then through acquiring the same number of stock, although the upside in both the cases is the same. This flexibility, as well as the underlying complexity of some of the exotic derivatives such as CDOs, makes derivatives risky for the overall economy. Warren Buffet famously referred to financial derivatives as time bombs and financial weapons of mass destruction. CDOs involving housing loans played a significant role in the economic crisis of 2008. (See, e.g., (Duffie, 2010)). The key reason being that while it was difficult to precisely measure the risk in a CDO (due extremal dependence amongst loan defaults), CDOs gave a false sense of confidence to the loan originators that since the risk was being diversified away to investors, it was being mitigated. This in turn prompted acquisition of far riskier sub-prime loans that had very little chance of repayment.

In the remaining paper we focus on developing underlying ideas for pricing relatively simple European options such as call or put options. In Section 2, we illustrate the no-arbitrage principle for a two-security-two-scenario-two-time-period Binomial tree toy model. In Section 3, we develop the pricing methodology for pricing European options in more realistic continuous-time-continuous-state framework. From a probabilistic viewpoint, this requires concepts of Brownian motion, stochastic Ito integrals, stochastic differential equations, Ito's formula, martingale representation theorem and the Girsanov theorem. These concepts are briefly introduced and used to develop the pricing theory. They remain fundamental to the modern finance pricing theory. We end with a brief conclusion in Section 4.

Financial mathematics is a vast area involving interesting mathematics in areas such as risk management (see, e.g., (McNeil *et al.*, 2005)) , calibration and estimation methodologies for financial models, econometric models for algorithmic trading as well as for forecasting and prediction (see, e.g., (Cambell *et al.*, 2006)), investment portfolio optimization (see, e.g., (Meucci, 2005)) and optimal stopping problem that arises in pricing American options (see, e.g., (Glasserman, 2003)). In this chapter, however, we restrict our focus to some of the fundamental derivative pricing ideas.

2   Binomial Tree Model

We now illustrate how the no-arbitrage principle helps price options in a simple Binomial-tree or the 'two-security-two scenario-two-time-period' setting. This approach to price options was first proposed by (Cox *et al.*, 1979) (see Shreve 2004 for an excellent comprehensive exposition of Binomial tree models).

Consider a simplified world consisting of two securities: The risky security or the stock and the risk free security or an investment in the safe money market. We observe these at time zero and at time Δt. Suppose that stock has value S at time zero. At time Δt two scenarios *up* and *down* can occur (see Figure 6 for a graphical illustration): Scenario up occurs with probability $p \in (0, 1)$ and scenario *down* with probability $1 - p$. The stock takes value $S \exp(u\Delta t)$ in scenario up and $S \exp(d\Delta t)$ otherwise, where

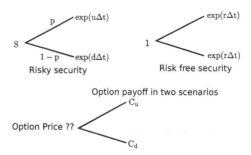

Figure 6: *World comprising a risky security, a risk-free security and an option that needs to be priced. This world evolves in future for one time period where the risky security and the option can take two possible values. No arbitrage principle that provides a unique price for the option that is independent of $p \in (0, 1)$, where $(p, 1 - p)$ denote the respective probabilities of the two scenarios.*

Color image of this figure appears in the color plate section at the end of the book.

$u > d$. Money market account has value 1 at time zero that increases to $\exp(r\Delta t)$ in both the scenarios at time Δt. Assume that any amount of both these assets can be borrowed or sold without any transaction costs.

First note that the no-arbitrage principle implies that $d < r < u$. Otherwise, if $r \leq d$, borrowing amount S from the money market and purchasing a stock with it, the investor earns at least $S \exp(d\Delta t)$ at time Δt where his liability is $S \exp(r\Delta t)$. Thus with zero investment he is guaranteed sure profit (at least with positive probability if $r = d$), violating the no-arbitrage condition. Similarly, if $r \geq u$, then by short selling the stock (borrowing from an owner of this stock and selling it with a promise to return the stock to the original owner at a later date) the investor gets amount S at time zero which he invests in the money market. At time Δt he gets $S \exp(r\Delta t)$ from this investment while his liability is at most $S \exp(u\Delta t)$ (the price at which he can buy back the stock to close the short position), thus leading to an arbitrage.

Now consider an option that pays C_u in the up scenario and C_d in the down scenario. For instance, consider a call option that allows its owner an option to purchase the underlying stock at the strike price K for some $K \in (S \exp(d\Delta t), S \exp(u\Delta t))$. In that case, $C_u = S - K$ denotes the benefit to option owner in this scenario, and $C_d = 0$ underscores the fact that option owner would not like to purchase a risky security at value K, when he can purchase it from the market at a lower price $S \exp(d\Delta t)$. Hence, in this scenario the option is worthless to its owner.

2.1 Pricing using no-arbitrage principle

The key question is to determine the fair price for such an option. A related problem is to ascertain how to cancel or hedge the risk that the seller of the option is exposed to by taking appropriate positions in the underlying securities. Naively, one may consider the expectation $pC_u + (1 - p)C_d$ suitably discounted (to account for time value of money and the risk involved) to be the correct value of the option. As we shall see, this approach may lead to an incorrect price. In fact, in this simple world, the answer based on the no-arbitrage principle is straightforward and is independent of the probability p. To see this, we construct a portfolio of the stock and the risk free security that exactly replicates the payoff of the option in the two scenarios. Then the value of this portfolio gives the correct option price.

Suppose we purchase $\alpha \in \Re$ number of stock and invest amount $\beta \in \Re$ in the money market, where α and β are chosen so that the resulting payoff matches the payoff from the option at time Δt in the two scenarios. That is,

$$\alpha S \exp(u\Delta t) + \beta \exp(r\Delta t) = C_u$$

and

$$\alpha S \exp(d\Delta t) + \beta \exp(r\Delta t) = C_d.$$

Thus,

$$\alpha = \frac{C_u - C_d}{S(\exp(u\Delta t) - \exp(d\Delta t))},$$

and

$$\beta = \frac{C_d \exp((u - r)\Delta t) - C_u \exp((d - r)\Delta t)}{\exp(u\Delta t) - \exp(d\Delta t)}.$$

Then, the portfolio comprising α number of risky security and amount β in risk free security exactly replicates the option payoff. The two should therefore have the same value, else an arbitrage can be created. Hence, the value of the option equals the value of this portfolio. That is,

$$\alpha S + \beta = \exp(-r\Delta t)\left[\frac{\exp(r\Delta t) - \exp(d\Delta t)}{\exp(u\Delta t) - \exp(d\Delta t)}C_u + \frac{\exp(u\Delta t) - \exp(r\Delta t)}{\exp(u\Delta t) - \exp(d\Delta t)}C_d\right].$$

$$(1)$$

Note that this price is independent of the value of the physical probability vector $(p, 1 - p)$ as we are matching the option pay-off over each probable scenario. Thus, p could be .01 or .99, it will not in any way affect the price of the option. If by using another methodology, a different price is reached for this model, say a price higher than (1), then an astute trader would be happy to sell options at that price and create an arbitrage for himself by exactly replicating his liabilities at a cheaper price.

2.1.1 Risk neutral pricing

Another interesting observation from this simple example is the following: Set $\hat{p} = \frac{\exp(r\Delta t) - \exp(d\Delta t)}{\exp(u\Delta t) - \exp(d\Delta t)}$, then, since $d < r < u$, $(\hat{p}, 1 - \hat{p})$ denotes a probability vector. The value of the option may be re-expressed as:

$$\exp(-r\Delta t) (\hat{p}C_u + (1 - \hat{p})C_d) = \hat{E}(\exp(-r\Delta t)C) \qquad (2)$$

where $\hat{E}$ denotes the expectation under the probability $(\hat{p}, 1 - \hat{p})$ and $\exp(-r\Delta t)C$ denotes the discounted value of the random pay-off from the option at time 1, discounted at the risk free rate. Interestingly, it can be checked that

$$S = \exp(-r\Delta t) (\hat{p} \exp(u\Delta t)S + (1 - \hat{p}) \exp(d\Delta t)S)$$

or $\hat{E}(S_1) = \exp(r\Delta t)S$, where S_1 denotes the random value of the stock at time 1. Hence, under the probability $(\hat{p}, 1 - \hat{p})$, stock earns an annualized continuously compounded rate of return r. The measure corresponding to these probabilities is referred (in more general set-ups that we discuss later) as the *risk neutral* or the *equivalent martingale* measure. Clearly, these are the probabilities the risk neutral investor would assign to the two scenarios in equilibrium (in equilibrium both securities should give the same rate of return to such an investor) and (2) denotes the price that the risk neutral investor would assign to the option (blissfully unaware of the no-arbitrage principle!). Thus, the no-arbitrage principle leads to a pricing strategy in this simple setting that the risk neutral investor would in any case follow. As we observe in Section 3, this result generalizes to far more mathematically complex models of asset price movement, where the price of an option equals the mathematical expectation of the payoff from the option discounted at the risk free rate under the risk neutral probability measure.

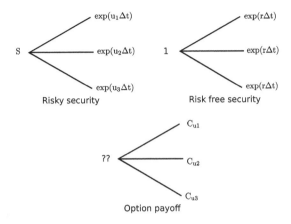

Figure 7: *World comprising a risky security, a risk-free security and an option that needs to be priced. This world evolves in future for one time period where the risky security and the option can take three possible values. No arbitrage principle typically does not provide a unique price for the option in this case.*

Color image of this figure appears in the color plate section at the end of the book.

2.2   Some extensions of the binomial model

Simple extensions of the binomial model provide insights into important issues directing the general theory. First, to build greater realism in the model, consider a two-security-three scenario-two-time-period model illustrated in Figure 7. In this setting it should be clear that one cannot replicate most options exactly without having a third security. Such a market where all options cannot be exactly replicated by available securities is referred to as *incomplete*. Analysis of incomplete markets is an important area in financial research as empirical data suggests that financial markets tend to be incomplete. (See, e.g., (Magill and Quinzii, 2002), (Fouque *et al.*, 2000)).

Another way to incorporate three scenarios is to increase the number of time periods to three. This is illustrated in Figure 8. The two securities are now observed at times zero, $0.5\Delta t$ and Δt. At times zero and at $0.5\Delta t$, the stock price can either go up by amount $\exp(0.5u\Delta t)$ or down by amount $\exp(0.5d\Delta t)$ in the next $0.5\Delta t$ time. At time Δt, the stock price can take three values. In addition, trading is allowed at time $0.5\Delta t$ that may

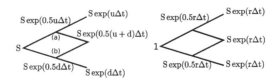

Figure 8: *World comprising a risky security, a risk-free security and an option that needs to be priced. This world evolves in future for two time periods 0.5Δt and Δt. Trading is allowed at time 0.5Δt that may depend upon the value of stock at that time. This allows exact replication and hence unique pricing of any option with payoffs at time Δt.*

Color image of this figure appears in the color plate section at the end of the book.

depend upon the value of stock at that time. This additional flexibility allows replication of any option with payoffs at time Δt that are a function of the stock price at that time (more generally, the payoff could be a function of the path followed by the stock price till time Δt). To see this, one can repeat the argument for the two-security-two scenario-two-time-period case to determine how much money is needed at node (a) in Figure 8 to replicate the option pay-off at time Δt. Similarly, one can determine the money needed at node (b). Using these two values, one can determine the amount needed at time zero to construct a replicating portfolio that exactly replicates the option payoff along every scenario at time Δt. This argument easily generalizes to arbitrary n time periods (see (Hull, 2008), (Shreve, 2004)). Cox *et al.* (1979) analyze this as n goes to infinity and show that the resultant risky security price process converges to geometric Brownian motion (exponential of Brownian motion; Brownian motion is discussed in section 3), a natural setting for more realistic analysis.

3 Continuous Time Models

We now discuss the European option pricing problem in the continuous-time-continuous-state settings which has emerged as the primary regime for modeling security prices. As discussed earlier, the key idea is to create a dynamic portfolio that through continuous trading in the underlying securities up to the option time to maturity, exactly replicates the option payoff in every possible scenario. In this presentation we de-emphasize technicalities to maintain focus on the key concepts used for pricing options and to keep the discussion accessible to a broad audience. We refer the reader to Shreve (2004), (Duffie, 1996), (Steele, 2001) for simple and engaging account of stochastic analysis for derivatives pricing; also see, (Fouque *et al.*, 2000)).

First in Section 3.1, we briefly introduce Brownian motion, perhaps the most fundamental continuous time stochastic process that is an essential ingredient in modeling security prices. We discuss how this is crucial in driving stochastic differential equations used to model security prices in Section 3.2. Then in Section 3.3, we briefly review the concepts of stochastic Ito integral, quadratic variation and Ito's formula, necessary to appreciate the stochastic differential equation model of security prices. In Section 3.4, we use a well known technique to arrive at a replicating portfolio and the Black Scholes partial differential equation for options with simple payoff structure in a two security set-up.

Modern approach to options pricing relies on two broad steps: First we determine a probability measure under which discounted security prices are martingales (martingales correspond to stochastic processes that on an average do not change. These are precisely defined later). This, as indicated earlier, is referred to as the equivalent martingale or the risk neutral measure. Thereafter, one uses this measure to arrive at the replicating portfolio for the option. First step utilizes the famous Girsanov theorem. Martingale representation theorem is crucial to the second step. We review this approach to options pricing in Section 3.5. Sections 3.4 and 3.5 restrict attention to two security framework: one risky and the other risk-free. In Section 3.6, we briefly discuss how this analysis generalizes to multiple risky assets.

3.1   Brownian motion

Louis Bachelier in his dissertation (Bachelier, 1900) was the first to use Brownian motion to model stock prices with the explicit purpose of evaluating option prices. This in fact was the first use advanced mathematics in finance. Samuelson (1965) much later proposed geometric Brownian motion (exponential of Brownian motion) to model stock prices. This, even today is the most basic model of asset prices.

The stochastic process $(W_t : t \geq 0)$ is referred to as standard Brownian motion (see, e.g., (Revuz and Yor, 1999)) if

1. $W_0 = 0$ almost surely.

2. For each $t > s$, the increment $W_t - W_s$ is independent of $(W_u : u \leq s)$. This implies that $W_{t_2} - W_{t_1}$ is independent of $W_{t_4} - W_{t_3}$ and so on whenever $t_1 < t_2 < t_3 < t_4$.

3. For each $t > s$, the increment $W_t - W_s$ has a Gaussian distribution with zero mean and variance $t - s$.

4. The sample paths of $(W_t : t \geq 0)$ are continuous almost surely.

Technically speaking, $(W_t : t \geq 0)$ is defined on a probability space

$$(\Omega, \mathcal{F}, \{\mathcal{F}_t\}_{t \geq 0}, \mathcal{P})$$

where $\{\mathcal{F}_t\}_{t \geq 0}$ is a filtration of $\mathcal{F}$, that is, an increasing sequence of sub-sigma algebras of $\mathcal{F}$. From an intuitive perspective, $\mathcal{F}_t$ denotes the information available at time t. The random variable W_t is $\mathcal{F}_t$ measurable for each t (i.e., the process $\{W_t\}_{t \geq 0}$ is adapted to $\{\mathcal{F}_t\}_{t \geq 0}$). Heuristically this means that the value of $(W_s : s \leq t)$ is known at time t. In this chapter we take $\mathcal{F}_t$ to denote the sigma algebra generated by $(W_s : s \leq t)$. This has ramifications for the martingale representation theorem stated later. In addition, the filtration $\{\mathcal{F}_t\}_{t \geq 0}$ satisfies the *usual conditions*. See, e.g., (Karatzas and Shreve, 1991).

The independent increments assumption is crucial to modeling risky security prices as it captures *the efficient market hypothesis* on security prices (Fama, 1970) , that is, any change in security prices is essentially due to arrival of new information (independent of what is known in past). All past information has already been incorporated in the market price.

3.2 Modeling Security Prices

In our model, the security price process is observed till $T > 0$, time to maturity of the option to be priced. In particular, in the Brownian motion defined above the time index is restricted to $[0, T]$. The stock price evolution is modeled as a stochastic process $\{S_t\}_{0 \leq t \leq T}$ defined on the probability space $(\Omega, \mathcal{F}, \{\mathcal{F}_t\}_{0 \leq t \leq T}, \mathcal{P})$ and is assumed to satisfy the stochastic differential equation (SDE)

$$dS_t = \mu_t S_t \, dt + \sigma_t S_t dW_t, \tag{3}$$

where μ_t and σ_t maybe deterministic functions of (t, S_t) satisfying technical conditions such that a unique solution to this SDE exists (see, e.g., (Oksendal, 1998), (Karatzas and Shreve, 1991), (Steele, 2001) for these conditions and a general introduction to SDEs). In (3), the simplest and popular case corresponds to both μ_t and σ_t being constants.

Note that $\frac{S_{t_1+t_2}-S_{t_1}}{S_{t_1}}$ denotes the return from the security over the time period $[t_1, t_2]$. Equation (3) may be best viewed through its discretized Euler's approximation at times t and $t + \delta t$:

$$\frac{S_{t+\delta t} - S_t}{S_t} = \mu_t \delta t + \sigma_t (W_{t+\delta t} - W_t).$$

This suggests that μ_t captures the drift in the instantaneous return from the security at time t. Similarly, σ_t captures the sensitivity to the independent noise $W_{t+\delta t} - W_t$ present in the instantaneous return at time t. Since, μ_t and σ_t maybe functions of (t, S_t), independent of the past security prices, and Brownian motion has independent increments, the process $\{S_t\}_{0 \leq t \leq T}$ is Markov. Heuristically, this means that, given the asset value S_s at time s, the future $\{S_t\}_{s \leq t \leq T}$ is independent of the past $\{S_t\}_{0 \leq t \leq s}$.

Equation (3) is a heuristic differential representation of an SDE. A rigorous representation is given by

$$S_t = S_0 + \int_0^t \mu_s S_s \, ds + \int_0^t \sigma_s S_s dW_s. \tag{4}$$

Here, while $\int_0^t \mu_s S_s \, ds$ is a standard Lebesgue integral defined path by path along the sample space, the integral $\int_0^t \sigma_s S_s dW_s$ is a stochastic integral known as Ito integral after its inventor (Ito, 1946). It fundamentally differs from Lebesgue integral as it can be seen that Brownian motion does not have bounded variation. We briefly discuss this and related relevant results in the subsection below.

3.3 Stochastic Calculus

Here we summarize some useful results related to stochastic integrals that are needed in our discussion. The reader is referred to (Karatzas and Shreve, 1991), (Protter, 1990) and (Revuz and Yor, 1999) for a comprehensive and rigorous analysis.

3.3.1 Stochastic integral

Suppose that $\phi : [0, T] \to \mathfrak{R}$ is a bounded continuous function. One can then define its integral $\int_0^T \phi(s)dA_s$ w.r.t. to a continuous process $(A_t : 0 \leq t \leq T)$, of finite first order variation, as the limit

$$\lim_{n \to \infty} \sum_{i=0}^{n-1} \phi(iT/n)(A_{(i+1)T/n} - A_{iT/n}).$$

(In our analysis, above and below, for notational simplification we have divided T into n equal intervals. Similar results are true if the intervals are allowed to be unequal with the caveat that the largest interval shrinks to zero as $n \to \infty$). The above approach to defining integrals fails when the integration is w.r.t. to Brownian motion $(W_t : 0 \leq t \leq T)$. To see this informally, note that in

$$\sum_{i=0}^{n-1} \left| W_{(i+1)T/n} - W_{iT/n} \right|$$

the terms $|W_{(i+1)T/n} - W_{iT/n}|$ are i.i.d. and each is distributed as $\sqrt{T/n}$ times $N(0, 1)$, a standard Gaussian random variable with mean zero and variance 1. This suggests that, due to the law of large numbers, this sum is close to $\sqrt{nT}E|N(0, 1)|$ as n becomes large. Hence, it diverges almost surely as $n \to \infty$. This makes definition of path by path integral w.r.t. Brownian motion difficult.

Ito defined the stochastic integral $\int_0^T \phi(s)dW_s$ as a limit in the L^2 sense[1]. Specifically, he considered adapted random processes $(\phi(t) : 0 \leq t \leq T)$ such that

$$E \int_0^T \phi(s)^2 \, ds < \infty.$$

For such functions, the following Ito's isometry is easily seen

$$E \left[\sum_{i=0}^{n-1} (\phi(iT/n)(W_{(i+1)T/n} - W_{iT/n}))^2 \right] = E \left(\sum_{i=0}^{n-1} \phi(iT/n)^2 T/n \right).$$

---

[1] A sequence of random variables $(X_n : n \geq 1)$ such that $EX_n^2 < \infty$ for all n is said to converge to random variable X (all defined on the same probability space) in the L^2 sense if $\lim_{n \to \infty} E(X_n - X)^2 = 0$.

To see this, note that ϕ is an adapted process, the Brownian motion has independent increments, so that $\phi(iT/n)$ is independent of $W_{(i+1)T/n} - W_{iT/n}$. Therefore, the expectation of the cross terms in the expansion of the square on the LHS can be seen to be zero. Further, $E[\phi(iT/n)^2(W_{(i+1)T/n} - W_{iT/n})^2]$ equals $E[\phi(iT/n)^2]T/n$.

This identity plays a fundamental role in defining $\int_0^T \phi(s)dW_s$ as an L^2 limit of the sequence of random variables $\sum_{i=0}^{n-1} \phi(iT/n)(W_{(i+1)T/n} - W_{iT/n})$ (see, e.g., (Steele, 2001), (Karatzas and Shreve, 1991)).

3.3.2 Martingale property and quadratic variation

As is well known, martingales capture the idea of a fair game in gambling and are important to our analysis. Technically, a stochastic process $(Y_t : 0 \le t \le T)$ on a probability space $(\Omega, \mathcal{F}, \{\mathcal{F}_t\}_{0 \le t \le T}, \mathcal{P})$ is a martingale if it is adapted to the filtration $\{\mathcal{F}_t\}_{0 \le t \le T}$, if $E|Y_t| < \infty$ for all t, and:

$$E[Y_t|\mathcal{F}_s] = Y_s$$

almost surely for all $0 \le s < t \le T$.

The process $(I(t) : 0 \le t \le T)$,

$$I(t) = \sum_{i=0}^{k-1} \phi(iT/n)(W_{(i+1)T/n} - W_{iT/n}) + \phi(kT/n)(W_t - W_{kT/n})$$

where k is such that $t \in [kT/n, (k+1)T/n)$, can be easily seen to be a continuous zero mean martingale using the key observation that for $t_1 < t_2 < t_3$,

$$E\left[\phi(t_2)(W_{t_3} - W_{t_2})|\mathcal{F}_{t_1}\right] = E\left(E[\phi(t_2)(W_{t_3} - W_{t_2})|\mathcal{F}_{t_2}]|\mathcal{F}_{t_1}\right)$$

and this equals zero since

$$E\left[\phi(t_2)(W_{t_3} - W_{t_2})|\mathcal{F}_{t_2}\right] = \phi(t_2)E\left[(W_{t_3} - W_{t_2})|\mathcal{F}_{t_2}\right] = 0.$$

Using this it can also be shown that the limiting process ($\int_0^t \phi(s)dW_s : 0 \le t \le T$) is a zero mean martingale if $(\phi(t) : 0 \le t \le T)$ is an adapted process and

$$E\int_0^T \phi(s)^2 \, ds < \infty.$$

Quadratic variation of any process $(X_t : 0 \le t \le T)$ may be defined as the L^2 limit of the sequence $\sum_{i=0}^{n-1}(X_{(i+1)T/n} - X_{iT/n})^2$ when it exists. This can be seen to equal $\int_0^T \phi(s)^2 \, ds$ for the process $(\int_0^t \phi(s)dW_s : 0 \le t \le T)$. In particular the quadratic variation of $(W_t : 0 \le t \le T)$ equals a constant T.

3.3.3   Ito's formula

Ito's formula provides a key identity that highlights the difference between Ito's integral and ordinary integral. It is the main tool for analysis of stochastic integrals. Suppose that $f : \mathcal{R} \to \mathcal{R}$ is twice continuously differentiable and $E \int_0^t f'(W_s)^2 \, ds < \infty$. Then, Ito's formula states that

$$f(W_t) = f(W_0) + \int_0^t f'(W_s)dW_s + \frac{1}{2}\int_0^t f''(W_s) \, ds. \tag{5}$$

Note that in integration with respect to processes with finite first order variation, the correction term $\frac{1}{2}\int_0^t f''(W_s) \, ds$ would be absent.

To see why this identity holds, re-express $f(W_t) - f(W_0)$ as

$$\sum_{i=0}^{n-1} \left(f(W_{(i+1)t/n}) - f(W_{it/n}) \right).$$

Expanding the summands using the Taylor series expansion and ignoring the remainder terms, we have

$$\sum_{i=0}^{n-1} f'(W_{it/n})(W_{(i+1)t/n} - W_{it/n}) + \frac{1}{2}\sum_{i=0}^{n-1} f''(W_{it/n})(W_{(i+1)t/n} - W_{it/n})^2.$$

The first term converges to $\int_0^t f'(W_s)dW_s$ in L^2 as $n \to \infty$. To see that the second term converges to $\frac{1}{2}\int_0^t f''(W_s) \, ds$ it suffices to note that

$$\sum_{i=0}^{n-1} f''(W_{it/n}) \left((W_{(i+1)t/n} - W_{it/n})^2 - t/n \right) \tag{6}$$

converges to zero as $n \to \infty$. This is easily seen when $|f''|$ is bounded as then (6) is bounded from above by

$$\sup_x |f''(x)| \sum_{i=0}^{n-1} \left((W_{(i+1)t/n} - W_{it/n})^2 - t/n \right).$$

The sum above converges to zero in L^2 as the quadratic variation of $(W(s) : 0 \le s \le t)$ equals t.

In the heuristic differential form (5) may be expressed as

$$df(W_t) = f'(W_t)dW_t + \frac{1}{2}f''(W_t)\,dt.$$

This form is usually more convenient to manipulate and correctly derive other equations and is preferred to the rigorous representation.

Similarly, for $f : [0, T] \times \Re \to \Re$ with continuous partial derivatives of second order, we can show that

$$df(t, W_t) = f_t(t, W_t)\,dt + f_x(t, W_t)dW_t + \frac{1}{2}f_{xx}(t, W_t)\,dt \qquad (7)$$

where f_t denotes the partial derivative w.r.t. the first argument of $f(\cdot, \cdot)$ and f_x and f_{xx} denote the first and the second order partial derivatives with respect to its second argument. Again, the rigorous representation for (7) is

$$f(t, W_t) = f(0, W_0) + \int_0^t f_t(s, W_s)\,ds$$
$$+ \int_0^t f_x(s, W_s)dW_s + \frac{1}{2}\int_0^t f_{xx}(s, W_s)\,ds.$$

3.3.4 Ito processes

The process $(X_t : 0 \le t \le T)$ of the (differential) form

$$dX_t = \alpha_t\,dt + \beta_t dW_t$$

where $(\alpha_t : 0 \le t \le T)$ and $(\beta_t : 0 \le t \le T)$ are adapted processes such that $E(\int_0^T \beta_t^2\,dt) < \infty$ and $E(\int_0^T |\alpha_t|\,dt) < \infty$, are referred to as Ito processes. Ito's formula can be generalized using essentially similar arguments to show that

$$df(t, X_t) = f_t(t, X_t)\,dt + f_x(t, X_t)dX_t + \frac{1}{2}f_{xx}(t, X_t)\sigma_t^2\,dt. \qquad (8)$$

3.4 Black Scholes partial differential equation

As in the Binomial setting, here too we consider two securities. The risky security or the stock price process satisfies (3). The money market $(R_t :$

$0 \le t \le T$) is governed by a *short rate* process ($r_t : 0 \le t \le T$) and satisfies the differential equation

$$dR_t = r_t R_t \, dt$$

with $R_0 = 1$. Here short rate r_t corresponds to instantaneous return on investment in the money market at time t. In particular, Rs. 1 invested at time zero in the money market equals $R_t = \exp(\int_0^t r_s \, ds)$ at time t. In general r_t may be random and the process ($r_t : 0 \le t \le T$) may be adapted to $\{\mathcal{F}_t\}_{0 \le t \le T}$, although typically when short time horizons are involved, a deterministic model of short rates is often used. In fact, it is common to assume that $r_t = r$, a constant, so that $R_t = \exp(rt)$. In our analysis, we assume that ($r_t : 0 \le t \le T$) is deterministic to obtain considerable simplification.

Now consider the problem of pricing an option in this market that pays a random amount $h(S_T)$ at time T. For instance, for a call option that matures at time T with strike price K, we have $h(S_T) = \max(S_T - K, 0)$. We now construct a replicating portfolio for this option. Consider the recipe process ($b_t : 0 \le t \le T$) for constructing a replicating portfolio. We start with amount P_0. At time t, let P_t denote the value of the portfolio. This is used to purchase b_t number of stock (we allow b_t to take non-integral values). The remaining amount $P_t - b_t S_t$ is invested in the money market. Then, the portfolio process evolves as:

$$dP_t = (P_t - b_t S_t) r_t \, dt + b_t dS_t. \tag{9}$$

Due to the Markov nature of the stock price process, and since the option payoff is a function of S_T, at time t, the option price can be seen to be a function of t and S_t. Denote this price by $c(t, S_t)$. By Ito's formula (8), (assuming $c(\cdot, \cdot)$ is sufficiently smooth):

$$dc(t, S_t) = c_t(t, S_t) \, dt + c_x(t, S_t) dS_t + \frac{1}{2} c_{xx}(t, S_t) \sigma_t^2 \, dt,$$

with $c(T, S_T) = h(S_T)$. This maybe re-expressed as:

$$dc(t, S_t) = \left(c_t(t, S_t) + c_x(t, S_t) S_t \mu_t + \frac{1}{2} c_{xx}(t, S_t) \sigma_t^2 S_t^2 \right) dt$$
$$+ c_x(t, S_t) \sigma_t S_t dW_t \tag{10}$$

Our aim is to select ($b_t : 0 \le t \le T$) so that P_t equals $c(t, S_t)$ for all (t, S_t). To this end, (9) can be re-expressed as

$$dP_t = ((P_t - b_tS_t)r_t + b_t\mu_tS_t)\, dt + b_t\sigma_tS_t dW_t. \tag{11}$$

To make $P_t = c(t, S_t)$ we equate the drift (terms corresponding to dt) as well as the diffusion terms (terms corresponding to dW_t) in (10) and (11). This requires that $b_t = c_x(t, S_t)$ and

$$c_t(t, S_t) + \frac{1}{2}c_{xx}(t, S_t)S_t^2\sigma_t^2 - c(t, S_t)r_t + c_x(t, S_t)S_tr_t = 0.$$

The above should hold for all values of $S_t \geq 0$. This specifies the famous Black-Scholes partial differential equation (pde) satisfied by the option price process:

$$c_t(t, x) + c_x(t, x)xr_t + \frac{1}{2}c_{xx}(t, x)x^2\sigma_t^2 - c(t, x)r_t = 0 \tag{12}$$

for $0 \leq t \leq T$, $x \geq 0$ and the the boundary condition $c(T, x) = h(x)$ for all x.

This is a parabolic pde that can be solved for the price process $c(t, x)$ for all $t \in [0, T)$ and $x \geq 0$. Once this is available, the replicating portfolio process is constructed as follows. $P_0 = c(0, S_0)$ denotes the initial amount needed. This is also the value of the option at time zero. At this time $c_x(0, S_0)$ number of risky security is purchased and the remaining amount $c(0, S_0) - c_x(0, S_0)S_0$ is invested in the money market. At any time t, the number of stocks held is adjusted to $c_x(t, S_t)$. Then, the value of the portfolio equals $c(t, S_t)$. The amount $c(t, S_t) - c_x(t, S_t)S_t$ is invested in the money market. These adjustments are made at each time $t \in [0, T)$. At time T then the portfolio value equals $c(T, S_T) = h(S_T)$ so that the option is perfectly replicated.

3.5 Equivalent martingale measure

As in the discrete case, in the continuous setting as well, under mild technical conditions, there exists another probability measure referred to as the risk neutral measure or the equivalent martingale measure under which the discounted stock price process is a martingale. This then makes the discounted replicating portfolio process a martingale, which in turn ensures that if an option can be replicated, then the discounted option price process is a martingale. The importance of this result is that it brings to bear the well developed and elegant theory of martingales to derivative pricing leading to deep insights into derivatives pricing and hedging (see Harrison and Kreps 1979 and Harrison and Pliska 1981 for seminal papers on this approach).

Martingale representation theorem and Girsanov theorem are two fundamental results from probability that are essential to this approach. We state them in a simple one dimensional setting.

Martingale Representation Theorem: If the stochastic process $(M_t : 0 \leq t \leq T)$ defined on $(\Omega, \mathcal{F}, \{\mathcal{F}_t\}_{0 \leq t \leq T}, \mathcal{P})$ is a martingale, then there exists an adapted process $(v(t) : 0 \leq t \leq T)$ such that $E(\int_0^T v(t)^2 dt) < \infty$ and

$$M_t = M_0 + \int_0^t v(s) dW_s$$

for $0 \leq t \leq T$.

As noted earlier, under mild conditions a stochastic integral process is a martingale. The above theorem states that the converse is also true. That is, on this probability space, every martingale is a stochastic integral.

Some preliminaries to help state the Girsanov theorem: Two probability measures $\mathcal{P}$ and $\mathcal{P}^*$ defined on the same space are said to be equivalent if they assign positive probability to same sets. Equivalently, they assign zero probability to same sets. Further if $\mathcal{P}$ and $\mathcal{P}^*$ are equivalent probability measures, then there exists an almost surely positive Radon-Nikodym derivative of $\mathcal{P}^*$ w.r.t. $\mathcal{P}$, call it Y, such that $\mathcal{P}^*(A) = E_{\mathcal{P}} YI(A)$ (here, the subscript on E denotes that the expectation is with respect to probability measure $\mathcal{P}$ and $I(\cdot)$ is an indicator function). Furthermore, if Z is a strictly positive random variable almost surely with $E_{\mathcal{P}} Z = 1$ then the set function $Q(A) = E_{\mathcal{P}} ZI(A)$ can be seen to be a probability measure that is equivalent to $\mathcal{P}$. Girsanov Theorem specifies the new distribution of the Brownian motion $(W_t : 0 \leq t \leq T)$ under probability measures equivalent to $\mathcal{P}$. Specifically, consider the process

$$Y_t = \exp\left(\int_0^t v_s dW_s - \frac{1}{2} \int_0^t v_s^2 ds\right).$$

Let $X_t = \int_0^t v_s dW_s - \frac{1}{2} \int_0^t v_s^2 ds$. Then $Y_t = \exp(X_t)$ so that using Ito's formula

$$dY_t = Y_t v_t dW_t,$$

or in its meaningful form

$$Y_t = 1 + \int_0^t Y_s v_s dW_s.$$

Under technical conditions the stochastic integral is a mean zero martingale so that $(Y_t : 0 \le t \le T)$ is a positive mean 1 martingale. Let $\mathcal{P}^v(A) = E_{\mathcal{P}}[Y_T I(A)]$.

Girsanov Theorem: Under the probability measure $\mathcal{P}^v$, under technical conditions on $(v_t : 0 \le t \le T)$, the process $(W_t^v : 0 \le t \le T)$ where

$$W_t^v = W_t - \int_0^t v_s \, ds$$

(or $dW_t^v = dW_t - v_t \, dt$ in the differential notation) is a standard Brownian motion. Equivalently, $(W_t : 0 \le t \le T)$ is a standard Brownian motion plus the drift process $(\int_0^t v_s \, ds : 0 \le t \le T)$.

3.5.1   Identifying the equivalent martingale measure

Armed with the above two powerful results we can now return to the process of finding the equivalent martingale measure for the stock price process and the replicating portfolio for the option price process. Recall that the stock price follows the SDE

$$dS_t = \mu_t S_t \, dt + \sigma_t S_t dW_t.$$

We now allow this to be a general Ito process, that is, $\{\mu_t\}$ and $\{\sigma_t\}$ are adapted processes (not just deterministic functions of t and S_t). The option payoff H is allowed to be an $\mathcal{F}_T$ measurable random variable. This means that it can be a function of $(S_t : 0 \le t \le T)$, not just of S_T.

Note that $R_t^{-1} S_t$ has the form $f(t, S_t)$ where S_t is an Ito's process. Therefore, using the Ito's formula, the discounted stock price process satisfies the relation

$$d(R_t^{-1} S_t) = -r_t R_t^{-1} S_t + R_t^{-1} dS_t = R_t^{-1} S_t \left((\mu_t - r_t) \, dt + \sigma_t dW_t \right). \tag{13}$$

It is now easy to see from Girsanov theorem that if $\sigma_t > 0$ almost surely, then under the probability measure $\mathcal{P}^v$ with

$$v_t = \frac{r_t - \mu_t}{\sigma_t}$$

the discounted stock price process satisfies the relation

$$d(R_t^{-1}S_t) = R_t^{-1}S_t\sigma_t dW_t^\nu$$

(this can be seen by replacing dW_t by $dW_t^\nu + \nu_t\,dt$ in (13). This operation in differential notation can be shown to be technically valid). This being a stochastic integral is a martingale (modulo technical conditions), so that $\mathcal{P}^\nu$ is the equivalent martingale measure.

It is easy to see by applying the Ito's formula (Note that $S_t = R_t X_t$ where $X_t = R_t^{-1}S_t$ satisfies the SDE above) that

$$dS_t = r_t S_t\,dt + S_t\sigma_t dW_t^\nu. \tag{14}$$

Hence, under the equivalent martingale measure $\mathcal{P}^\nu$, the drift of the stock price process changes from $\{\mu_t\}$ to $\{r_t\}$. Therefore, $\mathcal{P}^\nu$ is also referred to as the risk neutral measure.

3.5.2 Creating the replicating portfolio process

Now consider the problem of creating the replicating process for an option with pay-off H. Define V_t for $0 \le t \le T$ so that

$$R_t^{-1}V_t = E_{\mathcal{P}^\nu}[R_T^{-1}H|\mathcal{F}_t].$$

Note that $V_T = H$.

Our plan is to construct a replicating portfolio process $(P_t : 0 \le t \le T)$ such that $P_t = V_t$ for all t. Then, since $P_T = H$ we have replicated the option with this portfolio process and P_t then denotes the price of the option at time t, i.e.,

$$P_t = E_{\mathcal{P}^\nu}\left[\exp\left(-\int_t^T r_s\,ds\right)H|\mathcal{F}_t\right]. \tag{15}$$

Then, the option price is simply the conditional expectation of the discounted option payoff under the risk neutral or the equivalent martingale measure.

To this end, it is easily seen from the law of iterated conditional expectations that for $s < t$,

$$R_s^{-1}V_s = E_{\mathcal{P}^\nu}\left(E_{\mathcal{P}^\nu}[R_T^{-1}V_T|\mathcal{F}_t]|\mathcal{F}_s\right) = E_{\mathcal{P}^\nu}[R_t^{-1}V_t|\mathcal{F}_s],$$

that is, $(R_t^{-1}V_t : 0 \le t \le T)$ is a martingale. From the martingale representation theorem there exists an adapted process $(w_t : 0 \le t \le T)$ such that

$$d(R_t^{-1}V_t) = w_t dW_t^\nu. \tag{16}$$

Now consider a portfolio process $(P_t : 0 \le t \le T)$ with the associated recipe process $(b_t : 0 \le t \le T)$. Recall that this means that we start with wealth P_0. At any time t the portfolio value is denoted by P_t which is used to purchase b_t units of stock. The remaining amount $P_t - b_t S_t$ is invested in the money market. Then, the portfolio process evolves as in (9). The discounted portfolio process

$$d(R_t^{-1}P_t) = R_t^{-1}b_t S_t((\mu_t - r_t)dt + \sigma_t dW_t) = R_t^{-1}b_t S_t \sigma_t dW_t^\nu. \tag{17}$$

Therefore, under technical conditions, the discounted portfolio process being a stochastic integral, is a martingale under $\mathcal{P}^\nu$.

From (16) and (17) its clear that if we set $P_0 = V_0$ and $b_t = \frac{w_t R_t}{S_t \sigma_t}$ then $P_t = V_t$ for all t. In particular we have constructed a replicating portfolio. In particular, under technical conditions, primarily that $\sigma_t > 0$ almost surely, for almost every t, every option can be replicated so that this market is complete.

3.5.3   Black Scholes pricing

Suppose that the stock price follows the SDE

$$dS_t = \mu S_t \, dt + \sigma S_t dW_t$$

where μ and $\sigma > 0$ are constant. Furthermore the short rate r_t equals a constant r. From above and (14), it follows that under the equivalent martingale measure $\mathcal{P}^\nu$

$$dS_t = rS_t \, dt + S_t \sigma dW_t^\nu. \tag{18}$$

Equivalently,

$$S_t = S_0 \exp\left(\sigma W_t^\nu + (r - \sigma^2/2)t\right). \tag{19}$$

The fact that S_t given by (19) satisfies (18) can be seen applying Ito's formula to (19). Since (18) has a unique solution, it is given by (19). Since,

$\{W_t^v\}$ is the standard Brownian motion under $\mathcal{P}^v$, it follows that $\{S_t\}$ is geometric Brownian motion and that for each t, $\log(S_t)$ has a Gaussian distribution with mean $(r - \sigma^2/2)t + \log S_0$ and variance $\sigma^2 t$.

Now suppose that we want to price a call option that matures at time T and has a strike price K. Then, $H = (S_T - K)^+$ (note that $(x)^+ = \max(x, 0)$). The option price equals

$$E\left(\left[\exp\left(N((r - \sigma^2/2)T + \log S_0, \sigma^2 T)\right) - K\right]^+\right) \tag{20}$$

where $N(a, b)$ denotes a Gaussian distributed random variable with mean a and variance b. (20) can be easily evaluated to give the price of the European call option. The option price at time t can be inferred from (15) to equal

$$E\left(\left[\exp\left(N((r - \sigma^2/2)(T - t) + \log S_t, \sigma^2(T - t))\right) - K\right]^+\right).$$

Call this $c(t, S_t)$. It can be shown that this function satisfies the Black Scholes pde (12) with $h(x) = (x - K)^+$.

3.6 Multiple assets

Now we extend the analysis to n risky assets or stocks driven by n independent sources of noise modeled by independent Brownian motions $(W_t^1, \ldots, W_t^n : 0 \le t \le T)$. Specifically, we assume that n assets $(S_t^1, \ldots, S_t^n : 0 \le t \le T)$ satisfy the SDE

$$dS_t^i = \mu_t^i S_t^i \, dt + \sum_{j=1}^{n} \sigma_t^{ij} S_t^i dW_t^j \tag{21}$$

for $i = 1, \ldots, n$. Here we assume that each μ^i and σ^{ij} is an adapted process and satisfy restrictions so that the integrals associated with (21) are well defined. In addition we let $R_t = \exp(\int_0^t r(s) \, ds)$ as before. Above, the number of Brownian motions may be taken to be different from the number of securities, however the results are more elegant when the two are equal. This is also a popular assumption in practice.

The key observation from essentially repeating the analysis as for the single risky asset is that for the equivalent martingale measure to exist, the matrix $\{\sigma_t^{ij}\}$ has to be invertible almost everywhere. This condition is also essential to create a replicating portfolio for any given option (that is, for the market to be complete).

It is easy to see that

$$dR_t^{-1}S_t^i = S_t^i\left[\left(\mu_t^i - r_t\right)dt + \sum_{j=1}^{n}\sigma_t^{ij}dW_t^j\right]$$

for all i. Now we look for conditions under which the equivalent martingale measure exists. Using the Girsanov theorem, it can be shown that if $\mathcal{P}^\nu$ is equivalent to $\mathcal{P}$ then, under it,

$$dW_t^{\nu,j} = dW_t^j - \nu_t^j\,dt$$

for each j, are independent standard Brownian motions for some adapted processes $(\nu^j : j \le n)$. Hence, for $\mathcal{P}^\nu$ to be an equivalent martingale measure (so that each discounted security process is a martingale under it) a necessary condition is

$$\mu_t^i - r_t = \sum_{j=1}^{n}\sigma_t^{ij}\nu_j$$

for each i everywhere except possibly on a set of measure zero. This means that the matrix $\{\sigma_t^{ij}\}$ has to be invertible everywhere except possibly on a set of measure zero.

Similarly, given an option with payoff H, we now look for a replicating portfolio process under the assumption that equivalent martingale measure $\mathcal{P}^\nu$ exists. Consider a portfolio process $(P_t : 0 \le t \le T)$ that at any time t purchases b_t^i number of security i. The remaining wealth $P_t - \sum_1^n b_t^i S_t^i$ is invested in the money market. Hence,

$$dP_t = \left(\left(P_t - \sum_1^n b_t^i S_t^i\right)r_t + \sum_1^n b_t^i \mu_t^i S_t^i\right)dt + \sum_{i=1}^{n}b_t^i S_t^i \sum_{j=1}^{n}\sigma_t^{ij}dW_t^j \quad (22)$$

and

$$dR_t^{-1}P_t = R_t^{-1}\sum_{i=1}^{n}b_t^i S_t^i \sum_{j=1}^{n}\sigma_t^{ij}dW_t^{\nu,j}.$$

Since this is a stochastic integral, under mild technical conditions it is a martingale.

Again, since the equivalent martingale measure $\mathcal{P}^v$ exists we can define V_t for $0 \leq t \leq T$ so that

$$R_t^{-1} V_t = E_{\mathcal{P}^v} \left[R_T^{-1} H | \mathcal{F}_t \right].$$

Note that $V_T = H$.

As before, $(R_t^{-1} V_t : 0 \leq t \leq T)$ is a martingale. Then, from a multi-dimensional version of the martingale representation theorem (see any standard text, e.g., (Steele, 2001)), the existence of an adapted process $(w_t \in \mathcal{R}^n : 0 \leq t \leq T)$ can be shown such that

$$d\left(R_t^{-1} V(t)\right) = \sum_{j=1}^{n} w_t^j dW_t^{v,j}. \tag{23}$$

Then, to be able to replicate the option with portfolio process $(P_t : 0 \leq t \leq T)$ we need that $P_0 = V_0$ and

$$w_t^j = R_t^{-1} \sum_{i=1}^{n} b_t^i S_t^i \sigma_t^{ij}$$

for each j almost everywhere. This again can be solved for $(b_t \in \mathcal{R}^n : 0 \leq t \leq T)$ if the transpose of the matrix $\{\sigma_t^{ij}\}$ is invertible for almost all t. We refer the reader to standard texts, e.g., (Karatzas and Shreve, 1998), (Duffie, 1996), (Shreve, 2004), (Steele, 2001) for a comprehensive treatment of options pricing in the multi-dimensional settings.

4   Conclusion

In this chapter we introduced derivatives pricing theory emphasizing its history and its importance to modern finance. We discussed some popular derivatives. We also described the no-arbitrage based pricing methodology, first in the simple binomial tree setting and then for continuous time models.

References

Baaquie, B.E. 2010. *Interest Rate and Coupon Bonds in Quantum Finance*, Cambridge.

Bachelier, L. 1900. *Theorie de la Speculation*, Gauthier-Villars.

Black, F. and Scholes, M. 1973. The Pricing of Options and Corporate Liabilities, *Journal of Political Economy*. **81(3)**:637–654.

Brigo, D. and Mercurio, F. 2006. *Interest Rate Models: Theory and Practice: with Smile, Inflation, and Credit*, Springer Verlag.

Cairns, A.J.G. 2004. *Interest Rate Models*, Princeton University Press.

Cambell, J.Y., Lo, A.W. and Mackinlay, A.C. 2006. *The Econometrics of Financial Markets*, Princeton University Press.

Cox, J.C., Ross, S.A. and Rubinstein, M. 1979. Option Pricing: A Simplified Approach, *Journal of Financial Economics*. **7**:229–263.

Duffie, D. 1996. *Dynamic Asset Pricing Theory*, Princeton University Press.

Duffie, D. 2010. *How Big Banks Fail: and What To Do About It*, Princeton University Press.

Duffie, D. and Singleton, K. 2003. *Credit Risk: Pricing, Measurement, and Management*, Princeton University Press.

Fouque, J.P., Papanicolaou, G. and Sircar, K.R. 2000. *Derivatives in Financial Markets with Stochastic Volatility*, Cambridge University Press.

Harrison, J.M. and Kreps, D. 1979. Martingales and Arbitrage in Multiperiod Security Markets, *Journal of Economic Theory*. **20**:381–408.

Harrison, J.M. and Pliska, S.R. 1981. *Martingales and Stochastic Integrals in the Theory of Continuous Trading, Stochastic Processes and Applications*, **11**:215–260.

Fama, E.F. 1970. Efficient Capital Markets: A Review of Theory and Empirical Work, *Journal of Finance*. **25**:2.

Glasserman, P. 2003. *Monte Carlo Methods in Financial Engineering*, Springer-Verlag, New York.

Hull, J.C. 2008. *Options, Futures and Other Derivative Securities*, Seventh Edition, Prentice Hall.

Ito, K. 1946. On a Stochastic Integral Equation, *Proc. Imperial Academy Tokyo*. **22**:32–35.

Karatzas, I. and Shreve, S. 1991. *Brownian Motion and Stochastic Calculus*, Springer, New York.

Karatzas, I. and Shreve, S. 1998. *Methods of Mathematical Finance*, Springer, New York.

Lando, D. 2005. *Credit Risk Modeling*, New Age International Publishers.

Magill, M. and Quinzii, M. 2002. *Theory of Incomplete Markets*, Volume 1, MIT Press.

McNeil, A.J., Frey, R. and Embrechts, P. 2005. *Quantitative Risk Management: Concepts, Techniques, and Tools*, Princeton University Press

Meucci, A. 2005. *Risk and Asset Allocation*. Springer.

Oksendal, B. 1998. *Stochastic Differential Equations: An Introduction with Applications*, 5th ed. Springer, New York.

Protter, P. 1990. *Stochastic Integration and Differential Equations: A New Approach*, Springer-Verlag.

Revuz, D. and Yor, M. 1999. *Continuous Martingales and Brownian motion*, 3rd Ed., Springer-Verlag.

Samuelson, P. 1965. Rational Theory of Warrant Pricing, *Industrial Management Review.* **6(2):**13–31.

Schonbucher, P.J. 2003. *Credit Derivative Pricing Models*, Wiley.

Shreve, S.E. 2004. *Stochastic Calculus for Finance I: The Binomial Asset Pricing Model*, Springer.

Shreve, S.E. 2004. *Stochastic Calculus for Finance II: Continuous-Time Models*, Springer.

Steele, J.M. 2001. *Stochastic Calculus and Financial Applications*, Springer.

Part III
Mathematics and Computer Science

Linear Time, Almost Linear Time, and Almost Always Linear Time: Snippets from the Work of Robert Endre Tarjan

Jaikumar Radhakrishnan

School of Technology and Computer Science,
Tata Institute of Fundamental Research,
Mumbai 400 005, India.
e-mail: jaikumar@tifr.res.in

Rober Endre Tarjan   *Robert Endre Tarjan has been a leading figure in the area of algorithms for four decades. It is said that he wanted to be an astronomer as a child, but changed his mind and took up mathematics when he was in the eighth grade, apparently under the influence of an excellent mathematics teacher. He obtained his PhD in 1972 from Stanford University working under the supervision of Robert Floyd and Donald Knuth. Tarjan has been recognised and honoured through numerous awards, including the Nevanlinna Prize (the first recipient of the award, 1983) and the Turing Award (jointly with John Hopcroft, 1986). He is at present a Professor in the Department of Computer Science at Princeton University, USA.*

Tarjan has made deep contributions in the areas of Algorithms and Data Structures. He provided breakthroughs in the design and analysis of highly

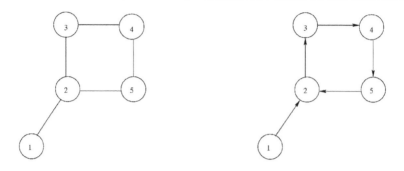

Figure 1: *An undirected graph and a directed graph*

efficient and very elegant algorithms, especially concerning problems on graphs. In this article, we present small parts taken from Tarjan's works, chosen so as to illustrate some of the important considerations that go into the design of algorithms.

1   Linear time algorithms

Hopcroft and Tarjan pioneered the use of depth-first search for designing linear-time algorithms for many natural problems on graphs. Crucial to all these innovations was a representation of graphs, known as the adjacency list representation. In this section, we will describe this representation.

Recall that a graph consists of a set of vertices and a set of edges: $G = (V, E)$, where $E \subseteq V \times V$. If E is symmetric $(u, v) \in E$ whenever $(v, u) \in E$, then we think of the graph as an undirected graph. Otherwise, we think of G as a directed graph. For this discussion we will assume that the graph has no self-loops: that is $(v, v) \notin E$ for all $v \in V$. Here are some examples of graphs, depicted as pictures. How are such graphs represented in a computer's memory? Perhaps, the most natural representation is its adjacency matrix: an $n \times n$ matrix of 0s and 1s, where rows and columns are indexed by elements of V, and the entry in position (v, w) is 1 if (v, w) is an edge of the graph, and is 0 otherwise. Given such a representation of a graph, one might expect the computer to answer several questions about it: Is the graph connected? Does it have a cycle? Is it planar? etc. The algorithm the computer uses must be efficient, that is, it must use as few basic steps as possible. It can be shown that for the problems mentioned above, one must examine $\binom{n}{2}$ entries of the adjacency matrix.

$$\begin{bmatrix} 0 & 1 & 0 & 0 & 0 \\ 1 & 0 & 1 & 0 & 1 \\ 0 & 1 & 0 & 1 & 0 \\ 0 & 0 & 1 & 0 & 1 \\ 0 & 1 & 0 & 1 & 0 \end{bmatrix} \qquad \begin{bmatrix} 0 & 1 & 0 & 0 & 0 \\ 0 & 0 & 1 & 0 & 0 \\ 0 & 0 & 0 & 1 & 0 \\ 0 & 0 & 0 & 0 & 1 \\ 0 & 1 & 0 & 0 & 0 \end{bmatrix}$$

Figure 2: *Adjacency matrices of the two graphs above*

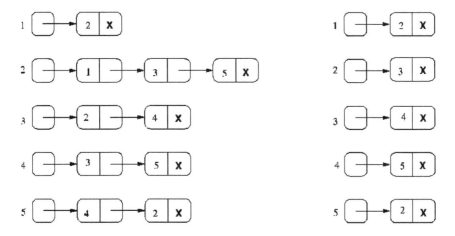

Figure 3: *The adjacency lists of the two graphs above*

In particular, to determine if a graph is planar, the time taken by the algorithm grows at least quadratically in the number of vertices, even though, we know that the size of a planar graph (the number of vertices plus the number of edges) is linear in the number of vertices. In practice, graphs are often sparse, and they have far fewer than $\binom{n}{2}$ edges. If we want very efficient linear-time algorithms, then it is infeasible to work with the adjacency matrix representation. To get around this fundamental bottleneck, the following *adjacency list representation* (apparently first explicitly defined in the work of (Tarjan, 1972)) is preferred. We will not formally define the representation; since it is natural enough, we will just illustrate how the graphs above are represented using adjacency lists. This representation is central to the analysis of algorithms for very sparse graphs. For example, suppose a large graph with n edges has only a linear number of edges (for example, planar graphs have this property) out of the $\binom{n}{2}$ possible edges. Given a vertex, we might want to know the set of its neighbours. With an adjacency matrix, this will involve scanning at least $n - 1$ entries, even though for a typical vertex the number of neighbours is going to be at most 6. For such tasks, clearly the adjacency list is better suited, for it lists the neighbours of every vertex sequentially. Important graph exploration algorithms, such as depth-first search, breadth-first search, topological sort, etc., exploit this convenience offered by the adjacency list representation. Through such exploration, several properties of graphs can be recognised in linear time. In particular, we have the following celebrated theorem of Hopcroft and Tarjan.

Theorem 1. *(Hopcroft and Tarjan, 1974) Given the adjacency list representation of a graph on n vertices, one can, in time $O(n)$, determine if it is planar.*

We will not be describing how this theorem is proved, nor any other algorithms based on adjacency lists. The reader is encouraged to consult standard texts, for e.g., (Kleinberg and Tardos, 2006), to learn more.

2 Almost linear time

In many algorithmic applications, it is required to use a certain procedure repeatedly. Such a need arises most commonly in the use of Data Structures, where information needs to be maintained and updated, and then queries about the data need to be answered. For example, we might maintain distances from the origin in a weighted graph as we explore a graph. Our estimate of the distances will constantly change. We might also want to explore the graph by moving to the next nearest vertex, etc., so the way we store the data must admit efficient retrieval of such information. Can we always store data so that every query can be answered efficiently? The answer unfortunately is 'No'. It is easy to show that in many cases, a long sequence of operations if responded to very efficiently, will make such a mess of the data that some request will need a lot of effort to process. This does not, however, make the situation hopeless. All we need is to show that the total cost of all operations put together remains small; after all we will be making many requests, and what counts is the total time. This pragmatic style of analysis, pioneered by Tarjan and his collaborators, is known

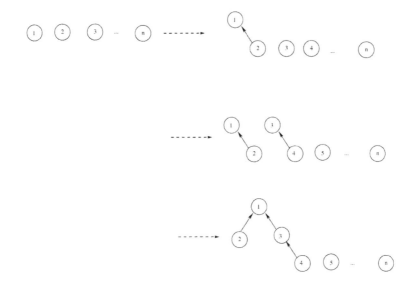

Figure 4: *The union-find data structure responds to three* Union *requests*

as *amortized analysis*. We will illustrate this beautiful idea by discussing
the problem of maintaining equivalence relations.

> We consider an evolving equivalence relation R on a universe
> of n elements, $[n] = \{1, 2, \ldots, n\}$. Initially, the relation is min-
> imal, containing only the essential ordered pairs: $(i, i) \in R$ for
> all $i \in [n]$. Thus, the number of equivalence classes is n. We
> need a method for processing two kinds of requests.
>
> Union(i, j): Add an element (i, j) to the relation. Note that
> this essentially means that the equivalence classes of i
> and j are merged into a common equivalence class.
>
> Find(i): Given an element i, return a representative of the
> equivalence class of i; the same representative must be
> returned for all elements in a common equivalence class
> at any point.

We need a scheme to serve an arbitrary sequence of Unions and Finds
over a universe of size $[n]$. One scheme for maintaining these equivalence
classes is shown in Figure 4. The computer represents the information in

the memory as a forest of trees, one for each equivalence class. Requests for unions are processed by linking the trees, making the root of one point to the root of the other. In the figure, first Union(1, 2) is requested, then Union(3, 4), and finally Union(1, 3).

It is now straightforward to process Find(i) requests. We just follow the arrows from i until we can proceed no further, and return the label on the last node on the path. Clearly, the answers returned by this scheme are valid. How efficient is it? If care is not taken, trees could become very lopsided and requests of Find(i) for an element i deep down in the tree could take n steps to trace. For typical applications, repeatedly taking n steps to process requests is not affordable. Note, that lopsided trees arise because we allow tall trees to be attached below shorter trees. If we disallow these, we obtain immediate benefits.

Rule I: When linking two trees, we have a choice: which of the two roots involved will be the root of the final tree. It is easy to see that making the root of the smaller tree point to the root of the bigger tree will ensure that no tree ever has depth more than $\log n$.

Proposition 1. *If Rule I is implemented, the equivalence classes can be maintained such that the* Union(i, j) *operations take constant time, and* Find*s take $O(\log n)$ time in the worst case.*

It is not hard to come up with instances where m Find(i) requests after $n - 1$ union instructions are issued in such a way that to serve them the algorithm needs $m \log n$ steps, that is, $\log n$ operations per Find(i). One way to work around this example is to move the element i all the way to the top whenever Find(i) is processed. The following rule goes even further, and compresses the entire path from i to the root, ensuring that all nodes on the path point to the root directly.

Rule II: After Find(i) traces the path to the root r, every node on this path is made to point to r. This has the advantage that if the request Find(i) is issued for some i far away from the root, it and many other nodes move close to the root, making it unlikely that future Find(j) requests will require a large number of operations to process. This is called *path compression*.

If Rule I and Rule II are used, one can show that averaged over a long sequence of operations, the cost of each Find(i) is almost constant.

Theorem 2. *A sequence of m* Union *and* Find *operations is always processed in $(m + n)\alpha(n)$ time, where $\alpha(n)$ is the inverse of the Ackermann*

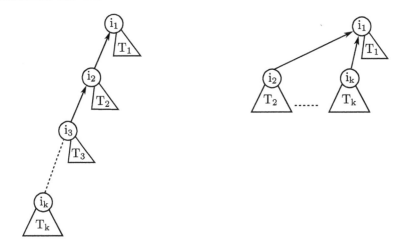

Figure 5: *Path compression, all nodes on the path from i_k to i_1 are made to point to i_1*

function, a very slowly growing function, which is defined as follows.

$$A_0(x) = x + 1;$$

$$A_{k+1}(x) = A_k^{(x)}(x),$$

where $A_k^{(i)}(x)$ is the result obtained by A_k i-times to x, that is $A_k^{(0)}(x) = x$, and $A_k^{(i+1)}(x) = A_k(A_k^{(i)}(x))$. Let $\alpha(n)$ be the least number k such that $A(k) \geq n$.

Note that different authors define the Ackermann function differently, but this hardly makes any difference to the definition of $\alpha(n)$. For most practical values of n, the $\alpha(n)$ is at most 4, and it is, therefore, reasonable to say that the union-find algorithm described above responds to a sequence of requests in almost linear time (in the length of the sequence), that is, constant time per request in the sequence.

2.1 Amortization

This theorem does not assure us that every instruction takes only $O(\alpha(n))$ steps. It is still possible that some Find(i) requires $\log n$ steps. However, these operations will be rare: the total cost divided by the total number of operations is $O(\alpha(n))$. It is important not to confuse the above result with solutions that guarantee efficiency assuming that the requests are drawn from specified probability distribution. The above result is true in the worst case; the solution is guaranteed to be efficient even for the worst sequence.

In the mathematical study of algorithms, as in many other areas of mathematics, one is not satisfied with showing some result. One also wonders if the result is the best possible. Now, Theorem 2 shows that a sequence of m operations requires only $O((m + n)\alpha(n))$ steps to process. Is this the best possible? Perhaps, a better analysis might yield a stronger result. Could it be that this scheme can support requests for Union and Find with constant number of operations per instruction. Unfortunately, as the following theorem shows, the answer is 'No'.

Theorem 3. *(Tarjan, 1975) There is a sequence of n instructions on a universe of n elements, where the above method (with Rule I and Rule II) takes $cn\alpha(n)$ steps.*

2.2 Unexpected dividends

Theorem 3 had a remarkable influence on some results in combinatorial geometry.

Definition 1 (Davenport-Schinzel sequences). A sequence $\langle a_1, a_2, \ldots, a_\ell \rangle$ is a Davenport-Schinzel sequence (DSS) of order s over $[n]$ if it satisfies the following conditions.

(i) Each $a_i \in [n]$.

(ii) No two adjacent a_i's are the same: $a_i \neq a_{i+1}$.

(iii) There is no subsequence $\langle a_{i_1}, a_{i_2}, \ldots, a_{i_{s+2}} \rangle$ such that $a_{i_1} \neq a_{i_2}$, and $a_{i_1} = a_{i_3} = \ldots$, and $a_{i_2} = a_{i_4} = a_{i_6} = \ldots$.

Let $\lambda_s(n)$ be the length of the longest DSS of order s over $[n]$.

DSS were defined by Davenport and Schinzel (1965) in their study of geometric problems from control theory. These sequences play an important role in combinatorial and computational geometry. The history and applications of these sequences are described in the book of (Sharir and Agarwal, 1995), and their survey articles (Agarwal and Sharir, 2000a,b). A readable summary of these results appears in Matousek's book (Matousek, 2002).

It is now known that $\lambda_3(n)$ is bounded above and below by some constant times $n\alpha(n)$: formally, $\lambda_3(n) = \theta(n\alpha(n))$. This result of Hart and Shamir was actually derived using the insights in the proof of Tarjan's Theorem 3. It is remarkable that a problem in discrete geometry was resolved using combinatorial insights from the analysis of computer algorithms.

3   Almost always in linear time

In this section, we discuss the randomized algorithm (Karger *et al.*, 1995) of Karger, Klein and Tarjan for computing minimum spanning trees in undirected graphs. Recall that a spanning tree in a connected undirected graph G is a connected subgraph of G with no cycles. If G has n vertices, the spanning tree has exactly $n - 1$ edges. Assume that each edge of G is assigned a distinct real weight. Then, we may define the weight of spanning tree to be the sum of the weights of its edges; we refer to the spanning tree of minimum weight as the minimum spanning tree (MST); note that our assumption that all edge weights are distinct implies that the MST is unique.

The problem of efficiently finding the minimum spanning tree has been studied for a long time. These studies resulted in the application of a large number of algorithmic and data structure tools. The best deterministic algorithm at present takes slightly superlinear time: the current champion, the algorithm of (Gabow *et al.*, 1984, 1986) on graphs with n vertices and m edges, takes time $O(m \log \beta(m, n))$, where β is a very slowly growing function, related to the inverse of the Ackermann function.

In the early 90s, through a sequence of remarkable papers, (Karger, 1993) and (Klein and Tarjan, 1994), showed a randomized procedure for identifying the MST; this procedure terminates in linear time with high probability.

Theorem 4. *(Karger et al., 1995) The minimum spanning tree can be computed using a randomized algorithm that always correctly computes the minimum weight spanning tree, and terminates in $O(m)$ time with probability $1 - \exp(-\Omega(m))$.*

In this section, we review the key ideas involved in the proof of the above theorem.

As noted earlier, the problem of finding the Minimum Spanning Tree has been studied for a long time. Most of the algorithms exploit the following insights (see Tarjan (1983) for details).

Blue rule: Let $S \subseteq V$ and let $E(S, \overline{S})$ denote the edges that connect S to $\overline{S}$. Then, the minimum weight edge in $E(S, \overline{S})$ must be in the MST.

Red rule: Let C be a cycle in G. Then, the maximum weight edge in G cannot be in any MST.

The randomized algorithm proceeds as follows. In the key randomized step, it picks (randomly) a subset of roughly half the edges and builds a minimum weight spanning tree T' recursively using these edges. This tree T' may not be the best spanning tree in the original graph. In fact, the edges produced by the sample may not even induce a connected graph, so in general, T' would be a collection of trees, one for each connected component. Nevertheless, T' will be handy for applying the *red rule*. If some edge of G is not the minimum weight edge of the cycle it induces when added to T', then it can be eliminated, for it will not be part of the final tree. How many edges can we expect to be eliminated in this process? Answer: all but $O(n)$. How long will it take to find the edges that need to be eliminated? Answer: linear time; this is not straightforward, but turns out to have been shown in earlier works (King, 1997). Roughly, the answer to these two questions will allow us to decompose (we omit a crucial application of the Blue Rule) the problem as follows.

$$\mathsf{MST}(m, n) = \mathsf{MST}\left(\frac{m}{2}, \frac{n}{4}\right) + \mathsf{MST}\left(2n, \frac{n}{4}\right) + O(m),$$

where the subproblems of smaller size are solved recursively using the same idea. It can be shown that Theorem 4, follows from this. We will not prove this here but just invite the reader to consult the beautiful paper of (Karger *et al.*, 1995). We will, however, comment a little more on one aspect: The key to this randomized algorithm is the sampling step. It is analysed using a striking but elementary argument, which we reproduce below.

Lemma 1. *Let H be a subgraph obtained from G by including each edge independently with probability $\frac{1}{2}$. If any edge of G is not the minimum weight edge of the cycle it induces with the MST of H, then it is discarded* *(using the Red Rule). The expected number (averaged over the choice of the random subgraph H) of edges that survive after this pruning is (on an average) at most 2n.*

Proof. Consider the edges of G in the increasing order of their weight. We do this for the analysis, and are not assuming that the algorithm sorts the edges. In our analysis we will imagine that the forest T' is built using the edges as they are included in the subgraph H. When a coin is tossed to determine if the edge e is to be part of H, we have one of the following two cases.

I. The two end points of e are in different components of H at that moment. In this case, if the coin toss determines that e belong to H, then the number of connected components in H reduces by 1.

II. The two end points of e lie in the same component of H; in particular, there is a cycle in H containing e, of which e is the heaviest edge; in this case, e will definitely be eliminated in the pruning step.

The number edges that survive is thus the expected number of times we encounter Case I. Thus, to estimate the number of edges that survive after the pruning step of the algorithm, we need to understand how many times the above procedure passes through Case I. Note that each time Case I is encountered, with probability $\frac{1}{2}$, the number of components reduces by 1; after the number of components reduces to one, Case I will never be encountered again; and all subsequent edges will in fact end up being eliminated by the algorithm. The expected number of times we might encounter Case I is thus at most the expected number of coin tosses needed to observe $n - 1$ heads. This is easily shown to be $2n - 2$.   □

4   Conclusion

We wished to introduce the reader to some of the key concerns that arise in the theoretical study of algorithms. The three works presented were chosen to demonstrate the roles of various mathematical processes involved in this study. They highlight three important concerns encountered when designing efficient algorithms: the input representation, the measure of efficiency, and the model of computation. The reader, I hope, will be encouraged to look up the texts (Kleinberg and Tardos, 2006; Tarjan, 1983;

Mitzenmacher and Upfal, 2005; Matousek, 2002) listed in the references and study this relatively young mathematical area, where increasingly sophisticated mathematics is now starting to be employed.

Acknowledgments

I thank Mohit Garg for helping me make the figures, and to Mohit Garg and Girish Varma for proofreading this article.

References

Agarwal, P.K. and Sharir, M. 2000a. *Davenport-Schinzel sequences and their geometric applications.* In J.-R. Sack and J. Urrita, editors, Handbook of Computational Geometry, 1–47, North-Holland, Amsterdam.

Agarwal, P.K. and Sharir, M. 2000b *Arrangements and their applications.* In J.-R. Sack and J. Urrita, editors, Handbook of Computational Geometry, 49–119, North-Holland, Amsterdam.

Gabow, H.N., Galil, Z. and Spencer, T.H. 1984. Efficient implementation of graph algorithms using contractions. *Proceedings of the 25th annual IEEE symposium on Foundations of Computer Science*, 347–357.

Gabow, H.N., Galil, Z., Spencer, T.H. and Tarjan, R.E. 1986. Efficient algorithms for finding minimum spanning trees in undirected and directed graphs. *Combinatorica*, vol. 6, 109–122.

Karger, D. 1993. Random sampling in matroids and applications to graph connectivity and minimum spanning trees. *Proceedings of the 34th annual IEEE symposium on Foundations of Computer Science*, 84–93.

Hopcroft, J. and Tarjan, R.E. 1974. Efficient planarity testing, *Journal of the ACM*, vol. 21, no. 4, 549–568.

Karger, D.R., Klein, P.N. and Tarjan, R.E. 1995. A randomized linear-time algorithm for finding minimum spanning trees. *Journal of the ACM*, vol. 42, no. 2, 321–328.

Kleinberg, J. and Tardos, E. 2006. *Algorithm Design.* Pearson.

Klein, P.N. and Tarjan, R.E. 1994. A randomized linear time algorithm for finding minimum spanning trees. *Proceedings of the 26th annual ACM Symposium on Theory of Computing*, 9–15.

King, V. 1997. A simpler minimum spanning tree verification algorithm. *Algorithmica*, vol. 18, no. 2, 263–270.

Matousek, J. 2002. *Lectures on discrete geometry.* Springer.

Mitzenmacher, M. and Upfal, E. 2005. *Probability and computing: randomized algorithms and probabilistic analysis.* Cambridge University Press, 2005.

Tarjan, R.E. 1972. Depth-first search and liner graph algorithms, *SIAM Journal on Computing*, vol. 1, no. 2, 146–160.

Tarjan, R.E. 1975. Efficiency of a good but not linear set union algorithm. *Journal of the ACM*, vol. 22, 215–225, 1975.

Tarjan, R.E. 1983. *Data structures and network algorithms.* SIAM.

Sharir, M. and Agarwal, P.K. 1995. *Davenport-Schinzel sequences and their geometric applications.* Cambridge University Press, Cambridge.

CHAPTER 15

Group Representations and Algorithmic Complexity

K.V. Subrahmanyam

Chennai Mathematical Institute,
Chennai 603103 India.
e-mail: kv@cmi.ac.in

1 Introduction

Let M be an $n \times n$ matrix with real entries. Two fundamental quantities associated with an $n \times n$ matrices are the determinant and permanent, whose definition we recall below.

$$det(M) = \sum_{\sigma \in S_n} \Pi_{i=1}^{i=n} sgn(\sigma) M_{i,\sigma(i)}$$

$$perm(M) = \sum_{\sigma \in S_n} \Pi_{i=1}^{i=n} M_{i,\sigma(i)}$$

Here S_n is the group of permutations of the set $[n] = \{1, 2, \ldots, n\}$. For $\sigma \in S_n$, $sgn(\sigma)$ is its signature - this is 1 if the set $\{(i, j) | 1 \le i < j \le n, \sigma(i) > \sigma(j)\}$ has even cardinality and it is -1 otherwise.

Now, computing the determinant is not very hard. For example, one can use the standard Gaussian elimination algorithm to compute it. In fact we will show that the number of multiplication and addition operations needed to compute the determinant of a $n \times n$ matrix is $O(n^3)$ - recall that this means that there is a constant k, such that the number of addition and multiplication operations needed to compute the determinant of an $n \times n$

matrix, is less than or equal to kn^3. The best known algorithm for computing the determinant of a matrix uses $O(n^{2.38})$ arithmetic operations. A long standing open problem is if it can be done with $O(n^2)$ operations. Scientists believe that it may be so, but nobody knows how to do this.

On the other hand, the permanent of a matrix is extremely hard to compute. And scientists believe that the number of multiplication and addition operations needed to compute the permanent of a matrix should be super-polynomial in n - that is, there is no constant c such that the permanent of an arbitrary $n \times n$ matrix can be computed using $O(n^c)$ addition and multiplication operations. However the proof of this has eluded scientists for almost forty years and is expected to be very hard.

The above problems are typical of the kind of questions that algorithm designers and complexity theorists ask. What is the best solution to a problem, under a certain resource measure? In the above problems the resource measure is the number of arithmetic operations. Is there an $O(n^2)$ *upper bound* to the determinant problem? Is there a super-polynomial *lower bound* to the permanent problem.

Complexity theory is concerned with classifying problems based on the resources needed to solve a problem. Problem are attacked from both ends—shave off a resource here and there, to get a better upper bound, or improve existing lower bounds and assert that you cannot do better. The famous *P vs NP* problem is also one such problem - the reader is encouraged to look up the internet for a discussion on this!

The representation theory of groups occupies an important position, at the interface of mathematics and physics. The subject has a rich history of about 150 years. There are some wonderful textbooks on the subject, some of which are alluded to in the references.

It now appears that the representation theory of groups may actually help us in solving the determinant upper bound problem, and the permanent lower bound problem as well. Though, how far we are from actually solving these problems, is anybody's guess.

In 2000, (Mulmuley and Sohoni, 2001) proposed an approach to lower bound problems in complexity theory, via the representation theory of groups. In their approach the representations of certain groups will serve as a proof, that the permanent will require super-polynomial number of arithmetic operations. They further argue that not only will these representations serve as a lower bound proof, there will exist efficient algorithms to reason about these representations.

Remarkably, the representation theory of finite groups is expected to play a decisive role in showing, that there may be way of computing the determinant of a $n \times n$ matrix using only $O(n^2)$ arithmetic operations. This is work of (Cohn and Umans, 2003).

In this article I will give a reasonably self-contained account of the work of (Cohn and Umans, 2003). I will also describe a construction which appears in a subsequent paper by (Cohn *et al.*, 2005). There will be a few gaps however, and I hope that the reader will be motivated enough to read the relevant literature and get a complete picture of their very beautiful work. Our exposition has been partly taken from lecture notes by (Boaz Barrack, 2008) available on the net.

Presenting Mulmuley and Sohoni's approach to lower bounds requires more sophisticated mathematics. And I will not be able to do any justice to this for lack of space. However I would encourage the reader to go and look up Mulmuley's article (Mulmuley, 2007), written for non-experts. There is also a very readable survey of this approach written by (Regan, 2002).

2 Computing the determinant, matrix multiplication

Throughout this exposition, arithmetic operations will mean addition and multiplication operations.

For a long time computer scientists have known that the number of arithmetic operations needed to compute the determinant of an $n \times n$ matrix is the same as the number of operations needed to compute the product of two $n \times n$ matrices, up to a constant factor - which means that if we can multiply two $n \times n$ matrices using $k \cdot n^c$ arithmetic operations for some constants k, c, then the determinant of an $n \times n$ matrix can also be computed using $k' \cdot n^c$ operations albeit for a different constant k' - note that the exponent of n in both cases, is the same constant c. We will not prove this statement here - a reference to this is (Cormen *et al.*, 1990). So instead of looking at the determinant upper bound problem, we will concentrate on an upper bound for finding the product of two $n \times n$ matrices.

So let A and B be two n by n matrices. Let C be the product of these two matrices. We know that the (i, j)-th entry in C is given by

$$C_{i,j} = \sum_{k=1}^{k=n} A_{i,k} B_{k,j}$$

To compute an entry of matrix C we can multiply n pairs of elements in A and B, and add them - so we perform a total of $n - 1$ additions and n multiplications per entry of C. Since there are n^2 entries in C, we perform a total of $O(n^3)$ arithmetic operations giving,

Theorem 1. *Two $n \times n$ matrices can be multiplied using $O(n^3)$ arithmetic operations.*

Now matrix multiplication is a fundamental operation, and is a basic subroutine in many computer algebra programs. Any improvement in the time it takes to multiply two matrices, will translate to an improvement in the running time of many such programs. So a natural question is if one can one multiply two matrices, using a fewer number of arithmetic operations than what the naive method suggests? Indeed, if there is such a procedure, it will surely translate to faster programs to multiply two matrices.

In a remarkable insight Strassen showed that this is possible (see, Strassen (1969), (Cormen *et al.*, 1990)).

3 Better matrix multiplication algorithms

Theorem 2. *Matrix multiplication can be done using $O(n^{2.81})$ arithmetic operations.*

We will give a proof of this in the sequel. But before that let us consider a different problem - how many arithmetic operations are needed to add two $n \times n$ matrices A, B. Now each entry in the sum of the two matrices, depends upon exactly two entries, one from A and one from B. So each entry of the sum can be computed using one addtion, so we perform a total of n^2 arithmetic operations to add two matrices.

Strassen's algorithm for multiplying two matrices is based on the above observation and another idea - divide the given matrices into small chunks, multiply these chunks, and then assemble the results obtained by multiplying these chunks, to solve the original problem. In computer science parlance this paradigm of solving problems is known as *divide and conquer*, and has been successfully employed to solve a number of problems. To achieve a better running time, Strassen trades multiplication of matrices for matrix additions. Since addition of matrices is inexpensive, everything works out! Let us see how.

3.1 The algorithm

For simplicity let us assume that n is a power of two. We divide A and B into four $n/2 \times n/2$ submatrices, as follows:

$$A = \begin{pmatrix} A_{11} & A_{12} \\ A_{21} & A_{22} \end{pmatrix}, \qquad B = \begin{pmatrix} B_{11} & B_{12} \\ B_{21} & B_{22} \end{pmatrix}$$

Dividing the product matrix, C, similarly, we have

$$C = \begin{pmatrix} C_{11} & C_{12} \\ C_{21} & C_{22} \end{pmatrix} = \begin{pmatrix} A_{11}B_{11} + A_{12}B_{21} & A_{11}B_{12} + A_{12}B_{22} \\ A_{21}B_{11} + A_{22}B_{21} & A_{21}B_{12} + A_{22}B_{22} \end{pmatrix}$$

Clearly one can compute C by computing each of the terms $C_{11}, C_{12}, C_{21}, C_{22}$. And each term can be computed by first multiplying two pairs of $n/2 \times n/2$ matrices, and then adding the resulting $n/2 \times n/2$ matrices - in all we perform 8 $n/2 \times n/2$ matrix multiplications, and 4 $n/2 \times n/2$ matrix additions. Now how do we compute the product of the $n/2 \times n/2$ matrices - are we back to square one? Well no, we have after all reduced the sizes of matrices involved! To compute the new products, we divide the matrices involved in a likewise fashion and proceed as we did - this technique is called recursion by algorithm designers. So we recurse, till we are down to matrices of size 2×2, which we multiply using 8 multiplications and 4 additions. Denoting by $T(n)$ the number of arithmetic operations needed for multiplying two $n \times n$ matrices, we have the formula

$$T(n) = 8T(n/2) + 4(n/2)^2$$

Using $T(2) = 12$, we solve the above equation to get $T(n) = O(n^3)$. This is the same bound we got by using the naive algorithm! So, indeed, we are back to square one!

However Strassen observed that by performing 7 intermediate $n/2 \times n/2$ matrix products instead of 8 as we did above, one can recover the terms $C_{11}, C_{12}, C_{21}, C_{22}$ Set

$$P_1 = A_{11}(B_{12} - B_{22})$$
$$P_2 = (A_{11} + A_{12})B_{22}$$
$$P_3 = (A_{21} + A_{22})B_{11}$$
$$P_4 = A_{22}(B_{21} - B_{11})$$
$$P_5 = (A_{11} + A_{22})(B_{11} + B_{22})$$

$$P_6 = (A_{12} - A_{22})(B_{21} + B_{22})$$
$$P_7 = (A_{11} - A_{21})(B_{11} + B_{12})$$

One can then check that C can be obtained as

$$C = \begin{pmatrix} P_5 + P_4 - P_2 + P_6 & P_1 + P_2 \\ P_3 + P_4 & P_5 + P_1 - P_3 - P_7 \end{pmatrix}$$

Note that Strassens algorithm performs many more additions - in fact 18 as compared to 4 that the earlier recursive procedure did. So it has effectively traded one matrix multiplication for 14 extra matrix additions. Asymptotically everything works out, because matrix additions is easier! For Strassen's algorithm we get the recurrence,

$$T(n) = 7 * T(n/2) + 18(n/2)^2$$

Plugging $T(2) = 12$ and solving, this time we get $T(n) = O(n^{2.81})$.

Essentially the same argument can be used to show that if for some constant k, we can multiply two $k \times k$ matrices using k^ω arithmetic operations, then two $n \times n$ matrices can be multiplied using n^ω arithmetic operations. This was the basis for further improvements algorithms for matrix multiplication. However these improvements were significantly more complicated than the above algorithm of Strassen. In 1990 Coppersmith and Winograd proposed a very complicated solution which results in $T(n) = O(n^{2.38})$ arithmetic operations. To date this remains the best upper bound for the number of arithmetic operations needed for matrix multiplication.

Denote by ω the least exponent for which there is an algorithm for multiplying two $n \times n$ matrices using n^ω arithmetic operations. Since C has n^2 entries, each of which needs to be computed, it is clear that $\omega \geq 2$. And we know that $\omega \leq 2.38$. So what is the real value of ω? Remarkably, after Strassen's breakthrough, and the series of improvements up to Coppersmith and Winograd's, it is now believed that ω may actually be equal to 2. However improvement beyond 2.38 has proven to be very difficult.

A new insight into the problem, proposed by Cohn and Umans (2003), offers fresh hope that there is an algorithm for $n \times n$ matrix multiplication using $O(n^2)$ arithmetic operations. The approach proposed by Cohn and Umans is based on group representation theory. It has the advantage that it offers a generic procedure to obtain matrix multiplication algorithms.

It provides us with concrete conjectures which, if true, would result in establishing $\omega = 2$.

We recall some notions from group representation theory. We will assume that the reader has some familiarity with the theory of groups, rings, algebras and vector spaces. Throughout our discussion we will restrict ourselves to vector spaces over the complex numbers $\mathbb{C}$, and our algebras will all be vector spaces over $\mathbb{C}$, with multiplication being associative.

4   Group representation theory basics

There are many outstanding books on the subject of group representations, some of which are alluded to in the list of references (Serre, 1977), (Goodman and Wallach, 1998).

Recall that a group is a set G with a binary operation $\cdot$ satisfying

- Closure. $\forall a, b \in G, a \cdot b \in G$.

- Associativity. $\forall a, b, c \in G, (a \cdot b) \cdot c = a \cdot (b \cdot c)$

- Existence of identity. $\exists id \in G, \forall a \in G, a \cdot id = id \cdot a = a$

- Existence of inverse. $\forall a \in G, \exists b \in G, a \cdot b = b \cdot a = id$

Examples

1 For example, $n \times n$ matrices with complex entries and non-zero determinant form a group, denoted $GL(n, \mathbb{C})$. The operation here is matrix multiplication. The identity matrix acts as the identity element of the group and the inverse of a matrix is its inverse in the group.

   Matrices having determinant 1 are closed under the above group operations for matrices, and so form a subgroup of $GL(n, \mathbb{C})$. We denote this as $SL(n, \mathbb{C})$.

2 The set of permutations of the set $\{1, 2, \ldots, n\}$ forms a group under composition, called the symmetric group on n letters.

Let V be a vector space over the complex numbers $\mathbb{C}$. Recall that a basis of a vector space is a set of linearly independent vectors spanning V. The dimension of the vector space, denoted $dim(V)$, is the number of elements in any basis.

Given two vector spaces V, W a linear transformation from V to W is a function $\phi : V \rightarrow W$ such that

$$\phi(v_1 + v_2) = \phi(v_1) + \phi(v_2)(v_1, v_2 \in V)$$
$$\phi(tv) = \phi(v)t \in \mathbb{C}, v \in V$$

We sometimes refer to a linear transformation as a linear map, or a linear homomorphism. The set of linear transformations from V to W is denoted by $Hom(V, W)$.

When $W = V$, we denote this as $End(V)$. Note that $End(V)$ is a $\mathbb{C}$ algebra with addition and multiplication defined as follows—the reader is encouraged to check that the other axioms of an algebra are satisfied by $End(V)$.

- For $\phi_1, \phi_2 \in End(V)$, $(\phi_1 + \phi_2)(v) = \phi_1(v) + \phi_2(v)$.

- For $\phi_1, \phi_2 \in End(V)$, $(\phi_1 \cdot \phi_2)(v) = \phi_1(\phi_2(v))$.

So multiplication in $End(V)$ is the composition of the two linear transformations involved.

The set of linear transformations in $End(V)$ which are bijective is denoted by $GL(V)$. One can easily check that $GL(V)$ is a group under the $\cdot$ (composition) operator.

If we choose a basis for V then each element of V can be identified with a vector of size $dim(V)$ with complex entries. And then $End(V)$ can be identified with the algebra of $dim(V) \times dim(V)$ matrices with complex entries. Likewise $GL(V)$ can be identified with matrices of size $dim(V) \times dim(V)$ with complex entries, and having non-zero determinant—using our earlier notation it is identified with $GL(dim(V), \mathbb{C})$.

Each such matrix m, gives us a transformation of V, by which a vector $v \in V$ is sent to $m \cdot v$, where $m \cdot v$, denotes multiplication of m with the column vector v. This motivates the following

Definition 1. We say V is a representation of G if we have a group homomorphism $\rho_V : G \rightarrow GL(V)$.

Such a representation then gives an action of G on V-an element g acts on V via the linear transformation $\rho_V(g)$-we denote the action of g on v by $g \cdot v$ instead of $\rho_V(g) \cdot v$, when the homomorphism ρ is clear from the context. When G acts on V, the following equations hold.

$$id \cdot v = v \qquad \forall v \in V \tag{1}$$

$$(gh) \cdot v = g \cdot (h \cdot v) \qquad \forall v \in V, \forall g, h \in G \qquad (2)$$
$$g \cdot (tv) = t(g \cdot v) \qquad \forall t \in \mathbb{C}, \forall g \in G \qquad (3)$$

- Homogeneous polynomials of degree n in the variables x, y form a vector space of dimension $n + 1$ and $x^n, x^{n-1}y, \ldots, y^n$ form a basis of this vector space. It can be checked that the following defines an action of $SL(2, \mathbb{C})$ on this space.

$$f(x, y) \cdot \begin{pmatrix} a & b \\ c & d \end{pmatrix} = f(ax + cy, bx + dy)$$

Definition 2. Given two representations V, W of a group G, we say a linear transformation from $\phi : V \to W$ is a G-morphism if for all $g \in G, v \in V$ we have $\rho_W(g) \cdot \phi(v) = \phi(\rho_V(g) \cdot v)$

Definition 3. Two representations V, W are said to be equivalent if there is a G linear isomorphism between them. Denoting the isomorphism between V, W by T, it is clear that for all g, the matrices $\rho_V(g)$ and $\rho_W(g)$ are related by

$$\rho_W(g) = T^{-1}\rho_V(g)T$$

Note that the kernel of a G-morphism ϕ is a G submodule of V. This is clear because if $\phi(v) = 0$ then $\phi(g \cdot v) = g\phi(v) = 0$, since ϕ is a G-morphism. Likewise the image $\phi(V)$ is a G submodule of W. We encourage the reader to check these details.

Given representations V, W of a group G, we can construct new representations of G. Recall that the dual space of V, V^*, is the space of all linear transformations from V to $\mathbb{C}$. We have a representation of G on V^* - given $f \in V^*$, the action of g on f is the linear transformation $g \cdot f$ defined by $g \cdot f(v) = f(g^{-1} \cdot v)$. We have an action of G on $V \otimes W$, the tensor product of V and W by $g \cdot (v \otimes w) = (g \cdot v \otimes g \cdot w)$.

Now let W be a subspace of V. We say W is a G-invariant subspace of V, if for all $g \in G, w \in W, g \cdot w \in W$. So in fact W is also a representation of G, with the homomorphism from G to $GL(W)$ being the restriction of of the image of ρ to W, $\rho(G)|_W$. Note that the zero vector in V is a subspace of V which is invariant under G. If this is the only proper subspace of V invariant under G we say V is an irreducible representation of G.

Now if V has a G-invariant subspace W, we can do the following - select a basis $e_1, e_2, \ldots, e_r$ for W, and complete it to a basis $e_1, e_2, \ldots, e_r$,

$e_{r+1}, \ldots, e_m$ for V. Then if we write down the images of elements in G under ρ as a matrix, it is of the form

$$
\begin{pmatrix}
* & \vdots & * \\
\cdots & \vdots & \cdots \\
0 & \vdots & *
\end{pmatrix}
$$

In fact we can do better - and that is the content of Mashke's theorem. In the rest of the article, unless otherwise specified, by a group G we will always mean a finite group G.

Theorem 3. *Let V be a representation of G and let W be a proper G-invariant subspace of V. Then there is a G-invariant complement of W i.e. there is a subspace W' of V, which is also G-invariant and $V = W \oplus W'$.*

Recall that $V = W \oplus W'$ means then each element of V can be uniquely written as the sum of an element in W and an element in W'.

Proof. Take a basis of $\{e_1, e_2, \ldots, e_r\}$ of W and complete it to a basis $\{e_1, e_2, \ldots, e_r, \ldots, e_m\}$ of V. Now every vector in $v \in V$ is expressible as a linear combination of the basis vectors e_i. Define a linear transformation $p : V \to V$ sending $v = \sum_{i=1}^{i=m} c_k e_k$ to the vector $\sum_{j=1}^{j=r} c_j e_j$. Clearly p is a projection onto the subspace W with $p(w) = w, w \in W$, $p \cdot p(v) = p(v), \forall v \in V$.

Now define a transformation $p_0 : V \to V$ as follows:

$$
p_0(v) = \frac{1}{|G|} \sum_{g \in G} \rho(g) \cdot p \cdot \rho(g)^{-1}(v)
$$

Now since p maps V onto W, the image of p_0 is also all of W. We show that p_0 is a G-morphism. Let $v \in V$ and $h \in G$. Then we have

$$
p_0(h \cdot v) =
$$

$$
p_0(\rho(h) \cdot v) = \frac{1}{|G|} \sum_{g \in G} \rho(g) \cdot p \cdot \rho(g)^{-1} \rho(h)(v)
$$

$$
= \frac{1}{|G|} \sum_{g \in G} \rho(h)(\rho(h)^{-1}\rho(g) \cdot p \cdot \rho(g)^{-1}\rho(h))(v)
$$

$$= \rho(h)\frac{1}{|G|}\sum_{g \in G}(\rho(h)^{-1}\rho(g) \cdot p \cdot \rho(g)^{-1}\rho(h))(v)$$

$$= \rho(h)p_0(v)$$

So the kernel of p_0 is a G-invariant subspace W' of V. Also using the fact that $p(w) = w, w \in W$, it is clear that that $p_0(w) = w, \forall w \in W$. Likewise it follows that $p_0 \cdot p_0(v) = p_0(v)$. So for every vector $v \in V$, $v - p_0(v) \in W'$. So every vector in V is the sum of the vector $p_0(v)$ in W, and the vector $v - p_0(v)$ in W'. Since both are G-invariant subspaces of V with no nonzero vector in their intersection, it follows that $V = W \oplus W'$. □

We can now choose a basis of V which consists of a basis of W and a basis of the subspace W', which the above theorem guarantees. It follows then that under the homomorphism ρ, the image of an element $g \in G$ is a matrix of the form

$$\begin{pmatrix} * & \vdots & 0 \\ \cdots & \vdots & \cdots \\ 0 & \vdots & * \end{pmatrix}$$

Repeatedly using Maschke's theorem we get

Theorem 4. *Let G be a finite group and V a representation of G over $\mathbb{C}$. Then V splits into a direct sum of irreducible representations $V_1 V_2, \ldots, V_k$.*

In matrix terms the theorem tells us that one can choose a basis of the representation V, in which the image of every group element looks like a block diagonal matrix with k blocks - the i-th block corresponding to the restriction of ρ to the irreducible representation V_i.

4.1 Group algebra

If G is a group, the group algebra over $\mathbb{C}$, denoted by $\mathbb{C}G$, consists of elements of the form $\Sigma_{g \in G} c_g g$, where c_g are complex numbers. If $a = \Sigma_{g \in G} a_g g$ and $b = \Sigma_{g \in G} b_g g$ then $a + b$ is defined to be the element $\Sigma_{g \in G}(a_g + b_g)g$. The product of a and b is the convolution product defined as

$$ab = \sum_{g \in G}\left(\sum_{h \in G} a_h b_{h^{-1}g}\right)g$$

With addition and multiplication defined this way, $\mathbb{C}G$, is an algebra. As a vector space over $\mathbb{C}$ it has dimension $|G|$.

We recall now the notion of a representation of a $\mathbb{C}$ algebra. In the sequel we will apply this notion to the group algebra of a finite group.

Definition 4. A representation of a $\mathbb{C}$ algebra A is a $\mathbb{C}$ algebra homomorphism ρ, from A to $End(W)$ for some $\mathbb{C}$ vector space W, mapping the identity element of A to the identity element in $End(W)$. So we have $\rho(ab) = \rho(a)\rho(b)$, $\rho(a + b) = \rho(a) + \rho(b)$, $\forall a, b \in A$. For an element $k \in \mathbb{C}$, we further require that $\rho(ka) = k\rho(a)$.

Like we did in the case of group representations, we think of the elements of the algebra acting on the space W, via their images in $End(W)$. The representation W of an algebra A is said to be irreducible if there is no proper non zero subspace S of W such that $\rho(a)S \subseteq S, \forall a \in A$.

Given any G representation V, we can extend it to a representation of $\mathbb{C}G$ on V by defining $\rho((\Sigma_{g \in G} c_g g))v = \Sigma_g c_g g \cdot v, \forall v \in V$. Via this extension, we get an algebra map from $\mathbb{C}G$ to $End(V)$ - the image of an element in $\mathbb{C}G$ may no longer be in $GL(V)$.

Now let G act on the vector space $\mathbb{C}G$ by $h \cdot \Sigma_{g \in G} c_g g = \Sigma_{hg \in G} c_g hg$. It can be easily checked that this gives a homomorphism of groups from G to $GL(\mathbb{C}G)$, so $\mathbb{C}G$ is a representation of G. So from the previous paragraph this extends to action of $\mathbb{C}G$ on itself. We get an algebra map from $\mathbb{C}G$ to $End(\mathbb{C}G)$. We will show that this is in fact an isomorphism from $\mathbb{C}G$ to a product of matrix algebras.

Towards this we need to understand what are the possible G-morphisms between two irreducible representations V, W of a group G. This is answered by the following lemma of Schur.

Lemma 1. *Let V, W be irreducible representations of G. Then every G-homomorphism ϕ from V to W is either zero or is an isomorphism. If ϕ is an isomorphism then there is an $a \in \mathbb{C}$ such that $\phi(v) = av$.*

Proof. Since the kernel is a G submodule of V and the image a G submodule of W, both must be either trivial or the whole space, Now if the kernel is zero and the image is all of W, it follows that we can identify W with V and ϕ must be an isomorphism.

Now let ϕ be an isomorphism from V to V. Let v be an eigenvector of ϕ with eigenvalue a. Such an a exists since $\mathbb{C}$ is algebraically closed. Then

$\phi - a * (Id)_V$ is a G linear transformation from V to V sending v to zero. Sine V is irreducible, the kernel must be all of V from the first part. And so $\phi(v) = a \cdot v$, for all $v \in V$, completing the proof. □

Consider the space of functions from $G \to \mathbb{C}$. We define a Hermitian inner product $\langle,\rangle$ on this space as follows.

$$\langle f_1, f_2 \rangle = \frac{1}{|G|} \sum_g \overline{f_1(g)} f_2(g)$$

where $\overline{f_1()}$ represents complex conjugation.

Given a representation V of G, we define a rather special function associated to V, the character of the representation, $\chi_V : G \to \mathbb{C}$ by $\chi_V(g) = Trace(\rho_V(g))$ - the trace on the right hand side of this equation is the trace of a matrix, which is the sum of its diagonal entries - note that this is independent of the choice of basis for V, since the trace of a matrix remains the same under conjugation. Note that $\chi_V(Id) = dim(V)$

A little linear algebra reveals a lot of information about characters. We proceed to do that.

Lemma 2. *For every representation V of G, and every $g \in G$, the matrix $\rho_V(g)$ can be diagonalized. If $\lambda_1, \lambda_2, \ldots, \lambda_n$ are the eigenvalues of $\rho_V(g)$ (including repetitions) then $\chi_V(g) = \sum_{i=1}^{i=n} \lambda_i$. Each λ_i is a root of unity.*

Proof. Since G is a finite group every element in G has finite order - i.e. there is a k (depending upon g) such that $g^k = id$. So for any representation V of G, the matrix $\rho_V(g)$ satisfies the polynomial equation $X^k = 1$. This implies that the minimal polynomial $f(X)$ of ρ_V, divides the polynomial $X^k - 1$. And so the minimal polynomial has distinct roots - which means that the matrix $\rho_V(g)$ can be diagonalized. So $\rho_V(g)$ is conjugate to the diagonal matrix with diagonal entries $\lambda_1, \lambda_2, \ldots, \lambda_n$. So $\chi_V(g) = \sum_{i=1}^{i=n} \lambda_i$, by definition of the trace of a diagonal matrix, and invariance of trace under conjugation. Furthermore all the eigenvalues of $\rho_V(g)$ are roots of unity, since each eigenvalue λ_i satisfies the equation $\lambda_i^k = 1$. □

Theorem 5. *If V and W are two irreducible representations of G, then $\langle \chi_V, \chi_W \rangle = 1$ if they are isomorphic representations and $\langle \chi_V, \chi_W \rangle = 0$ otherwise.*

Proof. Notice first that $\chi_{V^*}(g) = \chi_V(g^{-1})$ for the dual representation V^*. If the eigenvalues of $\rho_V(g)$ are $\lambda_1, \lambda_2, \ldots, \lambda_n$ (including possible repetitions) it is clear that the eigenvalues of $\rho_V(g^{-1})$ are $\lambda_1^{-1}, \lambda_2^{-1}, \ldots, \lambda_n^{-1}$. So $\chi_V(g^{-1}) = \sum_{i=1}^{i=n} \lambda_i^{-1}$ which is the same as $\sum_{i=1}^{i=n} \overline{\lambda_i} = \overline{\chi_V(g)}$, since each λ_i is a root of unity.

Now the character of the representation $W \otimes V^*$ is $\chi_W \cdot \chi_{V^*}$. So

$$\chi_{W \otimes V^*}(g) = \chi_W(g)\chi_{V^*}(g)$$

and this is equal to $\overline{\chi_V(g)}\chi_W(g)$. It follows then that the inner product of χ_V and χ_W is

$$\frac{1}{|G|} \sum_g \overline{\chi_V(g)}\chi_W(g) = \frac{1}{|G|} \sum_g \chi_{W \otimes V^*}(g) \tag{4}$$

Now suppose there is an element x in $W \otimes V^*$ which is such that $g \cdot x = x$ for all $g \in G$ - we call such an element an invariant for the action of G on $W \otimes V^*$. We could then take a basis of $W \otimes V^*$ containing x. And for all g, the matrix for g would then have a 1 at the diagonal position corresponding to basis element x. This will cortribute 1 to $\chi_{W \otimes V^*}$. In fact this argument tells us that the right hand side is the number of number of linearly independent invariants for the G action on $W \otimes V^*$. Now there is a natural identification of $W \otimes V^*$ with $Hom(V, W)$. Under this identification an element of the form $w \otimes v^*$ goes to the function $f_{w \otimes v^*}$ which sends an element u of V to $v^*(u)w$ in W. Under this identification it can be easily checked that a G invariant element maps to a G morphism from V to W and that every G morphism comes from a G invariant element. But V and W are irreduible representations of G, so Schur's lemma tells us that there the number of linearly independent G morphisms is either zero or 1. So the right hand side is 1 if V and W are isomorphic and zero otherwise. This completes the proof. □

We now use this to analyze the representation of G on $\mathbb{C}G$.

Theorem 6. *Every irreducible representation V of G occurs $\dim(V)$ times in $\mathbb{C}G$.*

Proof. The character of the representation $\chi_{\mathbb{C}G}G$ is easy to compute - it is zero on all elements of G except the identity on which it is $|G|$. To see this consider the basis of $\mathbb{C}G$ consisting of elements of g (thought of as a formal linear combination). Then the image of each non-identity group element is a permutation matrix, with no fixed points, so its trace is zero. From this we get,

$$\langle \chi_V, \chi_{CG} \rangle = \frac{1}{|G|} \overline{\chi_V(1)} \chi_{CG}(1) = \frac{1}{|G|} dim(V)|G| = dim(V)$$

□

Now assume that $V_1, V_2, \ldots, V_k$ are the irreducible representations of G and their dimensions are $d_1, d_2, \ldots, d_k$. From the above theorem it follows that there are d_i copies of the irreducible representation V_i in CG. In particular this tells us the dimension of CG is $\sum_i d_i^2$. Furthermore from the above discussion it follows that we can select a basis of CG, so that the image of every element in G, is a block diagonal matrix of size $|G| \times |G|$, with entries in C. The first d_1 blocks, each of size $d_1 \times d_1$, corresponding to the irreducible representations V_1 occuring in CG, the next d_2 blocks each of size $d_2 \times d_2$ corresponding to the irreducible representation V_2, and so on.

Since the d_i blocks correspond to equivalent representations of G, we may conjugate the d_i, $d_i \times d_i$ matrices (see definition 3), to ensure that for each $g \in G$, $\rho(g)$ has identical matrices in its d_i blocks, for each i.

Now since ρ is a homomorphism of groups from G to $GL(CG)$, this block decomposition respects multiplication in G. Let the image of $g_1 \in G$ be the block diagonal matrix $(A_1, \ldots, A_2, \ldots, \ldots, \ldots A_k)$, A_i being a $d_i \times d_i$ matrix, appearing d_i times in this expression. Likewise let the image of g_2 be be the block diagonal matrix $(B_1, \ldots, B_2, \ldots, \ldots, \ldots B_k)$. Then if $g = g_1 \cdot g_2$ in the group, its image must be the block diagonal matrix $(A_1 * B_1, \ldots, A_2 * B_2, \ldots, \ldots, \ldots A_k * B_k)$. The $A_i * B_i$ here refers to matrix multiplication of the $d_i \times d_i$ matrices A_i, B_i.

Now extend this to a representation of the group algebra CG. It is clear that the image of every element in CG is also a block diagonal matrix, with identical matrices in the d_i blocks corresponding to the irreducible representation V_i. This holds for each i. In fact, for each i, the image of CG is all of $End(V_i)$ - that is, every possible $d_i \times d_i$ matrix is the image of some element in CG. This follows from Burnsides theorem, which we state below without proof.

Theorem 7. *Let V be an irreducible representation of an algebra* A *given by* $\rho : A \mapsto End(V)$. *If* $dim(V)$ *is finite and* $\rho(A)$ *is not equal to zero, then* $\rho(A) = End(V)$.

It follows therefore that $\mathbb{C}G$, is isomorphic to a subalgebra of the product of the matrix algebras $\Pi_{i=1}^{i=k} M(d_i)$. However the image of $\mathbb{C}G$ is a vector subspace of $\Pi_{i=1}^{i=k} M(d_i)$. Since both these vectors space have the same dimension, $\sum_{i=1}^{i=k} d_i^2$, we get

Theorem 8. *There is an algebra isomorphism, $\mathbb{C}G \cong M(d_1) \times M(d_2) \cdots M(d_k)$, where $M(d_i)$ is the algebra of of $d_i \times d_i$ matrices with complex entries.*

5 Matrix multiplication using group representations

Note that matrix multiplication of block diagonal matrices is easy - just multiply the corresponding blocks in each matrix to get the product.

The main idea of Cohn and Umans for multiplying two matrices A, B is to embed the matrices as elements $\tilde{A}, \tilde{B}$, in the group algebra $\mathbb{C}G$, of a non-abelian finite group, multiply $\tilde{A}, \tilde{B}$ in the group algebra to get an element $\tilde{C}$, and read of the product matrix from $\tilde{C}$. For all this to work we need some special embeddings.

We fix a group G and subsets $S, T, U \subseteq G, |S| = |T| = |U| = n$. We index the rows of A by elements of S and the columns of A by elements in T. We index the rows in B by elements in T and the columns of B by the elements of U. The rows of resulting product matrix C will be indexed by elements in S and the columns by elements of U.

We embed A in the group algebra as the element $\tilde{A} = \sum_{s \in S, t \in T} A_{s,t} s^{-1} t$ and B as the element $\tilde{B} = \sum_{t' \in T, u \in U} B_{t',u} t'^{-1} u$.

Definition 5. We say the sets S, T, U satisfy the triple intersection product property, if $\forall s_1, s_2 \in S, t_1, t_2 \in T, u_1, u_2 \in U$ we have

$$s_1^{-1} t_1 t_2^{-1} u_1 = s_2^{-1} u_2 \Leftrightarrow s_1 = s_2, t_1 = t_2, u_1 = u_2$$

Theorem 9. *Let S, T, U be three subsets of size n satisfying the triple intersection product property. Embedding A, B as described above, C_{su} is the coefficient of $s^{-1}u$ in the product of $\tilde{A}\tilde{B}$.*

Proof.

$$\tilde{A}\tilde{B} = \left(\sum_{s \in S, t \in T} A_{s,t} s^{-1} t \right) \left(\sum_{t' \in T, u \in U} B_{t',u} t'^{-1} u \right)$$

$$= \sum_{s \in S, u \in U} \left(\sum_{t \in T} A_{s,t} B_{t,u} \right) s^{-1} u$$

The second equality follows from the definition of triple intersection product completing the proof. □

The above theorem then describes a convenient method to multiply two matrices. Find a group G and subsets having the triple intersection product. Embed the matrices A, B as described. As elements of the the group algebra of G, A and B look like block diagonal matrices. We multiply these block diagonal matrices. Now this resulting matrix has a unique expression as a linear combination of the block diagonal matrices coming from terms of the form $s^{-1}u$. We can use them to read of the $s^{-1}u$-th term of the product using one operation. And so the product C can be written down using $O(n^2)$ more operations.

Now we are almost done. If G has k irreducible representations V_i of size d_i, theorem 8 tells us that multiplication in the group algebra reduces to multiplication of k matrices of size $d_i \times d_i$, $1 \leq i \leq k$. If the exponent for matrix multiplication is ω, these k matrices can be multiplied using $\sum_i O(d_i^\omega)$ arithmetic operations. Since this is the time it takes to multiply two $n \times n$ matrices and we expect to do this using at most $O(n^\omega)$ operations, by the definition of ω, it is reasonable to expect that $n^\omega \leq \sum_i d_i^\omega$. (Cohn and Umans, 2003) show

Theorem 10. *Let S, T, U be three subsets of size n of a group G which satisfy the triple intersection product. Let the size of G be n^α for some constant α and let d_i denote ths dimension of the i-th irreducible representation V_i, of G. Then*

$$n^\omega \leq \sum_i d_i^\omega$$

So, does this approach yield better algorithms?

If we start with an Abelian group and follow Cohn and Umans strategy, it can be shown that we cannot beat the $O(n^3)$ bound one gets from the naive strategy. So, looking in non-abelian groups is essential. In (Cohn et al., 2005) the above philosophy is used, and examples are constructed which beat the $O(n^3)$ bound. One such construction yields $\omega = 2.91$. We present details of that construction below.

Let G be a group, and let H be another group acting on G- when H acts on G, we think of G as a set on which H acts, satisfying equations 1, 2.

We can then form the wreath product of G, H denoted $H \wr G$. As a set it consists of pairs of elements from G, H, with the group operation defined as follows:

$$(g, h) \cdot (g', h') = (g \cdot (h \cdot g'), h \cdot h')$$

The first term on the right hand side needs some explanation - we act h on g' to get an element of G, and we multiply this element on the left with the element g.

Let $\mathbb{Z}_n$ be the Abelian group of integers modulo n. Let $G = (\mathbb{Z}_n \times \mathbb{Z}_n \times \mathbb{Z}_n)^2$. We think of each element of G as a three by two row, the first row indexing the first of the three copies of $\mathbb{Z}_n$ and the second row indexing the second of the three copies of $\mathbb{Z}_n$:

Let H be the group $\{id, z\}$. For $g \in G$, define $id_h \cdot g = g$ and let

$$z \cdot \begin{array}{|c|c|c|} \hline a & b & c \\ \hline d & e & f \\ \hline \end{array} = \begin{array}{|c|c|c|} \hline d & e & f \\ \hline a & b & c \\ \hline \end{array}.$$

So the element $z \in H$ acts by swapping the two rows of an element of G.

The identity element in $H \wr G$ is the element $\left(\begin{array}{|c|c|c|} \hline 0 & 0 & 0 \\ \hline 0 & 0 & 0 \\ \hline \end{array}, 1 \right)$. It can be checked that the inverse of the element $\left(\begin{array}{|c|c|c|} \hline a & b & c \\ \hline d & e & f \\ \hline \end{array}, z \right)$ is the element $\left(\begin{array}{|c|c|c|} \hline -d & -e & -f \\ \hline -a & -b & -c \\ \hline \end{array}, z \right)$

The element z is identified with the element $\left(\begin{array}{|c|c|c|} \hline 0 & 0 & 0 \\ \hline 0 & 0 & 0 \\ \hline \end{array}, z \right)$, and it is its own inverse.

It is clear that $\left(\begin{array}{|c|c|c|} \hline a & b & c \\ \hline d & e & f \\ \hline \end{array}, z \right) = \left(\begin{array}{|c|c|c|} \hline a & b & c \\ \hline d & e & f \\ \hline \end{array}, 1 \right) \cdot z.$

Let us define three sets in the group $H \wr G$ as follows. For all the sets we need that neither g_1 nor g_2 is equal to zero.

$$S = \left\{ \left(\begin{array}{|c|c|c|} \hline g_1 & 0 & 0 \\ \hline 0 & g_2 & 0 \\ \hline \end{array}, z^j \right) \Big| g_1, g_2, \in \mathbb{Z}_n, j = 0, 1 \right\}$$

$$T = \left\{ \left(\begin{array}{|c|c|c|} \hline 0 & g_1 & 0 \\ \hline 0 & 0 & g_2 \\ \hline \end{array}, z^j \right) \Big| g_1, g_2, \in \mathbb{Z}_n, j = 0, 1 \right\}$$

$$U = \left\{ \left(\begin{array}{|c|c|c|} \hline 0 & 0 & g_1 \\ \hline g_2 & 0 & 0 \\ \hline \end{array}, z^j \right) \Big| g_1, g_2, \in \mathbb{Z}_n, j = 0, 1 \right\}$$

We claim that these three sets satisfy the triple intersection product property needed to apply theorem 10. So let us pick $s_1, s_2 \in S$, $t_1, t_2 \in T$ and $u_1, u_2 \in U$. We need to show that if $s_2 s_1^{-1} t_1 t_2^{-1} u_1 u_2^{-1}$ is equal to identity, then $s_1 = s_2$, $t_1 = t_2$ and $u_1 = u_2$.

Let us show that when $x, y \in S$ then xy^{-1} is either of the form $x' \cdot z \cdot y'$ or of the form $x' \cdot y'$ where $x' \in S, y' \in S$. When x and y have no z term this is clear by taking $x' = x, y' = y$. When $x = \left(\begin{array}{ccc} g_1 & 0 & 0 \\ 0 & g_2 & 0 \end{array}, z \right)$, we can write x as $\left(\begin{array}{ccc} g_1 & 0 & 0 \\ 0 & g_2 & 0 \end{array}, 1 \right) z$. Now if y has no z term then we are done. If y has a z term in it, say $y = \left(\begin{array}{ccc} g_1' & 0 & 0 \\ 0 & g_2' & 0 \end{array}, z \right)$ then clearly $y^{-1} = z \cdot \left(\begin{array}{ccc} -g_1' & 0 & 0 \\ 0 & -g_2' & 0 \end{array}, 1 \right)$. Since $z^2 = 1$ we get

$$ xy^{-1} = \left(\begin{array}{ccc} g_1 & 0 & 0 \\ 0 & g_2 & 0 \end{array}, 1 \right) \left(\begin{array}{ccc} -g_1' & 0 & 0 \\ 0 & -g_2' & 0 \end{array}, 1 \right). $$

When x has no z the proof is similar.

So the product $s_2 s_1^{-1} t_1 t_2^{-1} u_1 u_2^{-1}$ is of the form $s_2' z^{J_1} s_1' t_1' z^{J_2} t_2' u_1' z^{J_3} u_2'$. Here J_1, J_2, J_3 are either 0 Or 1.

Clearly this is the identity element in the wreath product only when there are an even number of z's in the expression. Now a simple case analysis, depending upon the number of $z's$ in the above expression completes the proof. When there are no z's at all, we look at the leftmost box in row 1 and the second box in row 2. The terms in these boxes are contributed by $s_2' s_1'$. It can be easily verified that this forces $s_1 = s_2$. The other equations are proved similarly by looking at other boxes in the identity element. The case when there are two z's can be similarly proved, without much difficulty. We omit the details.

Theorem 11. $\omega \le 2.91$.

Proof. We need to use the following fact - that if A is any Abelian subgroup of a group K, then in $\mathbb{C}K$, the dimension of each irreducible representations is at most $|K|/|A|$.

Let A, B be two matrices of size $2(n-1)^2$ which we wish to multiply. Let G, H as above, and set $K = H \wr G$. Clearly G is an Abelian subgroup of K. From the above fact it follows each irreducible representation of d_i of K has dimension at most 2.

Use the sets S, T, U constructed above to embed matrices A and B in the group algebra $\mathbb{C}K$. These sets satisfy the triple intersection product property and are of size $2(n-1)^2$, so we are in a position to apply theorem 10. Note that $|K| = 2n^6 = \sum_i d_i^2$.

If the exponent were 3 theorem 10 gives

$$\left(2(n-1)^2\right)^3 \leq \sum_i d_i^3$$

But $d_i \leq 2$. So the above inequality reads

$$\left(2(n-1)^2\right)^3 \leq \sum_i d_i^3 \leq 2 \sum_i d_i^2 \leq 2|K| \leq 4n^6$$

But this is false, for example when $n = 10$. This means the exponent ω must be less than 3. In fact we can work backwards and show that this approach yields $\omega \leq 2.91$. $\qquad\square$

6 Conclusions

Using a much more sophisticated construction, in (Cohn et al., 2005) the authors also show that their technique can be used to show that $\omega \leq 2.41$. What is interesting, and satisfying, is their observation that the Coppersmith and Winograd algorithm can also be interpreted in this group theoretic setting. They also state conjectures which will prove that $\omega \leq 2$.

So, perhaps, representation theory may become an indispensable tool, in the toolkit of algorithm designers and complexity theorists. Indeed, the very first explicit construction of expander graphs was obtained by (Margulis, 1988), using the representation theory the special linear group $SL(n)$ in characteristic p. The work of Mulmuley and Sohoni seems to suggest that representation theory is the correct kaleidoscope with which to view the entire panorama of lower bound problems in complexity theory. And the beautiful work of Cohn, Kleinberg, Szegedy and Umans, complements their insight wonderfully.

Acknowledgements

I thank R Sujatha for inviting me to write an article for this volume. I would also like to thank Preena Samuel for painstakingly going over a draft of this article, and suggesting many improvements.

References

Boaz Barrack 2008. *Lecture notes from the course, COS 598D, Mathematical Methods in Theoretical Computer Science*, Princeton University, 2008.

Cohn, H. and Umans, C. 2003. A Group-theoretic Approach to Fast Matrix Multiplication, *Proceedings of the 44th Annual IEEE Symposium on Foundations of Computer Science (FOCS)*, 2003, pp. 438–449.

Cohn, H., Kleinberg, R., Szegedy, B. and Umans, C. 2005. Group-theoretic Algorithms for Matrix Multiplication. *Proceedings of the 46th Annual IEEE Symposium on Foundations of Computer Science (FOCS)*, 379–388.

Cormen, T.H., Leiserson, C.E. and Rivest, R.L. 1990. *Introduction to Algorithms*, MIT Press.

Goodman, R. and Wallach, N.R. 1998. *Representations and Invariants of the Classical Groups*, Cambridge University Press.

Margulis, G. 1988. Explicit group-theoretic constructions of combinatorial schemes and their applications in the construction of expanders and concentrators, (Russian) *Problemy Peredachi Informatsii* **24(1):**51–60, (English) Problems Inform. Transmission 24, no. 1, 39–46.

Mulmuley, K. and Sohoni, M. 2001. Geometric complexity theory. I. An approach to the P vs. NP and related problems, *SIAM J. Comput.* **31(2):**496–526.

Mulmuley, K. 2007. On P vs. NP, Geometric Complexity Theory, and The Flip I: a high-level view, Technical Report TR-2007-13, Computer Science department, The University of Chicago,September 2007.

Regan, K. 2002. Understanding the Mulmuley-Sohoni Approach to P vs. NP, *Bulletin of the European Association for Theoretical Computer Science* 78, October 2002, pp. 86–97.

Serre, J.P. 1977. *Linear representations of Finite Groups*, Graduate Texts in Mathematics, 42, Springer Verlag.

Strassen, V. 1969. Gaussian elimination is not optimal, *Numerical Mathematics*. **13:**354–356.

CHAPTER 16

Regular Languages: From Automata to Logic and Back

Madhavan Mukund

Chennai Mathematical Institute,
Siruseri 603103, India.
Homepage: http://www.cmi.ac.in/~madhavan

1 Introduction

The mathematical foundations of computer science predate the development of the first electronic computers by a few decades. The early 20th century saw pioneering work by Alonzo Church, Alan Turing and others towards the formulation of an abstract definition of computability. This effort was in the context of one of David Hilbert's challenges, the *Entscheidungsproblem*, which asked whether all mathematical truths could be deduced "mechanically" from the underlying axioms. Church and Turing showed, independently, that it is impossible to decide algorithmically whether statements in arithmetic are true or false, and thus a general solution to the Entscheidungsproblem is impossible.

This early research on computability yielded the Turing machine, a compelling abstraction that captures the essence of an automatic computational device. A Turing machine has a finite set of control states that guide its behaviour, but it has access to an infinite "tape" on which it can record intermediate results during a computation. This makes a Turing machine unrealizable in practice, since it requires an external storage device with unbounded capacity.

If we restrict our computing devices to use an arbitrarily large, but fixed, amount of external storage, the situation changes drastically. The number of distinct configurations of such a machine—all combinations of control states and contents of the external storage—now becomes bounded. This yields a new abstract model, finite automata, whose properties were investigated by Stephen Cole Kleene, Michael Rabin, Dana Scott and others in the decades following Turing's work.

Initially, the focus was on characterizing the computational power of this restricted class of machines. However, starting in the late 1950's, Julius Richard Büchi and others identified a deep connection between these automata and mathematical logic, which has since played a central role in computer science.

The aim of this article is to provide a quick introduction to this elegant interplay between finite automata and mathematical logic. We begin with a self-contained presentation of finite automata. In Section 3, we describe regular expressions, an alternative notation for describing the computations that finite automata can perform. Some shortcomings of regular expressions motivate us to turn to logical specifications in monadic second order logic, introduced in Section 4. The next two sections prove the central theorem due to Büchi, Elgot and Trakhtenbrot showing that finite automata are expressively equivalent to specifications in monadic second order logic. We conclude with a discussion where we provide some pointers to the literature.

2 Finite automata

A finite automaton A consists of a finite set of states Q with a distinguished *initial* state $q_{in} \in Q$ and a set of *accepting* or *final* states $F \subseteq Q$.

An automaton A processes sequences of abstract actions drawn from a finite *alphabet* Σ. The operation of A is given by a *transition function* $\delta : Q \times \Sigma \to Q$. A transition of the form $\delta(q, a) = q'$ means that whenever A is in state q, the action a takes it to state q'.

Thus, an automaton A over Σ is fully described by its four components (Q, q_{in}, F, δ). A useful way to visualize an automaton is to represent it as a directed graph, where the vertices are the states and the labelled edges correspond to the transition function, as shown in Figure 1. In our pictures, we denote the initial state with an unlabelled incoming arrow and final states with unlabelled outgoing arrows, so $q_{in} = q_1$ and $F = \{q_1\}$ in this example.

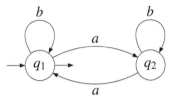

Figure 1: *A finite automaton*

The sequences of actions processed by finite automata are typically called *input words* and we usually say that an automaton *reads* such an input word. To process an input word $w = a_1a_2 \ldots a_n$, the automaton A starts in the initial state q_{in} and traces out a path labelled by w in the state-transition graph. More formally, a *run* of A on $w = a_1a_2 \ldots a_n$ is an alternating sequence of states and actions $q_0a_1q_1a_2q_2 \ldots a_nq_n$ such that $q_0 = q_{in}$ and for each $1 \le i \le n$, $\delta(q_{i-1}, a_i) = q_i$.

The automata we have defined are *deterministic*: for each state q and action a, $\delta(q, a)$ fixes uniquely the state to which Aut A moves on reading a in state q. This enforces that A has exactly one run on each input word. We will relax this condition later.

An automaton A *accepts* or *recognizes* an input w if it reaches a final state after processing w—in other words, the last state of the run of A on w belongs to the set F. The set of words accepted by A is called the *language* of A and is denoted $L(A)$.

Example 1. Consider the automaton in Figure 1. On input *aba*, the automaton goes through the sequence of states $q_1q_2q_2q_1$, while on input *babb*, the automaton goes through the sequence of states $q_1q_2q_2q_2$. Since $F = \{q_1\}$, the automaton accepts *aba* and does not accept *babb*. This automaton accepts all words with an even number of a's—we can show by induction that after reading any word with an even (respectively, odd) number of a's, the automaton will be in state q_1 (respectively, q_2).

2.1 Properties of regular languages

A set of words X is said to be a *regular language* if there exists a finite automaton A such that $X = L(A)$. Regular languages are closed with respect to boolean operations: complementation, union and intersection. Complementation is defined with respect to the set of all words over Σ—this set is normally denoted Σ^*. Thus, given a set of words X, its complement $\overline{X}$ is the set $\Sigma^* \setminus X$. Note that the set Σ^* includes the empty word, ε, of length zero.

Proposition 1. *Let X and Y be regular languages. Then $\overline{X}$, $X \cup Y$ and $X \cap Y$ are also regular languages.*

Proof. To show that $\overline{X}$ is regular, let A be an automaton that accepts X. If we interchange the final and non-final states of A we obtain an automaton B that accepts every input word w that A does not accept and vice versa. Thus $L(B) = \overline{X}$.

To show that $X \cap Y$ is regular, let $A_X = (Q_X, q_{in}^X, F_X, \delta_X)$ and $A_Y = (Q_Y, q_{in}^Y, F_Y, \delta_Y)$ be automata that accept X and Y respectively. We construct the direct product A_{XY} with states $Q_X \times Q_Y$, initial state (q_{in}^X, q_{in}^Y), final states $F_X \times F_Y$ and transition function δ_{XY}, such that for all $q_x \in Q_X$, $q_y \in Q_Y$ and $a \in \Sigma$, $\delta_{XY}((q_x, q_y), a) = (\delta_X(q_x, a), \delta_Y(q_y, a))$. The automaton A_{XY} simulates A_X and A_Y in parallel and accepts an input w provided both A_X and A_Y accept w. Hence, $L(A_{XY}) = X \cap Y$.

Since $X \cup Y = \overline{\overline{X} \cap \overline{Y}}$, it follows that $X \cup Y$ is also regular. We can also give a direct construction for this case by setting the final states of A_{XY} to $(F_X \times Q_Y) \cup (Q_X \times F_Y)$. □

2.2 Nondeterminism

An important generalization of our definition of finite automata is to relax the restriction that state transitions are deterministic. In a *nondeterministic* finite automaton, given a state q and an action a, the automaton may change to one of several alternative new states. Formally, a nondeterministic finite automaton (NFA) A is a tuple (Q, q_{in}, F, Δ) where Q, q_{in} and F are as before and the transition function $\Delta : Q \times \Sigma \rightarrow 2^Q$ maps each pair (q, a) to a subset of states. Notice that it is possible for $\Delta(q, a)$ to be empty, in which case A gets "stuck" on reading a in state q and cannot finish processing the input, and hence cannot accept it. We can still represent automata as directed graphs with labelled edges—an example of a nondeterministic automaton is shown in Figure 2. Here, for instance, $\Delta(q_1, b) = \{q_1, q_2\}$ and $\Delta(q_2, a) = \emptyset$.

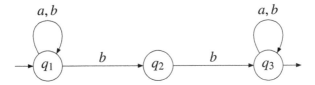

Figure 2: *A nondeterministic finite automaton*

As with deterministic automata, a nondeterministic automaton A processes an input w by tracing out a path labelled by w in the state transition graph: a *run* of A on $w = a_1 a_2 \ldots a_n$ is an alternating sequence of states and actions $q_0 a_1 q_1 a_2 q_2 \ldots a_n q_n$ such that $q_0 = q_{in}$ and for each $1 \leq i \leq n$, $q_i \in \Delta(q_{i-1}, a_i)$. Notice that a run *must* read the entire input.

Clearly, nondeterminism permits an automaton to have multiple runs (or even no run at all!) on an input. To accept an input, it is sufficient to have one good run: an automaton A accepts an input w if there exists a run on which A reaches a final state after processing w.

Example 2. Consider the automaton in Figure 2. This automaton starts in q_1 and must reach q_3 to accept an input. The automaton can stay in states q_1 and q_3 indefinitely on any sequence of a's and b's. To get from q_1 to q_3, the automaton must read two consecutive b's. Thus, this automaton accepts any word w of the form $w_1 bb w_2$ with two consecutive b's. On such an input, the automaton "guesses" after w_1 that it should move to q_2 on the next b, and eventually ends up in q_3 where it can remain till the end of w_2. If w does not have this form, it cannot make such a guess: after any b that takes the automaton from q_1 to q_2, the next input is an a, on which there is no transition from q_2, so the automaton gets "stuck" and there is no accepting run.

3   Regular Expressions

As we have seen, the overall behaviour of an automaton is defined in terms of the language that it accepts. Thus, two automata that accept the same language are essentially equivalent from our point of view, though they may vary greatly in their internal structure.

Since languages are the basic notion of behaviour, it is useful to be able to abstractly specify them, rather than having to present concrete implementations in terms of automata. For instance, it would be useful to have a precise notation to write descriptions of languages corresponding to the informal descriptions we have seen such as "all words with an even number of a's" or "all words with two consecutive b's"

Ideally, the specifications that we write in this manner should be realizable—that is, they should describe regular languages and there should be an effective algorithm to construct an automaton from such a specification. Even better would be to show that every regular language can be described within such a specification language.

One way of abstractly defining languages is to use algebraic expressions. The set of *regular expressions* over an alphabet Σ is defined inductively as follows:

- The symbols $\emptyset$ and ε are regular expressions.

- Every letter $a \in \Sigma$ is a regular expression.

- If e and f are regular expressions, so are $e + f$, $e \cdot f$ and e^*.

The operator $+$ denotes union while $\cdot$ denotes concatenation. We usually omit $\cdot$ and write ef for $e \cdot f$. The operation * represents iteration and is called the Kleene star.

Each regular expression corresponds to a language. This association is built up inductively, for which we need to define some operations on languages.

Let X and Y be languages. Then:

- $XY = \{xy \mid x \in X, y \in Y\}$, where xy represents concatenation of words: if $x = a_1 a_2 \ldots a_m$ and $y = b_1 b_2 \ldots b_n$ then $xy = a_1 a_2 \ldots a_m b_1 b_2 \ldots b_n$.

- $X^* = \bigcup_{i \geq 0} X^i$, where $X^0 = \{\varepsilon\}$ and for $i > 0$, $X^i = XX^{i-1}$.

We can now associate languages to regular expressions inductively, as promised. For an expression e, let $L(e)$ denote the language associated with e. We then have the following:

- $L(\emptyset) = \emptyset$ and $L(\varepsilon) = \{\varepsilon\}$.

- For each $a \in \Sigma$, $L(a) = \{a\}$.

- $L(e + f) = L(e) \cup L(f)$, $L(ef) = L(e)L(f)$ and $L(e^*) = (L(e))^*$.

It is not difficult to show that if X and Y are regular languages, then XY and X^* are also regular. Thus, every regular expression describes a regular language.

The following fundamental result shows that finite automata are, in fact, equivalent to regular expressions.

Theorem 1 (Kleene). *A language is regular L if and only if it is described by a regular expression.*

As we have seen, we can use the inductive structure of regular expressions to map them to regular languages. The translation from automata to expressions is harder: finite automata do not have any natural inductive structure and the global behaviour of an automaton cannot easily be decomposed into properties of individual states and transitions. We will not prove this result here: the final section lists standard references where the details can be found.

Example 3. Recall the automaton in Figure 1 that accepts words with an even number of a's. Any word w with an even number of a's can be decomposed into blocks $w_1 w_2 \ldots w_k$ where each w_i has exactly two a's. The expression $b^* a b^* a b^*$ describes all words with exactly two a's: before the first a, between the two a's and after the second a we can have zero or more b's, which is captured by b^*. We use the Kleene star to allow an arbitrary number of blocks of this form, resulting in the expression $(b^* a b^* a b^*)^*$ for this language.

The regular expression for the automaton in Figure 2 is $(a + b)^* bb (a + b)^*$: any word in the language starts and ends with an arbitrary sequence of a's and b's and contains the sequence bb in between.

We have seen that regular languages are closed under boolean operations such as negation and conjunction. Hence, regular expressions should also be able to cope with these operations. However, for instance, the process of going from a regular expression for a language L to an expression describing the complement of L is far from transparent. Complementation involves a switch of quantifiers: for example, the language of Figure 2 asks if *there exist* two consecutive positions labelled *b* while the complement language would demand *for every* position labelled *b*, the positions before and after this position are not labelled *b*. In general, regular expressions are convenient for expressing *existential* properties but are clumsy for describing *universal* properties—as another example, consider a variant of the language from Figure 2 where we ask that *every* position labelled *b* occurs as part of a pair of adjacent positions labelled *b*.

4 Logical Specifications

Another approach to describing sets of words over an alphabet is to use mathematical logic. We will write formulas that are interpreted over words: in other words, the formulas we write can be assigned either the value true or the value false when applied to a given word. For instance, if we write a formula that expresses the fact that *"the first letter in the word is a"*, then this formula will evaluate to true on words of the form *abab* or *aaaa* or *abbbb*, but will be false for words such as *ε*, *bbba* or *baaaa* that do not start with an *a*. Thus, a formula φ defines in a natural way a language of words $L(\varphi)$: those words over which φ evaluates to true. This gives us a way to specify languages using logic. Our main goal is to identify a formal logical notation whose formulas capture precisely the regular languages.

4.1 Formulating the logic

We begin with an informal presentation of our logical language. The basic entities we deal with in our logic are positions in words. Thus, the variables that we use in our logic, such as x and y, denote positions. We use predicates to describe the letters that appear at positions. For each $a \in \Sigma$, we have a predicate P_a such that $P_a(x)$ is true in a word w if and only if the letter at position x in w is a. We can compare positions: $x < y$ is true if the position denoted by x is earlier than the position denoted by y.

We use the usual logical connectives $\neg$ (negation), $\vee$ (disjunction) and $\wedge$ (conjunction) to combine formulas, as also derived connectives like $\rightarrow$ (implication), where $\varphi \rightarrow \psi \equiv (\neg\varphi \vee \psi)$ and $\leftrightarrow$ (if and only if), where $\varphi \leftrightarrow \psi \equiv (\varphi \rightarrow \psi) \wedge (\psi \rightarrow \varphi)$. To these logical connectives, we add, as usual, the quantifiers $\exists$ (there exists) and $\forall$ (for all).

For example, consider the language described by the automaton in Figure 2. A formula for this language is

$$\exists x. \exists y. P_b(x) \wedge P_b(y) \wedge next(x, y),$$

$$\text{where } next(x, y) \stackrel{\triangle}{=} x < y \wedge \neg \exists z. (x < z \wedge z < y),$$

which asserts that there are two consecutive positions labelled b in the word—the formula $next(x, y)$ expresses that position y occurs immediately after x by demanding that there are no other positions strictly between the two.

The first example we saw, Figure 1, is more complicated. We need to assert that the number of positions labelled a is even. Any formula we write has a fixed set of variables $x_1, x_2, \ldots, x_k$ and can only describe the relationship between k positions in the word at a time. Intuitively speaking, for words that have more than k a's, we cannot expect this formula to assert a property about *all* positions labelled a. This argument can be formalized, but the net result is that we have to enrich our language to directly refer to *sets of positions*.

We introduce a new class of variables, written $X, Y, \ldots$ that represent sets of positions. We write $X(y)$ to denote that the position y belongs to the set X. We are allowed to use the quantifiers $\exists$ and $\forall$ for set variables, just as we do for individual variables.

Let us consider a simpler language than the one in Figure 1—let L_e be the set of all words of even length. To describe this language, we can use a set variable X and restrict the members of X to precisely the odd positions. To do this, we demand that the minimum position be in X and, for any pair of consecutive positions x and y, x is in X if and only if y is not in X. This captures the fact that X contains every alternate (odd) position, starting from the first one.

Notice that we can use our logical language to define a constant *min* to refer to the minimum position, for use in such a formula. A formula $\varphi(min)$ that asserts some property involving the minimum position *min* can be rewritten as $\exists x. \varphi(x) \wedge (\neg \exists z. z < x)$.

Finally, to say that the overall numbers of positions is even, we just assert that the last position is not in X—like *min*, we can define *max*: the position x such that $\neg \exists z. x < z$. Putting this together, we have the following formula,

$$\exists X. X(min) \wedge \forall y, z. (next(y, z) \rightarrow (X(y) \leftrightarrow \neg X(z))) \wedge \neg X(max),$$

where we use the abbreviations *min*, *max* and *next* defined earlier.

From this, it is only a small step to describing the language in Figure 1. We relativize *min*, *max* and *next* to positions labelled a, so $min_a \stackrel{\Delta}{=} P_a(x) \wedge \neg(\exists z.z < x \wedge P_a(z))$ is the smallest a-labelled position, $max_a \stackrel{\Delta}{=} P_a(x) \wedge \neg(\exists z.x < z \wedge P_a(z))$ is the largest a-labelled position and $next_a(x, y) \stackrel{\Delta}{=} P_a(x) \wedge P_a(y) \wedge \neg(\exists z.x < z \wedge z < y \wedge P_a(z))$ describes when two a-labelled positions are consecutive. We can then modify the formula above to use min_a, max_a and $next_a$ in place of *min*, *max* and *next* and capture the fact that the number of a-labelled positions in a word is even.

To illustrate why logical specifications are more convenient than regular expressions, let us revisit some of the examples from the end of the previous section. Complementing logical specifications is easy—we just negate the specification: if φ describes a set of words, $\neg\varphi$ describes the complement. How about languages that implicitly incorporate universal quantifiers, such as "*every b* occurs as part of a pair of adjacent b's"? Since we have quantifiers at our disposal, we can describe this quite easily, as follows.

$$\forall x. \Big(P_b(x) \rightarrow \exists y. \big[(next(y, x) \wedge P_b(y)) \vee (next(x, y) \wedge P_b(y))\big]\Big)$$

This formula asserts that every position labelled b is either immediately preceded by or followed by another position labelled b.

4.2   Monadic second order logic

Formally, the logic we have looked at is called monadic second order logic, or MSO. Second order refers to the fact that we can quantify over predicates, not just individual elements, and monadic says that we restrict our attention to 1-place predicates, which are equivalent to sets.

As we have seen, our formulas are built from the following components.

- Individual variables $x, y, \ldots$ and set variables $X, Y, \ldots$ denoting positions and sets of positions, respectively. Recall that we have derived constants *min* and *max* that can be used wherever we use individual variables such as $x, y, \ldots$.

- Atomic formulas

- $x = y$ and $x < y$ over positions, with the natural interpretation.
- $X(x)$, denoting that position x belongs to set X
- $P_a(x)$ for each $a \in \Sigma$, denoting that position x is labelled a.

- Complex formulas are built using the usual boolean connectives $\neg$, $\vee$, $\wedge$, $\rightarrow$ and $\leftrightarrow$, together with the quantifiers $\exists$ and $\forall$.

Each (nonempty) word $w = b_1 b_2 \ldots b_m$ over the alphabet $\Sigma = \{a_1, a_w, \ldots, a_n\}$ provides us a *word model*

$$\langle w \rangle = \left(\{1, 2, \ldots, m\}, <^w, P_{a_1}^w, P_{a_2}^w, \ldots, P_{a_n}^w \right)$$

where

- $\mathrm{pos}(w) = \{1, 2, \ldots, m\}$ is the set of positions in w,

- $<^w$ is the (natural) ordering on the positions in $\mathrm{pos}(w)$,

- for each $a_i \in \Sigma$, $P_{a_i}^w = \{j \in \mathrm{pos}(w) \mid b_j = a_i\}$ describes the positions where a_i occurs.

We restrict our attention to nonempty words to avoid dealing with some boundary cases that occur when $\mathrm{pos}(w) = \emptyset$.

A variable x or X that occurs within the scope of a quantifier is said to be *bound*. Variables that are not bound are said to be *free*. We write

$$\varphi(x_1, x_2, \ldots, x_m, X_1, X_2, \ldots, X_n)$$

to denote that the set of free variables in φ is at most $\{x_1, x_2, \ldots, x_m, X_1, X_2, \ldots, X_n\}$.

The meaning of a bound variable is determined by its binding quantifier: for instance, if we write $\forall x. \varphi(x)$, this formula is true if $\varphi(j)$ evaluates to true for all possible choices $j \in \mathrm{pos}(w)$ for x. However, for free variables we need to additionally fix how the variable is to be interpreted. Hence, to give meaning to a formula $\varphi(x_1, x_2, \ldots, x_m, X_1, X_2, \ldots, X_n)$, we need

- A word model $\langle w \rangle$.

- Positions $k_1, k_2, \ldots, k_m$ as interpretations of $x_1, x_2, \ldots, x_m$.

- Sets $K_1, K_2, \ldots, K_n$ as interpretations of $X_1, X_2, \ldots, X_n$.

We write

$$(\langle w \rangle, k_1, k_2, \ldots, k_m, K_1, K_2, \ldots, K_n) \models \varphi(x_1, x_2, \ldots, x_m, X_1, X_2, \ldots, X_n)$$

to express the fact that φ holds in $\langle w \rangle$ if each x_i is interpreted as k_i and each X_j as K_j. As an abbreviation, we write $\langle w \rangle \models \varphi[k_1, k_2, \ldots, k_m, K_1, K_2, \ldots, K_n]$. We also write $\langle w \rangle \not\models \varphi[k_1, k_2, \ldots, k_m, K_1, K_2, \ldots, K_n]$ to indicate that a formula evaluates to false under an interpretation.

Example 4. Let $w = babaa$ with $\Sigma = \{a, b\}$ and

1. $\varphi_1(x_1, X_1) \triangleq P_b(x_1) \wedge \forall y.(x_1 < y \wedge X_1(y) \rightarrow P_a(y))$.

2. $\varphi_2 \triangleq \exists x.\exists y.(P_b(x) \wedge P_b(y) \wedge next(x, y))$

Then

1. $\langle w \rangle \models \varphi_1(1, \{2, 4, 5\})$.

2. $\langle w \rangle \not\models \varphi_1(1, \{2, 3, 4, 5\})$.

3. $\langle w \rangle \not\models \varphi_2$.

Notice that φ_2 has no free variables, so it can be interpreted directly on a word model $\langle w \rangle$.

A formula with no free variables is called a *sentence*. Sentences can be interpreted directly on word models. We can thus associate with each sentence φ, a language of words $L(\varphi) = \{w \mid \langle w \rangle \models \varphi\}$. Such a language is said to be MSO-definable.

If we restrict our formulas to not use set variables or set quantifiers, the corresponding logic is called first-order logic, or FO. For example φ_2 in Example 4 is an FO sentence. A language described by an FO sentence is said to be FO-definable.

The main result we are interested in is that MSO-definability is equivalent to regularity.

Theorem 2 (Büchi, Elgot, Trakhtenbrot). *A language L of nonempty words over Σ is regular if and only if it is MSO-definable. The transformation in both directions is effective.*

To prove this result, we need to extend our definitions so that we associate languages with arbitrary formulas, not just sentences. Let $\varphi(x_1, \ldots, x_m, X_1, \ldots, X_n)$ be a formula with free variables. We have seen that models for such formulas are of the form $(\langle w \rangle, k_1, k_2, \ldots, k_m, K_1, K_2, \ldots, K_n)$. For each position $j \in \text{pos}(w)$, we record whether $j = k_1, j = k_2, \ldots, j = k_m, j \in K_1, j \in K_2, \ldots, j \in K_n$. This gives us a bit vector of length $m + n$ for each position j. Formally, we can represent $(\langle w \rangle, k_1, k_2, \ldots, k_m, K_1, K_2, \ldots, K_n)$ as a word over the alphabet $\Sigma \times \{0, 1\}^{m+n}$. Clearly, for a sentence, $m = n = 0$, so the word associated with a model is just the word itself.

Example 5. Let $w = babaa$, $k_1 = 1$, $K_1 = \{3\}$ and $K_2 = \{1, 2, 4, 5\}$. Then, we represent $(\langle w \rangle, k_1, K_1, K_2)$ as a word over $\Sigma \times \{0, 1\}^3$ as follows.

w	b	a	b	a	a
k_1	1	0	0	0	0
K_1	0	0	1	0	0
K_2	1	1	0	1	1

5   From automata to logic ...

One half of proving the Büchi-Elgot-Trakhtenbrot theorem consists of associating an MSO-formula with each regular language. In general, let us assume that our regular language is described by an NFA. Given an NFA A, we have to construct a sentence φ_A such that $\langle w \rangle \models \varphi_A$ if and only if $w \in L(A)$. To do this, we write a formula that describes successful runs of A.

As a concrete example, let us consider the automaton Figure 2 that accepts all words with two consecutive b's. This automaton has three states $\{q_1, q_2, q_3\}$. For each state q_i, we associate a set variable X_i denoting the positions in which a run of the automaton assumes state q_i. For instance, on the inputs $abbab$, the accepting run $q_1, a, q_1, b, q_2, b, q_3, a, q_3, b, q_3$ would be encoded as follows, omitting the last state.

w	a	b	b	a	b
X_1	1	1	0	0	0
X_1	0	0	1	0	0
X_2	0	0	0	1	1

In general, if an automaton Aut A as k states, we use k set variables $X_1, X_2, \ldots, X_k$ to describe runs of Aut A. We can describe in MSO the constraints that $X_1, X_2, \ldots, X_k$ should satisfy to constitute an accepting run.

(i) The X_i's partition the set of positions—at each position the automaton is in exactly one state.

(ii) At the first position, the automaton is in an initial state.

(iii) For each pair of consecutive positions x, y, the states at x and y and the letter at x describe a valid transition.

(iv) At the final position, there is a transition leading to an accepting state.

Concretely, here is how we write these properties in MSO for the automaton in Figure 2.

(i) $\exists X_1, X_2, X_3.[\forall x.(X_1(x) \lor X_2(x) \lor X_3(x)) \quad \land \neg \exists x.(X_1(x) \land X_2(x))$

$\land \neg \exists x.(X_1(x) \land X_3(x))$

$\land \neg \exists x.(X_2(x) \land X_3(x))]$

In general, we would write

$$\exists X_1, X_2, \ldots, X_k. \left[\forall x.(X_1(x) \lor \cdots \lor X_k(x)) \land \bigwedge_{i \neq j} \neg \exists x.(X_i(x) \land X_j(x)) \right].$$

(ii) $X_1(min)$

(iii) $\forall x, y.next(x, y) \rightarrow \quad X_1(x) \land P_a(x) \land X_1(y) \lor X_1(x) \land P_b(x) \land X_1(y)$

$X_1(x) \land P_b(x) \land X_2(y) \lor X_2(x) \land P_b(x) \land X_3(y)$

$X_3(x) \land P_a(x) \land X_3(y) \lor X_3(x) \land P_b(x) \land X_3(y)$

In general, we would write

$$\forall x, y.next(x, y) \rightarrow \bigvee_{(q_i, a, q_j) \in \Delta} (X_i(x) \land P_a(x) \land X_j(y)).$$

(iv) $(X_2(max) \land P_b(max)) \lor (X_3(max) \land P_a(max)) \lor (X_3(max) \land P_b(max)).$

In general, we would write

$$\bigvee_{(q_i, a, q_j) \in \Delta, q_j \in F} X_i(max) \land P_a(max).$$

This establishes the first half of the Büchi-Elgot-Trakhtenbrot theorem. For every NFA A over Σ, we can write a sentence φ_A that describes all successful runs of A such that $L(A) = L(\varphi_A)$. In other words, A accepts w if and only if $\langle w \rangle \models \varphi_A$.

6 ...and back

In the converse direction, for each formula $\varphi(x_1, x_2, \ldots, x_m, X_1, X_2, \ldots, X_n)$, we have to construct an automaton A_φ over $\Sigma \times \{0, 1\}^{m+n}$ such that $L(\varphi) = L(A_\varphi)$. We build up A_φ by induction on the structure of φ.

To reduce the number of cases we have to consider, we define a subset $\text{MSO}_{\min}$ of MSO that is expressively equivalent to the original language. In $\text{MSO}_{\min}$, we eliminate individual variables, so we only have set variables and set quantifiers. An individual position x will be represented as a singleton set $\{x\}$. The atomic formulas in $\text{MSO}_{\min}$ are as follows.

- $X \subseteq Y$ and $X \subseteq P_a$.

- $singleton(X)$—"X is a singleton set"

- $X < Y$—"For each $x \in X$, $y \in Y$, $x < y$"

Once again, complex formulas are built using the boolean connectives $\neg, \vee, \wedge, \rightarrow$ and $\leftrightarrow$ and the quantifiers $\exists$ and $\forall$.

It is not difficult to see show that we can express every MSO formula in $\text{MSO}_{\min}$.

- $x = y$ translates as $singleton(X) \wedge singleton(Y) \wedge X \subseteq Y \wedge Y \subseteq X$.

- $x < y$ translates as $singleton(X) \wedge singleton(Y) \wedge X < Y$.

- $X(y)$ translates as $singleton(Y) \wedge Y \subseteq X$

- $P_a(x)$ translates as $singleton(X) \wedge X \subseteq P_a$

Since we no longer have individual variables, an $\text{MSO}_{\min}$ formula is of the form $\varphi(X_1, X_2, \ldots, X_n)$ and a word model of φ can be encoded as a word over $\Sigma \times \{0, 1\}^n$.

Recall that our aim is to associate with each such formula $\varphi(X_1, X_2, \ldots, X_n)$, an automaton A_φ over $\Sigma \times \{0, 1\}^n$ such that $L(\varphi) = L(A)$, where A_φ is built up inductively, based on the structure of φ.

We begin with automata corresponding to the atomic formulas in MSO_{min}.

- $X \subseteq Y$

  Given a word w over $\Sigma \times \{0, 1\}^n$, our automaton checks that whenever a position belongs to X, it also belongs to Y. Without loss of generality, let $X = X_1$ and $Y = X_2$. Our automaton checks that whenever the X_1 component is 1, so is the X_2 component.

$$\rightarrow \; \boxed{q_1} \; \rightarrow \quad \begin{bmatrix} \# \\ 0 \\ 0 \\ * \end{bmatrix}, \begin{bmatrix} \# \\ 0 \\ 1 \\ * \end{bmatrix}, \begin{bmatrix} \# \\ 1 \\ 1 \\ * \end{bmatrix}$$

  Here # denotes any letter in Σ and $*$ denotes an arbitrary bit vector in $\{0, 1\}^{n-2}$. This automaton accepts an input provided it does not contain $\begin{bmatrix} \# \\ 1 \\ 0 \\ * \end{bmatrix}$, which would denote a position in the input that belongs to X_1 but not X_2.

- $X \subseteq P_a$

  Again, without loss of generality, let $X = X_1$. Our automaton then looks like this.

$$\rightarrow \; \boxed{q_1} \; \rightarrow \quad \begin{bmatrix} \# \\ 0 \\ * \end{bmatrix}, \begin{bmatrix} a \\ 1 \\ * \end{bmatrix}$$

  This automaton checks that whenever $X_1 = 1$, the letter read is a. If $X_1 = 0$, the letter read is irrelevant.

- *singleton*(X)

  As before, we assume that $X = X_1$. We check that there is exactly one 1 along the component corresponding to X_1.

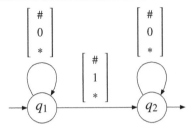

- $X < Y$

  As usual, let $X = X_1$ and $Y = X_2$. We demand that each $x \in X$ comes before each $y \in Y$. For this, we check that all 1's in the X_1 component of the input come before any 1 in the X_2 component. We loop in q_1, insisting that the X_2 component remain 0. After "guessing" when we have seen the last 1 in the X_1 component, we make a transition to q_2, at which point we insist that we see no more 1's in the X_1 component. In this automaton, ? denotes 0 or 1.

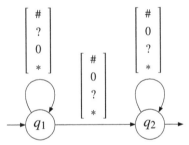

Having translated the atomic formulas, we must look at the connectives used to build up complex formulas. We have already seen that regular languages are closed with respect to Boolean operations, so given automata corresponding to φ and ψ, we can build automata for $\neg\varphi$ and $\varphi \vee \psi$, and hence for all boolean combinations of φ and ψ. Thus, we only need to describe how to handle the quantifiers $\exists$ and $\forall$. Further, since $\forall X.\varphi(X) \equiv \neg\exists X.\neg\varphi(X)$, it suffices to consider just $\exists$.

Consider a formula of the form $\psi(X_2, X_3, \ldots, X_n) \equiv \exists X_1.\varphi(X_1, X_2, \ldots, X_n)$. This formula is true if we can associate some set of positions with X_1 such that $\varphi(X_1, X_2, \ldots, X_n)$ is true. By the induction hypothesis, we already have an automaton $A_{\varphi(X_1, X_2, \ldots, X_n)}$ over $\Sigma \times \{0, 1\}^n$ whose language is the same as that of $\varphi(X_1, X_2, \ldots, X_n)$. From this we must construct, over $\Sigma \times \{0, 1\}^{n-1}$, an automaton $A_{\psi(X_2, X_3, \ldots, X_n)}$ whose language is the same as that of $\psi(X_2, X_3, \ldots, X_n)$.

Let $A_{\varphi(X_1, X_2, \ldots, X_n)} = (Q, q_{in}, F, \Delta)$. For convenience, we will represent input letters from $\Sigma \times \{0, 1\}^k$ as row vectors rather than as column vectors.

We retain (Q, q_{in}, F) and build our new automaton

$$A_{\psi(X_2, X_3, \ldots, X_n)} = (Q, q_{in}, F, \Delta')$$

such that $(q, [a, \bar{x}], q') \in \Delta'$ if and only if either $(q, [a, 0, \bar{x}], q') \in \Delta$ or $(q, [a, 1, \bar{x}], q') \in \Delta$. In other words, the new automaton "guesses" a value for the X_1 component at each step. It is not difficult to show that there is an accepting run $q_{in}, [a_1, \bar{x}_1], q_1, [a_2, \bar{x}_2], q_2, \ldots, q_{n-1}, [a_n, \bar{x}_n], q_n$ of $A_{\psi(X_2, X_3, \ldots, X_n)}$, where $q_n \in F$, if and only if there is some accepting run $q_{in}, [a_1, b_1, \bar{x}_1], q_1, [a_2, b_2, \bar{x}_2], q_2, \ldots, q_{n-1}, [a_n, b_n, \bar{x}_n], q_n$ of $A_{\varphi(X_1, X_2, \ldots, X_n)}$, where each $b_j \in \{0, 1\}$. The sequence $b_1 b_2 \ldots b_n$ describes the structure of the "witness" X_1 that is required to make $\psi(X_2, X_3, \ldots, X_n)$ true.

This establishes the second half of the Büchi-Elgot-Trakhtenbrot theorem — every MSO formula $\varphi(x_1, x_2, \ldots, x_m, X_1, X_2, \ldots, X_n)$ can be effectively transformed into an automaton A_φ over $\Sigma \times \{0, 1\}^{m+n}$ such that $L(\varphi) = L(A_\varphi)$.

7 Discussion

We have seen that monadic second order logic provides an appealing way to abstractly specify regular languages. Though MSO offers many advantages with respect to other expressively equivalent notations such as regular expressions, there is one glaring disadvantage: the cost of translating MSO formulas to automata. Stockmeyer has shown that there is an unbounded family of MSO formulas such that each formula with n symbols in the family blows up into an automaton with $2^{2^{\cdot^{\cdot^2}}}$ states where the height of the tower of exponentials is proportional to n.

Despite this, the connection between MSO and automata has laid the foundation for an important area of computer science called automated verification. The idea is to use logical specifications to describe abstract properties of computational systems and then use automata theory to check if a proposed implementation conforms to the specification. If the specification is given by a formula φ, we can use the theory developed in the chapter to construct an automaton A_φ that accepts $L(\varphi)$, the set of *all* words consistent with φ. Given a potential implementation, described as another automaton B, to check that B exhibits only "legal" behaviours with respect to φ, it suffices to verify that $L(B) \subseteq L(A_\varphi)$. In practice, we move from MSO to other logical formalisms such as temporal logic for which the translation from logic to automata is more tractable.

Further reading

We do not provide pointers to the original papers in these areas by Turing, Kleene, Rabin and Scott, Büchi and others. The original references are standard and the material is so classical by now that it appears in a number of textbooks and surveys.

All standard textbooks on automata theory cover finite automata and their connection to regular expressions in full detail—examples include (Hopcroft, 2007) and (Kozen, 1997). Textbooks that cover the connection between MSO and finite automata are more rare. However, there are excellent survey articles such as (Thomas, 1990), which present this connection between MSO and finite automata in a richer context, over infinite words. There are now numerous books that deal with automated verification in the context that we have discussed here, including (Clarke, 2000). For a more concise and accessible introduction, see (Merz, 2001).

Acknowledgment

This article borrows heavily from Wolfgang Thomas's lecture notes on *Applied Automata Theory*, informally available from various sources on the Internet.

References

Clarke, Edmund M., Orna Grumberg and Doron A. Peled: *Model Checking*, MIT Press, 2000.

Hopcroft, John E., Rajeev Motwani and Jeffrey D. Ullman: *Introduction to Automata Theory, Languages and Computation*, Addison-Wesley, 2007.

Kozen, Dexter: *Automata and Computability*, Springer-Verlag, 1997.

Merz, Stephan: Model Checking: A Tutorial Overview, in *Modeling and Verification of Parallel Processes*, Lecture Notes in Computer Science, Volume 2067, Springer-Verlag. 3–38, 2001.

Thomas, Wolfgang: Automata on infinite objects, in Jan van Leeuwen (ed.), *Handbook of Theoretical Computer Science*, Volume B, North-Holland, Amsterdam. 133–191, 1990.

Part IV
Mathematics and Physics

CHAPTER 17

Variational Calculus in Mathematics and Physics

Rajesh Gopakumar

Harish-Chandra Research Institute,
Allahabad 211019, India.
e-mail: gopakumr@hri.res.in

1   Introduction

The search for unifying principles has been an integral part of the development of physical laws of nature. One of the most fruitful principles of this kind has been the so-called variational principle which has appeared in a number of different guises, most famously as the *Principle of Least Action*. In this article we will discuss some aspects of this principle, its applications in physics and some of the mathematics underlying it.

Mathematically, the variational principle arises in questions related to extremisation. The extremisation which is involved here is more general than the extremisation of functions. However, to explain the more general class of questions, let us start by considering the latter case which is simpler. In many circumstances involving maximisation or minimisation, the question can be reduced to one that can be addressed by differential calculus. For instance, consider the following simple question. Among all rectangles of fixed perimeter $4L$, which is the one of maximum area? This is easy to answer since we can take the sides of any such rectangle to be of length x and $2L - x$. The area of such a rectangle is then $A(x) = x(2L - x)$.

The one with the maximum area is the one for which the function $A(x)$ is extremised i.e.

$$\frac{dA(x)}{dx} = 0 \Rightarrow 2(L - x) = 0 \Rightarrow x = L. \tag{1}$$

This is easily checked to be a maximum i.e. $\frac{d^2A}{dx^2}\big|_{x=L} < 0$ Thus the rectangle which maximises the area is the square with all sides of equal length L.

There are many such questions which can be reduced to extremisation of functions. One quickly realises that one needs to consider functions of *many variables* and extremise with respect to all the variables. Thus, for instance, a simple generalisation of the above problem can be the following. Among all parallelograms with a fixed perimeter $4L$, which is the one of maximal area? Now one needs to consider also the variable angle α between two adjacent sides of the parallelogram, in addition to the length x of one of the sides. Thus the area of all such parallelograms can be expressed as a function of *two* variables

$$A(x, \alpha) = x(2L - x) \sin \alpha. \tag{2}$$

We need to extremise with respect to both variables to obtain the figure of maximum area. In other words we must have

$$\frac{\partial A(x, \alpha)}{\partial x}\bigg|_\alpha = 0 = \frac{\partial A(x, \alpha)}{\partial \alpha}\bigg|_x \Rightarrow 2(L - x) = 0, \cos \alpha = 0 \Rightarrow x = L, \alpha = \frac{\pi}{2}. \tag{3}$$

Thus one ends up again with a square of side L.

In general, the problem of extremising a function $F(x_1, x_2, \ldots, x_n)$ with respect to its variables is one that involves solving the set of n simultaneous equations

$$\frac{\partial F(x_1, x_2, \ldots, x_n)}{\partial x_i}\bigg|_{\{x_j\}} = 0 \qquad (j \neq i). \tag{4}$$

If these equations admit solutions then one can go on to to check whether these are maxima or minima (or in some cases, points of inflexion). And if there are many such, one can also address questions of global maxima/minima as opposed to local maxima/minima. Many problems in mathematical economics and optimisation in engineering can be formulated in such a way.

[**Question:** Generalise our problem to one of extremising amongst quadrilaterals of a fixed total perimeter. How many independent variables is the area now a function of?]

2   Variational Problems

However, there are many problems of extremisation that cannot be framed in terms of maximising or minimising a function of *finitely* many variables. Indeed the extremisation problem and its variants that we considered above are simple special cases of a more general problem known as the *Isoperimetric Problem*. This asks the following question. Amongst all closed curves one could draw on a plane, of fixed total perimeter, which is the one of maximal area. We can see that by no longer restricting oneself to a restricted class of curves (namely, rectangles/parallelograms/quadrilaterals) we have significantly complicated the question. In the former case we had a finite set of variables which characterised the most general member of the class. Hence the problem of extremising the area was one of computing the area as a function of these variables and taking partial derivatives and setting them equal to zero. The solution of the resulting equations gave the answer to the question.

However, the general isoperimetric problem cannot be so easily solved. *A general planar curve cannot be specified by giving a finite number of parameters.* It is easy to see this by thinking of approximating a planar curve by a series of polygons i.e. replacing short segments of arcs of the curve by the corresponding chords. For a polygon of n sides the area will depend on a finite number of parameters but the number grows at least as fast as n (**Question:** Work out the exact number). But the number n has to be arbitrarily large since we need more and more sides to approximate a general curve better and better. Hence the number of parameters to specify the area of a general curve must be infinite. Thus our approach of trying to write a function and minimise it seems quite hopeless.

In this particular case one can still arrive at the solution of this problem in a couple of steps. Essentially one first argues that amongst polygons of fixed number of sides, the regular one has maximal area for fixed perimeter (just as we saw for the square in the case $n = 4$). Then one compares two regular curves (of fixed number of sides m and n respectively) with the same perimeter, and finds that if $m > n$ then the area of the regular m-sided polygon is greater than that of the n-sided polygon (Compare the cases $n = 3$ and $n = 4$). In other words the area can be made to increase (for fixed perimeter) by keeping on increasing the number of sides of the regular polygon. The limiting curve is then a circle and this is indeed the solution of the isoperimetric problem.

The isoperimetric problem is however only one member of a set of problems on extremisation which are not addressed by the usual differential calculus simply because one has to extremise amongst an infinite parameter family (all planar curves in the above example). Another way to phrase this is that the variation has to performed not just over a space of parameters but more generally over a space of functions. Thus a closed planar (convex) curve can be described by a function $r = f(\theta)$ and we are asked to find the function $f(\theta)$ which extremises a certain quantity (the area under the curve in the above example) that depends on the function in a complicated way. Here (r, θ) are the usual polar coordinates on the plane. The quantity one is minimising is not quite a function itself, since its value depends not on a finite set of variables (like a normal function would) but on an entire function itself. Such a quantity is called a *functional* and it depends on a function. Thus we can speak of the *area functional* $A[f(\theta)]$ (often denoted simply as $A[f]$) which can be expressed in the above parameterisation of the curve in polar coordinates by

$$A[f] = \frac{1}{2} \int_0^{2\pi} f^2(\theta)\, d\theta. \tag{5}$$

Thus a functional is a generalisation of a function in that it takes as its input (argument) a function and its output (value) is a number. In the case of the isoperimetric problem we are asked to minimise the functional $A[f]$ subject to the constraint that the total perimeter

$$P[f] = \int_0^{2\pi} \sqrt{f^2(\theta) + \left(\frac{df}{d\theta}\right)^2}\, d\theta = L. \tag{6}$$

Notice that the perimeter is also a functional depending on the curve $f(\theta)$. We also see that generally a functional $F[f]$ may depend not just on the values of the function f but also on its derivatives f', f'' etc. Typically one needs to only consider functionals that depend on *finitely* many derivatives of f.

Another example of a problem which involves extremising a functional is one where one is asked to calculate the curve (between two fixed points in space) along which if a frictionless bead were to slide under gravity, it would take the least amount of time. This is what is known as the *Brachistochrone problem*. Again one can write down the total time taken by the bead as a functional $T[f]$ depending on the function specifying the curve

$y = f(x)$. The problem is again one of extremising this functional to find the specific curve. The answer was shown by Jakob Bernoulli to be an arc of a curve known as the cycloid. To tackle problems of this kind in a systematic way one needs to develop ways of extremising functionals just as one learnt to extremise functions through differential calculus. For more on the Brachistochrone and the Isoperimetric problems, as well as for an accessible introduction to the variational calculus we are going to describe, see the book, by Simmons (1972) mentioned in the bibliography.

3   Variational Calculus

The general method to obtain the extrema of functionals is through what is known as the *Variational Calculus*. To get an idea of how this is done let us review the extremisation of functions of finitely many variables. Given $f(x_1, x_2, \ldots, x_n)$ we can consider an infinitesimal variation of the arguments $x_i \rightarrow x_i + \delta x_i$. Under this variation we have

$$
\begin{aligned}
f(x_1, x_2, \ldots, x_n) \quad &\rightarrow \quad f(x_1 + \delta x_1, x_2 + \delta x_2, \ldots, x_n + \delta x_n) \\
&\approx \quad f(x_1, x_2, \ldots, x_n) + \sum_i \delta x_i \frac{\partial f}{\partial x_i}. \quad (7)
\end{aligned}
$$

Here we have neglected terms of $O(\delta x^2)$. Thus the variation

$$
\delta f(x_1, x_2, \ldots, x_n) \approx \sum_i \delta x_i \frac{\partial f}{\partial x_i}. \quad (8)
$$

For the function to be a (local) extremum at some value of the x_i, we must have $\delta f = 0$ for arbitrary variations δx_i about that point. Thus we recover the familiar condition for extremisation that

$$
\delta f(x_1, x_2, \ldots, x_n) = 0 \rightarrow \frac{\partial f}{\partial x_i} = 0 \qquad (all\ i). \quad (9)
$$

The basic idea with the extremisation of functionals is to start by approximating the function it depends on by its *values* at a finite (discrete) set of points. The functional is then approximated by evaluating it at these finitely many points. We thus reduce it to an object that depends on only finitely many variables, namely the values of the function at the finitely many points. We then finally take the limit $n \rightarrow \infty$ which makes this approximation arbitrarily good and thus exact. This is similar in spirit to what was done in the isoperimetric problem where we could consider polygons of a large but fixed number of sides n and find the extremal solution by taking the limit $n \to \infty$.

We can consider the variation of a functional approximated as described above and perform the extremisation by finally taking the limit $n \to \infty$. Thus we can write the discretisation as

$$F[f(x)] \approx F(f(x_1), f(x_2), \ldots, f(x_n)) \qquad (10)$$

And then consider the variation

$$F[f(x)] + \delta F[f(x)] = F[f(x) + \delta f(x)]$$
$$\approx F(f(x_1) + \delta f(x_1), f(x_2)$$
$$+ \delta f(x_2), \ldots, f(x_n) + \delta f(x_n)) \qquad (11)$$

In other words we approximate the variation of the function denoted by δf by considering an arbitrary variation of its values $f(x_i) \to f(x_i) + \delta f(x_i)$ at the discrete set of points $(x_1, x_2, \ldots x_n)$. Note that in the variation of a functional it is the values of the function it depends on that is being varied and not the argument x of the function itself. For convenience of notation we will denote $f(x_i) = f_i$ and $\delta f(x_i) = \delta f_i$ thus also indicating that it is the value of the function which is important in this context.

We can then write using the usual rule for variation of a function of many variable

$$F[f(x) + \delta f(x)] \approx F(f_1 + \delta f_1, f_2 + \delta f_2, \ldots, f_n + \delta f_n))$$
$$\approx F(f_1, f_2 \ldots, f_n) + \sum_{i=1}^{n} \delta f_i \frac{\partial F}{\partial f_i} + O(\delta f^2). \qquad (12)$$

Here in the last step we have assumed that the variations δf_i are infinitesimal and thus ignore terms of $O(\delta f^2)$. This is true for all n and thus we can proceed to take the limit $n \to \infty$. This can be written as

$$F[f(x) + \delta f(x)] = F[f(x)] + \int dx \delta f(x) \frac{\delta F[f]}{\delta f(x)} + O(\delta f^2). \qquad (13)$$

Essentially we have replaced the sum over i in (12) by a continuous integral as we take $n \to \infty$. We have also denoted the continuous version of $\frac{\partial F}{\partial f_i}$ by $\frac{\delta F[f]}{\delta f(x)}$ which is meant to indicate the variation of the functional

$F[f]$ with respect to the value of the function $f(x)$, i.e. a variation of the function $f(x)$ such that only its value at the point x is changed keeping the value unchanged at all other points. This limiting notion of a variation of a functional with respect to a function is called a functional derivative generalising the usual notion of the derivative of a function.

The condition for an extremum is then easy to state, either by considering a limit of (12) as $n \to \infty$ or directly from (13), in both cases requiring stationarity of the functional with respect to arbitrary variations δf. We obtain

$$\frac{\delta F[f]}{\delta f(x)} = 0 \qquad (14)$$

as the condition for extremising the functional. This is an example of a functional equation and typically can be written in terms of a *differential equation* which $f(x)$ has to satisfy to be an extremum of the functional.

4 Variational Principles in Physics

One of the very fruitful principles which played an important role in the development of classical mechanics and geometrical optics was the *Principle of Least Action* (which in the context of optics became known as the *Principle of Least Time* or *Fermat's principle*). As we will see these can be formulated as variational principles (see Feynman *et al.* 1964).

It is easiest to explain this by looking at a couple of elementary questions in optics. First, consider a reflecting mirror and a point-like light source and a light detector placed at fixed points above the reflecting plane of the mirror. The question is: what is the geometrical trajectory that a ray of light will take in going from the source to the detector through a reflection by the mirror? We know the answer to the question from our school textbooks. It is composed of two straight lines – an incident ray and a reflected ray which impinge on the mirror on such a way that the angle of incidence is equal to the angle of reflection (both measured with respect to the normal to the mirror). Is there a simple way to characterise this particular trajectory amongst all the possible ones that one could draw between the source and the detector. The answer is yes. Assuming light has a fixed speed in vacuum (or air), this trajectory is the one for which light takes the *least time*. Since the speed is a constant this is equivalent in this case to the trajectory of shortest length amongst all possible trajectories (between the source and detector and including at least one incidence on the mirror). It can be easily verified using a geometrical construction that the above trajectory is indeed the shortest one.

[**Exercise:** Verify this statement.]

We can go onto ask a slightly more involved question. Consider two media in which light travels at different speeds (e.g. air and water). Assume that there is a planar interface between them as in the surface of a glass of water. Now consider a source in the water and a detector outside in the air both held at fixed locations. What is the trajectory that light takes to go from the source to the detector passing through the interface? Again our schoolbooks enshrine the answer in terms of Snell's law which says that the trajectory is one in which an incident straight ray (from the source in the water) impinges on the interface at an angle θ_i and is *refracted* into another straight ray (at an angle θ_r) which continues to the detector such that θ_i and θ_r bear a fixed relationship. Namely that given by Snell's law

$$\frac{\sin \theta_i}{\sin \theta_r} = \frac{1}{n} = \frac{v_{water}}{v_{air}}. \tag{15}$$

Here $n > 1$ is the refractive index of water with respect to air and is given by the ratio of the speeds of light in the two respective media as indicated.

We can again ask what it is that characterises this particular trajectory. This time it is clear that the answer is not that of shortest length (which would have been a straight line instead of the bent refracted ray). In fact, one can easily argue that this is the trajectory of shortest time given that light would take different amounts of time in air and water to cover the same length.

[**Exercise:** Verify this statement. Hint: first argue that in each medium the rays would have to be straight lines. Then amongst the possible straight line segments in each medium argue that the one satisfying Snell's law will take the least time.]

In fact, this principle can be generalised to describe all of geometrical optics and it is a powerful way of obtaining the trajectories of rays of light in complicated media, perhaps with continuously varying refractive properties. We can state it in the variational language described above by constructing a functional $T[x]$ where $x(s)$ denotes a geometrical trajectory and T is the time for light to traverse that trajectory (subject to various constraints such as imposed in the above problems). The principle of least time asks for extremising (minimising) the functional $T[x]$ over all allowed trajectories. The trajectory $x(s)$ satisfying

$$\frac{\delta T}{\delta x(s)} = 0 \qquad (16)$$

will be the one that light takes. Principles such as this enable us therefore to see the inherent unifying element amongst the many different situations of geometrical optics. In fact, it was understood much later that the principle of least time followed in an appropriate limit from Maxwell's understanding of light as an electromagnetic wave. The limit of geometrical optics suffices to describe the propagation of light and holds when the wavelength of light is small compared to other length scales in the problem.

Emboldened by the success of such a principle for the propagation of light, physicists and mathematicians looked for a similar principle for the propagation of ordinary particles subject to Newton's laws of motion. It turns out that there is indeed a variational principle or extremisation of a functional which yields Newton's equations of motion. In the simplest case of a particle of mass m moving in a potential $V(\vec{x})$, one defines the *Lagrangian*

$$L = \frac{1}{2}m\left(\frac{d\vec{x}}{dt}\right)^2 - V(\vec{x}) \qquad (17)$$

as the *difference* between the kinetic and potential energies. (The total energy would have been the sum of the two). We can now define a functional which is traditionally called the action which is nothing other than the integral of the lagrangian over time

$$S[\vec{x}] = \int_{t_1}^{t_2} L\,dt = \int_{t_1}^{t_2} dt\left[\frac{1}{2}m\left(\frac{d\vec{x}}{dt}\right)^2 - V(\vec{x})\right]. \qquad (18)$$

The action functional $S[\vec{x}]$ is defined on the set of all trajectories $\vec{x}(t)$ which begin and end at some specified points $\vec{x}_1, \vec{x}_2$ at times t_1 and t_2 respectively. The claim is that Newton's equations for the classical trajectory

$$m\frac{d^2\vec{x}}{dt^2} = -\vec{\nabla}V(\vec{x}) \qquad (19)$$

then follows from considering the extremisation of this functional. In other words the condition

$$\frac{\delta S[\vec{x}]}{\delta \vec{x}(t)} = 0 \qquad (20)$$

gives rise to (19). This gives a new perspective on the classical equations of motion. In fact, it turns out that most classical equations of motion can be obtained as the variation of some suitably chosen action functional. This holds for the Maxwell equations of electrodynamics and the Einstein equations for gravity. Thus the variational principle of least action gives a unified way of viewing various different classical equations.

Furthermore it turns out that when one goes beyond classical mechanics to the world of quantum mechanics, the action functional continues to play an important role, as was first understood by Dirac and Feynman. The passage from classical mechanics to quantum mechanics turns out to have a close analogy to the passage from geometrical optics to wave optics. This gives a relatively more intuitive picture of quantum mechanics in which one actually samples over all the trajectories (and not just the extremal or classical one) with an appropriate weight which is simply $e^{i\frac{S[\vec{x}]}{\hbar}}$ where $\hbar$ is a fundamental constant of nature known as Planck's constant which governs quantum behaviour.

5 Conclusion

Hopefully the reader will be motivated enough by now to further explore the fascinating interplay of physics and mathematics that played a role in the development of variational calculus. The two references in the bibliography give a possible starting point to some of these topics.

References

Feynman, R.P., Leighton, R.B. and Sands, M. 1964. *The Feynman Lectures on Physics*, **Vol. 2**, Chapter 19 Addison-Wesley, New York.

Simmons, George F. 1972. *Ordinary Differential Equations and their Applications*, Chapter 9, TataMcGraw-Hill, New Delhi.

CHAPTER 18

The Forces of Nature

Sudarshan Ananth

Indian Institute of Science
Education and Research,
Pune 411021, India.
e-mail: ananth@iiserpune.ac.in

1 Fundamental forces

Gravity, although the oldest force known to man, is the force we understand the least. The weakness of the gravitational force precludes the possibility of performing simple experiments that will teach us more about this omnipresent force. Apart from gravity there are three other *fundamental* forces in Nature: the electromagnetic force, the weak force and the strong force. Fundamental here means that *all* forces we encounter may be attributed to one of these four forces. For example, the force of tension, seen frequently in mechanics problems, is electromagnetic in origin. Three of the fundamental forces are described by quantum Yang-Mills theories and theoretical predictions have, with stunning accuracy, matched experimental observations. Gravity, the fourth fundamental force, has a well tested description: the Einstein-Hilbert theory. However, this is a "classical" theory, meaning that it does not offer an accurate microscopic description. Such a description would require a quantum theory of gravity - something researchers have been working towards for several decades (without success). A problem that looms large in this search is the very structure of gravity. In this self-contained article we will attempt to explain this problem.

2   The principle of least action

We start with a review of the principle of least action. We are familiar
with the concept of "total energy" for a simple mechanical system: it is the
sum of kinetic and potential energies. For example, for a simple harmonic
oscillator, the potential energy is

$$V = \frac{1}{2}kx^2, \tag{1}$$

where x represents the displacement of the spring from its equilibrium (relaxed) position and k, the spring constant. The kinetic energy for this system is

$$T = \frac{1}{2}m\dot{x}^2, \tag{2}$$

where $\dot{x} = \frac{dx}{dt}$ represents the rate of change of position of the mass 'm'
attached to the spring. The total energy is

$$H = T + V = \frac{1}{2}m\dot{x}^2 + \frac{1}{2}kx^2. \tag{3}$$

Surprisingly, although the total energy is a useful quantity, it is not essential
to our understanding of classical mechanics. Instead, the really important
object in mechanics is the Lagrangian (Wikipedia: Lagrange) defined by

$$L = T - V. \tag{4}$$

Explicitly, for the simple harmonic oscillator, this reads

$$L(x, \dot{x}, t) = \frac{1}{2}m\dot{x}^2 - \frac{1}{2}kx^2. \tag{5}$$

The reason the Lagrangian is such an important object is because of the
"principle of least action" which we now state.
Consider a classical system at position x_1 at time t_1 and at position x_2 at
time t_2. This system will *always* move from x_1 to x_2 in a manner such that
the integral

$$S = \int_{t_1}^{t_2} L(x, \dot{x}, t)\, dt, \tag{6}$$

takes the least possible value[1]. This is referred to as the principle of least action (S being the action) and may be expressed as the following variation

$$\delta S = \delta \int_{t_1}^{t_2} L(x, \dot{x}, t)\, dt = 0. \tag{7}$$

It is amazing that all of classical mechanics (including laws like the conservation of energy) follows from this beautiful[2] principle.

The entire field of mathematics, called the "Calculus of variations", is devoted to extremizing functionals as in (7). Physics is littered with principles governed by the calculus of variations - a wonderful example of the key role mathematics plays in the physical sciences (Wikipedia: Calculus_of_variations).

Performing the variation in (7) produces the Euler-Lagrange equations of motion (Landau and Lifshitz, 1976)

$$\frac{d}{dt}\left(\frac{\partial L}{\partial \dot{x}}\right) - \frac{\partial L}{\partial x} = 0, \tag{8}$$

which for the simple harmonic oscillator implies

$$m\ddot{x} = -kx, \tag{9}$$

which matches what we obtain by using Newton's laws.

Each of the forces of Nature is also decribed by a Lagrangian. Conveniently, three of the four forces are described by the Yang-Mills Lagrangian. Gravity, on the other hand, is described by the Einstein-Hilbert Lagrangian. In the next section, we will briefly describe these two Lagrangians and then be in a position to highlight the problem with gravity.

3   Lagrangians

The Lagrangian for the simple harmonic oscillator involves two terms

$$L(x, \dot{x}, t) = \frac{1}{2}m\dot{x}^2 - \frac{1}{2}kx^2 . \tag{10}$$

---

[1]The integral is actually an extremum but we will ignore this point.

[2]Beauty here refers to the fact that one simple principle produces all the laws of mechanics.

The dynamical variable for this system is the position variable, x which appears in both terms (as x and $\dot{x}$).

The dynamical variable in Yang-Mills theory is called a "gauge field". We digress at this point and briefly introduce the reader to the concept of a field.

3.1 From classical mechanics to classical field theory

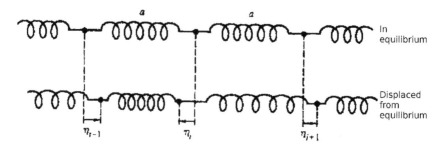

The figure above shows a system of an infinite number of mass-spring pairs in one-dimension (Goldstein, 1980). In equilibrium, the masses m are separated by distance a. Suppose the system is displaced - the displacement from rest for the ith mass being η_i. The dynamical variables for this system are then the η_i's. The net kinetic energy is the sum of all the individual kinetic energies

$$T = \frac{1}{2} \sum_{i=-\infty}^{+\infty} m \dot{\eta}_i^2, \tag{11}$$

and the net potential energy is the sum of all the potential energies (due to elongation or compression)

$$V = \frac{1}{2} \sum_{i=-\infty}^{+\infty} k (\eta_{i+1} - \eta_i)^2. \tag{12}$$

The Lagrangian for this system is

$$L = \frac{1}{2} \sum_{i=-\infty}^{+\infty} \left[m \dot{\eta}_i^2 - k (\eta_{i+1} - \eta_i)^2 \right], \tag{13}$$

which can be rewritten as

$$L = \frac{1}{2} \sum_{i=-\infty}^{+\infty} a \left[\frac{m}{a} \dot{\eta}_i^2 - ka \left(\frac{\eta_{i+1} - \eta_i}{a} \right)^2 \right] = \sum_{i=-\infty}^{+\infty} a L_i. \tag{14}$$

In the continuum limit, when the separation between adjacent masses becomes extremely small, $a \to 0$. The integer index i identifying a particular mass needs to be replaced by a continuous index, x. As a consequence

$$\lim_{a \to 0} \frac{\eta_{i+1} - \eta_i}{a} = \lim_{a \to 0} \frac{\eta(x + a) - \eta(x)}{a} = \frac{d\eta}{dx}, \tag{15}$$

$$\lim_{a \to 0} \sum_{i=-\infty}^{+\infty} aL_i = \int_{-\infty}^{+\infty} dx\, L(x). \tag{16}$$

So the Lagrangian, in the continuum limit, becomes

$$L = \frac{1}{2} \int_{-\infty}^{+\infty} \left[\mu \dot{\eta}^2 - Y \left(\frac{d\eta}{dx} \right)^2 \right] dx, \tag{17}$$

where $\mu = \frac{m}{a}$ is the mass density of the system and $Y = ka$ is the Young's modulus. We now have a field theory Lagrangian where the dynamical variable is the displacement (now a field) and the coordinate x is reduced to a mere label—adopting the role played earlier by the discrete variable i.

3.2   Yang-Mills and gravity

The field η encountered above is a single component object and is called a scalar field. The dynamical variable for Yang-Mills theory happens to have two components[3] A and $\bar{A}$ and is referred to as a gauge field. The Yang-Mills Lagrangian in terms of these variables has the following structure

$$L_{\text{YM}} \sim \bar{A}A + g\bar{A}AA + g\bar{A}\bar{A}A + g^2 \bar{A}\bar{A}AA. \tag{18}$$

Derivatives are not shown in the above equation since they have little relevance to the present discussion. The Yang-Mills Lagrangian thus involves four terms. The first term is referred to as the kinetic term while the remaining terms are called "interaction vertices". The quantity g, referred to as the coupling constant, describes the strength of the interaction.

We now turn to gravity which is also described by two field variables h and $\bar{h}$. In terms of these variables, the gravity Lagrangian is

$$L_{\text{g}} \sim \bar{h}h + \kappa \bar{h}hh + \kappa \bar{h}\bar{h}h + \kappa^2 \bar{h}\bar{h}hh + \kappa^3 \bar{h}\bar{h}\bar{h}hh + \kappa^3 \bar{h}\bar{h}hhh + \cdots \tag{19}$$

---

[3]Technical comment: this is in four dimensions, in light-cone gauge.

where the ... refers to an infinite number of interaction vertices each involving different numbers of the two fields h and $\bar{h}$. Unfortunately, the Lagrangian for gravity contains infinitely many "interaction vertices" making gravity quite different from other theories encountered previously. Any calculation involving the gravity Lagrangian must take into account infinitely many vertices thus becoming intractable.

A lot of current research today is aimed at finding ways around this problem. Recently, there have been some interesting developments in this direction: a new form for the Yang-Mills Lagrangian was found by using a clever change-of-variables (Gorsky and Rosly, 2006; Mansfield, 2006). This mapped the Yang-Mills Lagrangian, involving four terms, into a Lagrangian with infinitely many vertices *exactly* like the gravity Lagrangian. This immediately leads us to ask: is it possible that a similar judicious change-of-variables will allow us to map the infinitely long gravity Lagrangian into a Lagrangian with only a finite number of interaction vertices[4]?

Acknowledgments

This work is supported by the Department of Science and Technology, Government of India through a Ramanujan Fellowship and by the Max Planck Society, Germany through the Max Planck Partner Group in Quantum Field Theory.

References

Goldstein, H. 1980. *Classical Mechanics* Edition 3, Addison Wesley.

Gorsky, A. and Rosly, A. 2006. *Journal of High Energy Physics (JHEP)* 0601, Page 101, arXiv.org e-Print: hep-th/0510111.

Landau, L.D. and Lifshitz, E.M. 1976. *Mechanics, Course of Theoretical Physics* Volume 1, Edition 3, Butterworth-Heinemann.

Mansfield, P. 2006. *Journal of High Energy Physics (JHEP)* 0603, Page 037, arXiv.org e-Print: hep-th/0511264.

Wikipedia: http://en.wikipedia.org/wiki/Lagrange

Wikipedia: http://en.wikipedia.org/wiki/Calculus_of_variations

---

[4]Unfortunately, the answer to this question is very likely to be NO.

CHAPTER 19

The Reciprocal Interaction Between Mathematics and Natural Law

Sunil Mukhi

Tata Institute of Fundamental Research,
Mumbai 400 005, India.
e-mail: mukhi@tifr.res.in

1 Introduction

In ancient times most mathematicians were also physicists and vice versa, Newton being only the most famous example. But starting with the twentieth century, there have been fewer examples of a single person straddling both fields. The modern university usually has well-defined and separate mathematics and physics departments which consider themselves to be in pursuit of very distinct goals.

From this time, mathematics has typically been seen by natural scientists as a language in which their observations can be usefully codified. In his famous article *The Unreasonable Effectiveness of Mathematics in the Natural Sciences*[1], Wigner put it as follows: *The miracle of the appropriateness of the language of mathematics for the formulation of the laws of physics is a wonderful gift.* ... Many developments have taken place in the four and a half decades since Wigner made this observation, and the miracle to which he was referring remains alive and well today. But there are also indications that the situation is starting to evolve in an interesting way.

---

[1]http://www.dartmouth.edu/~matc/MathDrama/reading/Wigner.html

As Wigner indicated, the new physical notions that came up during the twentieth century readily found a setting in established areas of mathematics. One such example is the theory of continuous groups, known as *Lie groups*, and their associated algebras. It was realised that rotations in space form a group, the orthogonal group in three dimensions, $O(3)$, whose representations play an especially important role in quantum mechanics. In special relativity, the rotation group gets subsumed into the Lorentz group, $O(3, 1)$, which contains besides rotations, the "boosts" between space and time. In the 1960's it became necessary to invoke a new class of groups, the unitary groups, to classify fundamental particles. The experimentally observed particles conveniently fell into representations, or multiplets, of these unitary groups. It was just as well that orthogonal and unitary groups had already been discovered and classified. Otherwise one can imagine that the discoverers of particle spectra would have been perplexed by the observed regularities that heralded the existence of unitary symmetry, and would have been at a loss for an explanation.

From the mathematicians' viewpoint, however, it is fair to say that the induction of Lie groups into physics discourse did not add significant conceptual material to what was already known in the mathematics literature. To be sure, physicists did discover new and important facts about Lie groups, but these typically did not have the degree of beauty and depth that would appeal to mathematicians in a big way. Hence it is not surprising that the enthusiasm of physicists to learn mathematics has usually been greater than that of mathematicians to learn physics. The latter might simply feel they have nothing much to gain.

Many other twentieth-century developments, such as the introduction of differential geometry in the study of gravity, of infinite-dimensional vector spaces in quantum mechanics, and of fibre bundles in gauge theory, exemplify the same story. The role of mathematics in physics in these cases was limited in two ways. First, the core mathematical results needed had already been obtained, and second, mathematicians did not learn very significant new things after these areas came to be used in physics.

This is of course a value judgement, and heavily depends on what we mean by "very significant". Even in the twentieth century a considerable amount of mathematical research was done by physicists, and Wigner himself wrote a textbook entitled *Group Theory* (Wigner, 1959). But then, the subtitle of his book was *and its Application to the Quantum Mechanics of Atomic Spectra* and the primary audience for the book is made up of physicists. So the above picture of a one-way relationship between mathematics and physics does seem valid to a reasonable degree.

However, during the 1980's, this situation started to evolve. The evolution is hard to understand fully because we are still in the middle of it, or maybe close to the beginning, and it is the main focus of this article, whose main observation will be that the interface between mathematics and physics has turned into a two-way relationship of an unprecedented nature. Not only does mathematics provide physics with a formulation it can use, but physics has started to provide mathematics with methods and results that it needs, desires and appreciates.

It would be tempting to describe this new situation in terms of *The Unreasonable Effectiveness of Physics in Mathematics*, and an internet search reveals that over the last couple of years there have been articles and colloquia with precisely this title, by a number of distinguished mathematicians and physicists. It seems likely that this reversal (or perhaps equalising) of the traditional roles is wide-ranging in its scope and that there is much more to come. The newly two-sided relationship between mathematics and physics seems likely to occupy a part of centre-stage in both fields for the next few decades.

2   Consistency and Experiment

Before turning to the new paradigms thrown up in the 1980's, I would like to briefly touch upon an extra ingredient in the mathematics-physics relationship - the remarkable role played by consistency. In the twentieth century it happened, more than once, that mathematics did not merely act as a tool for quantification, but appeared to have predictive power over nature. Merely by requiring mathematical consistency, one was forced to believe in new, as yet undiscovered, physical phenomena. The phenomena in question were then later confirmed by experiment.

The classic example was Dirac's equation for the electron, which he postulated in 1928. It was soon realised that while the equation worked well for electrons, it also predicted the existence of another particle with the same mass as an electron but opposite charge. This prediction arose because for a charged particle of spin one-half, a Lorentz-invariant equation necessarily has at least two kinds of solutions. So it was mathematical consistency that required the new particle and no way could be found to get rid of it from the equation. Instead, in 1932 the new particle, the *positron*, was experimentally detected (see for example Frank Close, 2009).

As is widely known, Dirac later commented: *it is more important to have beauty in one's equations that to have them fit experiment*, a statement that appears bold, controversial, and even downright unscientific. What would Dirac, or anyone else, have done if the positron had not been discovered? Or was he saying that the positron had no choice but to be discovered? We do not know. But the rest of Dirac's comment, which is less widely publicised, gives a better picture of what he had in mind. *If there is not complete agreement between the results of one's work and experiment, one should not allow oneself to be too discouraged, because the discrepancy may well be due to minor features that are not properly taken into account and that will get cleared up with further development of the theory.* He was only saying that mathematics is a more reliable guide to nature than one might expect, but not that it is an ultimate arbiter of natural law.

In 1931, Dirac proposed the existence of magnetic monopoles from similar considerations of mathematical consistency. He argued that monopoles were consistent with all known facts about quantum mechanics, therefore they ought to exist. Fifty years later, during which period no monopole had been discovered, he retracted his proposal in the most abjectly empiricist language. He wrote to Abdus Salam in 1981: *I am inclined now to believe that monopoles do not exist. So many years have gone by without any encouragement from the experimental side* (see for example Sunil Mukhi, 2003).

Taken as a composite, the quotations above give a more accurate picture of the situation. Dirac's views about the role of mathematics were far from extremist. He simply believed that beauty in the equations was a more reliable guide than most people had hitherto believed. One should interpret this as a slight strengthening of Wigner's observations on the effectiveness of mathematics in the natural sciences.

3 Mathematics and Quantum Field Theory

The major influence in creating a different paradigm for the mathematics-physics relationship was undoubtedly Edward Witten. Despite being a physicist, he received the highest honour in mathematics, the Fields Medal, in 1990 for his contributions to mathematics using tools of physics. In the citation for the award[2], Ludwig Faddeev said that *Physics was always a source of stimulus and inspiration for Mathematics... In classical time[s] its connection with mathematics was mostly via Analysis, in particular through Partial Differential Equations. However, [the] quantum era gradually brought a new life. Now Algebra, Geometry and Topology, Complex Analysis and Algebraic Geometry enter naturally into Mathematical Physics and get new insights from it.*

Putting it as simply as possible, Witten's strategy was to rephrase mathematical concepts in the language of Quantum Field Theory, the framework for the description of fundamental particles and their interactions. He would then use insights and tools native to quantum field theory, such as symmetries and invariances, and the (notoriously hard to define) Feynman-Kac path integral. This would lead to a natural setting for the mathematics problem where not only a solution could be found but also an enormous variety of generalisations.

In an article[3] originally intended to be Witten's Fields citation but which he was unable to actually deliver at the award ceremony, Michael Atiyah notes 11 papers of Witten that justify his being awarded this prestigious prize. To give the reader a flavour of what was done, I will briefly touch upon a few of them.

3.1 Knot Theory

The study of knots in three-dimensional space was initiated in the 9th century by, of all people, the physicist Lord Kelvin. He had a theory of *vortex atoms*, according to which the different atoms that occur in nature are made of the same "substance" but knotted in different ways. According to this theory, the classification of atoms could be reduced to the classification of knots, and Kelvin accordingly embarked on the latter problem.

Over the ensuing decades, considerable progress on classifying knots was made by mathematicians. Two knots are distinct if they cannot be continuously deformed into each other. But for complicated configurations, it is hard to establish whether or not this is possible just by looking at presentations of the knots. Two presentations can look quite different from each

---

[2]www.mathunion.org/o/General/Prizes/Fields/1990/Witten/page1.html
[3]www.mathunion.org/o/General/Prizes/Fields/1990/Witten2/page1.html

other, but there might be some "moves" that smoothly deform one of them to the other - then they would describe the same knot. The idea emerged of associating an "invariant" to a knot, a number or function that is calculable starting from any picture of the knot and remains invariant under smooth deformations of the knot. Now if starting from two presentations we get two different invariants, then we can be sure the knots are distinct. The converse is not in general true – a weak knot invariant may return the same value for many different knots, but a powerful invariant will be successful in distinguishing many knots from each other.

Knot invariants were extensively developed, and the most powerful ones (some of which had been found by the mathematician Vaughn Jones) turned out to be polynomials in one or more parameters. Witten realised that in a certain quantum field theory – a *gauge theory* in three spatial dimensions, the natural observable is a *Wilson line* (a line of generalised electric flux) along an arbitrary closed contour, the same thing as a knot. Being an observable, the Wilson line can have an expectation value in the vacuum state of the quantum theory. When computed using techniques of quantum field theory, this expectation value turns out to be a polynomial in some parameters related to the coupling constant of the theory and the type of charges flowing through the Wilson line.

This approach beautifully reproduced the Jones polynomials. For knot theory it was a dazzling illumination: it gave an interpretation to the polynomials, it explained (via a relation to conformal field theory) some of their properties and relations to the theory of quantum groups, it related the parameters in the knot polynomials to an expansion parameter, and it provided a series of generalisations of the polynomial, for example by changing the gauge group of the gauge theory from $SU(2)$ to an arbitrary Lie group. Moreover, as Atiyah points out, *it is the only intrinsically three-dimensional interpretation of the Jones invariants; all previous definitions employ a presentation of a knot by a plane diagram or by a braid.*

Now the gauge theory at the root of this brilliant discovery, the so-called *Chern-Simons theory*, is a cousin of Yang-Mills theory that describes three of the four fundamental interactions in nature: the two nuclear interactions (weak and strong) and electrodynamics. Yang-Mills theory, in turn, was a generalisation of Maxwell's theory of electromagnetism – a relativistic field theory with a gauge symmetry, under which wave functions change by a local phase. So it was because of the whole circle of theoretical ideas and experiments relating to relativity and electromagnetism that

Witten found a new way to describe knot invariants, and a major mathematical illumination was brought about (Witten, 1989).

3.2 Morse Theory

Morse theory studies differentiable manifolds by relating the critical points of some suitable function defined on them to their homology, roughly the study of nontrivial *cycles* that can be embedded in them. One can learn about the topology of a manifold by knowing about its cycles and, via Morse theory, one gets this from the study of critical points. Witten provided a proof of certain inequalities, the *Morse inequalities* by setting up an equivalence between *supersymmetric quantum mechanics* and the cohomology/homology of the manifold.

This is harder to understand intuitively as compared to knot polynomials. But here too, we can see where the physics input came from. Starting around 1974, theoretical physicists made a breakthrough in their search for a symmetry that unifies bosons and fermions, the two different kinds of particles that occur in nature. The new symmetry was dubbed *supersymmetry* and there are compelling reasons to believe that, although it is hidden from view in everyday life, nature makes use of it in some subtle way. Indeed the reasons are so compelling that the Large Hadron Collider at CERN, Geneva has as one of its highest priorities the search for supersymmetric particles. Witten's supersymmetric quantum mechanics is a highly simplified version of the physical theories whose experimental confirmation is being sought, but uses the same basic structure. By mapping it onto the problem of Morse theory, Witten was able to use properties of the supersymmetric path integral to make headway into pure mathematics (Witten, 1982).

3.3 The Moduli Spaces of Riemann Surfaces

To a lay person, a Riemann surface is just a two dimensional surface, examples being a sphere, a torus (like a bicycle tyre) and multi-handled generalisations. Mathematicians like to think of them as *lines* of one complex dimension, and want to know how they can be smoothly varied and in what way their mathematical properties change as we vary them. The *moduli space* of such a surface is the space of parameters that label this variation. Sophisticated results due to Mumford and Morita and others, had given considerable insight into the global structure (topology) of these moduli spaces.

Witten was inspired to address this by considering a problem in gravitation. However, his theory of gravity was not the one discovered by Einstein, but a simpler version of it where spacetime is two-dimensional. From a physical point of view the theory has very little dynamics in this setup, but Witten analysed the operators that made up the observables of this theory, and asked what was the correlation function between them – a measure of how fluctuations are interlinked in a quantum field theory. What he discovered was that these correlation functions were identical to the Mumford-Morita classes on the moduli space of Riemann surfaces. As usual, this approach also gave rise to important generalisations of the known mathematical results.

This time the physical inspiration originated in the structure of gravitation as elucidated by Einstein. Though Witten's gravity theory was very different, it is possible that if there were no gravitational force in nature (and assuming we still existed) this insightful approach to moduli spaces would never have been discovered and some profound results about the abstract manifolds called Riemann surfaces would not have been known to mathematicians.

Atiyah once made a comment about Witten's contribution to mathematics on the following lines. Usually, physics provides an intuitive notion, or a way of thinking about things, and then mathematicians prove a corresponding rigorous result. This is the classic form in which physics has influenced mathematics over the centuries. But in the areas of Witten's work, many rigorous results in mathematics already existed and Witten subsequently provided the intuitive explanation for them. In this sense, the mathematicians were ahead of the game, though to say this in no way undermines the contribution of this work. As already indicated, the physical re-interpretation led to generalisations of the mathematics which would have been almost impossible to guess if one only knew the rigorous methods. And by providing a new setting for old ideas, they have provided a number of pointers towards new mathematical directions (Witten, 1990).

3.4 Mathematics and String Theory

I would now like to finally address an area in which not only is physics in the process of influencing mathematics, but, for the moment at least, the physics approach seems to be ahead of the game.

We only need to know some basic facts about string theory for the purpose of this discussion (see for example Sunil Mukhi, 1999). String theory is the most serious candidate known for a consistent quantum theory of gravity. It proposes that the fundamental excitations in nature are not point-like particles but extended (though tiny) objects called strings. This theory has not received direct experimental confirmation, but it is built on many experimental facts: the number of dimensions we live in, the presence in nature of both fermions and bosons, and the existence of gauge symmetry as in Yang-Mills theory and of general coordinate invariance as in Einstein's theory of General Relativity.

Einstein taught us that gravity is the geometry of space-time. But he was talking about classical gravity, not the quantum version. In his life-time, no quantum theory of gravity was known. Classical gravitation used a great deal of mathematics from the realm of Riemannian geometry, and at this level space-time is a (pseudo)-Riemannian manifold.

Even in Einstein's time, there were some theories of gravity that re-quired not only a physically observable spacetime, but also an extra hidden space that was too small to observe experimentally. It affected our world only in indirect ways. This notion was proposed by Kaluza and Klein the early in twentieth century. The hidden space was a Riemannian manifold and the possibilities for what it could be were governed by standard geo-metric considerations well-known to mathematicians.

But in string theory, we have a quantum version of gravity. And like the Kaluza-Klein theories, string theory too requires a hidden space in addition to the observable space-time. But the manifolds describing the hidden space are no longer purely classical objects, and therefore need not be in correspondence with conventional Riemannian geometry except in the *classical limit* – the limit in which quantum effects are negligible. To what sort of geometry do they correspond in general?

The answer is not completely known, but there is by now considerable evidence that there is an enormous geometrical structure, dubbed *stringy quantum geometry* for lack of a better word, which generalises the usual mathematical notions of geometry in an exceedingly strange way. Here I will highlight three properties of the quantum geometry associated to string theory.

3.4.1 Target Space Duality

In conventional Riemannian geometry, one has the notion of a metric which defines the distance between two points. This is a very appealing notion to a physicist, and it plays a key role in general relativity where the metric of spacetime is itself the fundamental dynamical variable. In conventional geometry, a distance can take any value from zero to infinity.

Now let us apply this to an internal spatial direction that is compactified into a circle. Classically this circle has a radius, given by the minimum distance one traverses in making a complete tour of the circle and returning to the starting point. In the limit of large radius, we call the direction *non-compact* and roughly that is what the three familiar spatial dimensions in the real world look like.

Suppose now that we consider a finite value of the radius, in some units, and then vary it so that it becomes smaller and smaller. In classical geometry this process never ends, the circle simply continues to shrink. If we place a quantum mechanical particle on this space, it will have difficulty exciting itself into a mode that can propagate on the small circle. The reason is that a well-defined wave on a tiny circle needs to be rapidly varying and therefore has a high wave number, or momentum. Therefore to probe this circle, the particle must have a high energy. This is precisely what we mean by saying that a compact internal dimension is physically unobservable in the limit of small radius.

Now replace the particle with a string. Physically it is clear that at low energies a string behaves very much like a point particle, and it will have the same difficulty entering the small internal space unless we give it a large energy. However, the string can do something else that also probes the space. It can wind itself around this circular direction. This is a classical configuration and it carries an energy proportional to the length of the direction. The energy required to excite this *stringy* mode actually decreases as the circle shrinks. Therefore on small circles, the string has a spectrum of heavy *momentum modes* and light *winding modes*.

Now suppose we replace the circle of a given radius by one of the inverse radius (in appropriate units). In this way a small-radius circle gets mapped to a large-radius one. In conventional geometry this is a major change. But a string on this space again has a spectrum of light modes (now they are the ones in which it propagates with momentum on the large circle) and a spectrum of heavy modes (when it winds over the circle).

In other words, the spectrum of a string remains invariant on replacing a tiny direction with a huge one. This is not a mere coincidence, rather the inversion of the radius has been to be an exact symmetry in string theory. It goes by the name of *target-space duality* or *T-duality*.

What is the consequence of this for mathematics? Geometry, the study of shapes of objects, clearly originated with something or someone being able to actually probe the object. That object is the macroscopic-sized human being, or perhaps the point particles out of which we are made. But with the introduction of a new object, the string, the probe is very different and the geometry seen is correspondingly different. In this new geometry, there would be a minimum length scale and the geometry would be invariant under inversion of a compact direction.

It is too early to say what are the implications of this result within mathematics. Perhaps one day, schoolchildren will be taught that a circle of a given radius is the same as a circle of the inverse radius, even though classical, pointlike objects would not be appropriate probes to demonstrate this fact. More likely it will enter the lore in a different way.

3.4.2   Mirror Symmetry

In string theory we believe that the internal space hidden from our view is at least in some approximation a 6-dimensional geometrical manifold. The number 6 arises because strings propagate consistently in 10 dimensions, while we live in only 4. This means the remaining 6 dimensions must be invisible, and are therefore assumed to be compact and small. String theory requires them to be special manifolds called *Calabi-Yau* spaces (previously discovered by the mathematicians E. Calabi and S.-T. Yau), with a definite set of properties.

In the 1980's some groups of string theorists discovered that such 6-dimensional manifolds come in pairs. Geometrically, each member of the pair has little or no resemblance to the other. However, strings propagate in precisely the same way on both members of the pair, and the interchange of two paired manifolds is an exact symmetry of string theory. In other words, in the domain of stringy quantum geometry, these manifolds would be completely indistinguishable from each other. They are called *mirror pairs* (Hori *et al.*, 2003).

Here is a regularity that was unexpected at the outset. The first thing a physicist would do is to enquire if mathematicians understand this regularity in terms of conventional geometry. And here is the remarkable surprise: they do not. Just like a small circle and a large circle, the two Calabi-Yau spaces that form a mirror pair appear to be completely different from each other, but string theory recognises them as being the same. So one may guess that there is a new branch of mathematics, *stringy quantum geometry*, within which such pairs are manifestly the same. This new branch of mathematics, which arose from string theory, is being investigated by leading mathematicians such as Maxim Kontsevich and others, who have been working to formalise and prove the mirror symmetry conjecture in several cases[4].

3.4.3 The Geometric Langlands Programme

The Langlands programme has been described as a collection of conjectures that relate results in disparate fields of mathematics, primarily number theory and the theory of group representations. Among other things it has connections to Andrew Wiles's proof of Fermat's last theorem. A variant of it called the *geometric Langlands programme* describes curves over complex numbers (Riemann surfaces) or over finite fields. About five years ago, Kapustin and Witten initiated the study of the geometric Langlands programme for Riemann surfaces via a physical theory.

In this case they use a quantum field theory called *maximally supersymmetric Yang-Mills (SYM) theory*. SYM is a close relative of a different Yang-Mills theory called Quantum Chromodynamics or QCD, that is accepted as the theory of the strong nuclear interactions. However at high energies the behaviour of this SYM differs from that of QCD. In the latter, the coupling strength decreases with increasing energy or shorter distance. This has the physical implicaton that quarks interact more weakly with each other when they are close by than when they are far apart, a fact confirmed by experiment. However the SYM of interest here, which does not directly correspond to a system observed in nature, is *scale-invariant* so its coupling strength does not change with distance.

Yang-Mills theories depend on a non-Abelian Lie group G for their definition. Excitations of the theory can be electric or magnetic by analogy with similar properties for electrodynamics, where the group G is just $U(1)$. A key property of maximal SYM is that it has exact electric-

---

[4]http://en.wikipedia.org/wiki/Homological_mirror_symmetry

magnetic duality, namely it remains unchanged if we trade the coupling constant for its inverse and exchange the lie group G with its dual group LG. The duality holds because the role and properties of the electric and magnetic excitations get interchanged in this process. Scale invariance of the theory is important in making this duality precise.

The reason Kapustin and Witten considered this theory, which was already being studied since a couple of decades, is that it provides new insights into the geometric Langlands programme. Briefly, they consider the SYM on a four-manifold that is a direct product of a Riemann surface with the Euclidean plane. They also introduce certain boundary conditions that define objects familiar in string theory called *branes*. They are then able to re-interpret the geometric Langlands programme as defining maps from branes to branes.

The key result to be proved in the geometric Langlands programme is a correspondence involving on one side flat LG bundles over the Riemann surface, and on the other side holomorphic G-bundles. In the field-theory setting, this is mapped to a relation between electric and magnetic branes. Now the two branes are interchanged by electric-magnetic duality, which also interchanges the group G with its dual LG, giving the desired result (Witten, 2005).

References

Close, Frank. (See for example) 2009. Paul Dirac: A Physicist of Few Words, *Nature*. **459**:326–327

Hori, K., Katz, S., Klemm, A., Pandharipande, R., Thomas, R., Vafa, C., Vakil, R. and Zaslow, E. 2003. *Mirror Symmetry*, Clay Mathematics Monographs. 1, Providence, USA, AMS.

http://en.wikipedia.org/wiki/Homological-mirror-symmetry.

http://www.dartmouth.edu/matc/MathDrama/reading/Wigner.html.

Mukhi, Sunil. (See for example) 2003. Dirac's conception of the magnetic monopole, and its modern avatars, *Resonance* 8, no. 8, 17–26

Mukhi, Sunil. (See for example) 1999. The Theory of Strings: A Detailed Introduction, http://theory.tifr.res.in/mukhi/Physics/string2.html.

Wigner, E. 1959. *Group Theory and its Application to the Quantum Mechanics of Atomic Spectra*, Academic Press (New York).

Witten, E. 1982. Supersymmetry and Morse theory, *J. Diff. Geom.* **17**:661.

Witten, E. 1989. Quantum field theory and the Jones polynomial, *Commun. Math. Phys.* **121**:351.

Witten, E. 1990. On The Structure of The Topological Phase of Two-Dimensional Gravity, *Nucl. Phys. B.* **340**:281.

Witten, E. Gauge theory and the geometric Langlands program, `http://insti.physics.sunysb.edu/ITP/conf/simonswork3/talks/Witten.pdf`.

Note: Revised and expanded version of an article by the author that appeared in the Special Section *Explaining the Obvious: The place of mathematics in the natural sciences*, Current Science Vol. 88 No. 3, 10 February 2005. Reproduced with permission.

Part V
Mathematics and Biology

CHAPTER 20

An Introduction to Mathematical Biology

Pranay Goel

Indian Institute of Science Education and Research (IISER)
Pune, Maharashtra 411 021, India.
e-mail: pgoel@iiserpune.ac.in

1 Prelude

Imagine a science fantasy world in the somewhat distant future. Humans have, of course, by now evolved a remarkable technology. Societies, advised by their Science Counselors, decided a long time ago to invest heavily in the fields of cybernetics, bionics and genetic engineering http://en.wikipedia.org/wiki/Biorobotics. Introductory courses in New Biology are now part of the standard curricula in schools, and with good reason: it is through this initiative that Mankind first resolved the conundrums of sustainable food and energy production, protected the environment and succeeded in securing human health http://www.nap.edu/catalog/12764.html. Indeed, for the first time in its history on the planet, Mankind has assured itself of a future for life on earth.

But alas, this Utopia is not ours to claim. Not yet anyway. We struggle with more immediate, more urgent problems: diseases that are epidemics, economies that offer little solace of justice, a planet with a climate that threatens, countries that bicker and war. Bravely, Science and Technology continues to battle on our side, indeed for our very survival.

In such circumstance it is altogether easy to crave a dream where technology has solved our problems. Dogs, for instance, have an exquisite sense of smell, nearly a 100 million times more sensitive than a human `http://en.wikipedia.org/wiki/Dog`; might we not electro-biochemically mimic a dog's nose[1] to counter terrorism (Beauchamp, 2004) for example. One almost begins to wonder: Why not? After all, why should it be so hard to invent all of this technology? For example, given that we have made such wondrous progress in electronics, it seems a short step to understanding the electrical activity in the nervous system. In the following I will argue to the contrary: That the electronics that will give us this technological Future is only just about beginning to emerge! Further, it turns out that to understand why the *technology* of Utopia is not yet within our grasp, one has also to look closely at our fundamental understanding of the *science* involved: There are a great many aspects of the biology that we do not yet understand; nor do we have very many fundamental laws on which to base theories of function[2]. To make progress in the twin directions of science and technology, mathematics[3] – together with (biological) experiments – seems to be the most powerful tool we have at our disposal.

To illustrate my arguments I will focus largely on electrical activity in neural tissues. In closing I will briefly touch upon other aspects of mathematical biology besides theoretical neuroscience; clearly even the enormous activity in theoretical and mathematical biology, especially over the last few decades, is but a tip of the iceberg for what will follow.

2 A primer of neuroscience

The nervous systems of animals `http://en.wikipedia.org/wiki/Neuron` are an intricate labyrinth of cells that "communicate" with each other

---

[1]The idea of essentially reproducing a nose electronically to achieve efficiency or safety is an easily imagined one: References to "electronic noses" (Gardner and Bartlett, 2000) can be found in Ray Bradbury's science fiction novel *Farenheit 451*, and in the movie *Predator 2* `www.technovelgy.com/ct/Science-Fiction-News.asp?NewsNum=2738`. Electronic sniffers seem to be very versatile, at least in principle: NASA, for example, is interested in a creating an e-nose to avoid poisoning and fires during space flight `http://science.nasa.gov/science-news/science-at-nasa/2004/06oct_enose/`.

[2]That is, there are few organizing principles in biology that can be used to construct theories of a general nature (Stevens, 2000). Biological systems are not derivable from a few simple, fundamental laws in the same way that classical physics understands kinematics through Newton's laws of motion (Reed, 2004).

[3]And especially, mathematical *modeling*.

through tiny electrical impulses passed between them. Nerve cells, or neurons, (i) pass signals from sensory organs, such as the skin, to the brain and spinal cord (sensory neurons), (ii) pass messages from the brain to muscles (motor neurons), or (iii) connect other neurons (interneurons), among other things. A typical neuron has three distinct parts: a cell body, or soma, that contains the nucleus, and dendrites and an axon, processes that emerge from the soma. The soma receives neural impulses from other neurons via dendrites, and axons transmit impulses processed by (or generated at) the soma to (dendrites of) other neurons. Electrical impulses travel down lengths of the slender axons in a wave-like fashion, and can reach great distances, comparable even to the length of the animal in some cases. The axons of neurons typically do not touch dendrites, but come very close at specialized junctions called synapses: the function of a synapse is to release chemicals (called neurotransmitters) when an electrical wave from the pre-synaptic neuron reaches it into the junctional cleft; these neurotransmitters are recognized by special receptors on the post-synaptic dendrite where they generate electrical activity[4]. In relaying signals thus in networks, neurons perfom tasks of astounding complexity.

2.1 The fundamental nature of biological electricity

It would seem that electricity in one system ought to be just the same as in another, be it neurons or domestic wiring. Yes, but not quite. Electrical conduction in biological tissues, especially nerves, is fundamentally different from that in, say, household wires or integrated circuits. In the common copper wire, for example, a potential difference applied across its ends generates an electric field along the length of the wire that stimulates electrons to flow along the potential gradient. These electrons typically have to negotiate the matrix of material's atoms (or molecules) and this gives rise to an electrical *resistance* to the flow of electrons: The typical rule that summarizes this behavior is therefore Ohm's Law: $V = IR$. In the case of electrical conduction in nerve axons an electrical potential difference exists as well, however this gradient exists *in the direction perpendicular to the length of the axon*. Further, currents that arise in response to this potential difference do not obey a simple Ohm's Law. Moreover, contrary to the case of the Ohmic wire where current is generated all along the length of a

---

[4]Another less common mode of transmission is via "electrical" synapses, where tiny proteins rivet together two neurons resulting in a direct electrical contact.

wire (whose ends are held at different potentials), currents in neurons are localized to short regions of the axons, and these *travel* along the length of the axon: a wave. Another question that is worth addressing here is: how is the potential difference generated in the first place? The nerve maintains an exquisite set of proteins on the cell membrane that are preferentially permeable to various ions, such as potassium (K^+) and sodium (Na^+) – called *channels* – as well as *pumps* that "separate" ions across the membrane against their natural dissipative gradients. By a concerted action of these transmembrane proteins, neurons maintain high levels of K^+ inside the cells and Na^+ outside: an electrical potential difference arises out of this separation of charge[5].

In 1952, Alan Lloyd Hodgkin and Andrew Huxley constructed a mathematical model describing the electric currents in the giant axon of the squid. In particular, they demonstrated that their equations reproduce the characteristic voltage transient observed in neurons, called an *action potential* (see below). They received the 1963 Nobel Prize in Physiology or Medicine for their work.

We can now discuss the essential difference between classical electrical circuit theory and electrical models of nerves: the "resistance" in the Hodgkin-Huxley model is not a resistor at all. For one, it is non-Ohmic. In fact, whatever it is, it is so special that it belongs in its own right to uniquely different class of circuit elements called *memristive devices*.

2.2 Memristive devices

In 1971, Leon Chua, in a remarkable insight recognized that a basic circuit element was "missing" from the list of the classical elements: the resistor, the capacitor and the inductor. On the basis of symmetry arguments and the fundamental equations of circuit theory, Chua pointed out that there ought to be a fourth element – which he called a memristor – that resembled a resistor, except in that the resistance ought to depend on history, i.e., it had a "memory". A few years later, Chua and Kang (Chua and Kang, 1976) generalized this to a class of elements they called *memristive devices*, that

---

[5]Although we do not describe this is detail, the situation is clearly a little bit more subtle than this: for example the positive charges on the two sides of the membrane surely do not "cancel each other out". A clear exposition of how this *electrochemical gradient* is maintained can be found in the textbook by Johnston and Wu (Daniel Johnston and Samuel Miao-Sin Wu, 1994).

obey the following equations:

$$I = g(w, V) \, V \tag{1}$$

$$\frac{dw}{dt} = f(w, V) \tag{2}$$

where V is the voltage difference across the element, I is the current through it, and w can be a set of state variables[6]. This element resembles a resistor in at least one respect (which also distinguishes it as a particular sort of dynamical system): that current through the element is zero if there is no potential difference across it.

It is precisely this element[7] that is found in the Hodgkin-Huxley model.

Despite the early success, for nearly forty years after Chua postulated the fourth element, no memristive device could ever be fabricated! Chua and Kang (Chua and Kang, 1976) had recognized that the mathematical equations of memristive devices could be used to model several prominent systems, including thermistors, vacuum tubes, Josephson junctions, and indeed, the Hodgkin-Huxley neuron itself: What remained experimentally elusive, however, was a standalone practical physical device capable of behaving as an isolated memristive element. In 2008, Strukov, Snider, Stewart and Williams (Strukov *et al.*, 2008) devised a layered platinum–titanium-oxide–platinum nanocell and demonstrated memristance in their system[8]. Since then several examples of memristive devices have been demonstrated.

I show next that the characteristic properties of neurons that arise from memristances in the membrane can be understood well using a geometric analysis of the dynamics of the Hodgkin-Huxley model. But first, I describe briefly the model equations that are arrived at by examining the electrical properties of the cell membrane through an equivalent circuit description; the complete set of the equations are compiled in the Appendix.

---

[6]And in the general version of the theory, g and f can also be explicitly dependent on time.

[7]Strictly speaking, a *memristor* is a particular case of memrisitve elements such that $V = R(q(t))I$ with $dq/dt = I$; I shall use the simple term *memristance* to denote the memristive devices of Equations (1) and (2).

[8]Devices with behavior similar to a memristor were invented before 2008, but the connection to Chua's theoretical element was first made explicit only in the work of (Strukov *et al.*, 2008).

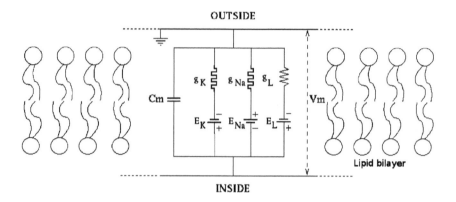

Figure 1: *The equivalent circuit model of the membrane electrical activity due to Hodgkin and Huxley: the lipid bilayer behaves as a dielectric layer with a capacitance, and the transmembrane proteins in the bilayer, including ion channels, behave as voltage sources and offer resistance as well. Note especially the memristances, g_K and g_{Na}.*

2.3   The Hodgkin-Huxley model

Hodgkin and Huxley sought to model electrical activity in the cell using the familiar circuit elements: capacitors, resistors and voltage sources. The circuit equivalent of the cell (membrane) is shown in Figure 1.

First, it is a good approximation to regard the fluidic intracellular media as well as the extracellular space as each being isopotentials; there is thus a net potential difference

$$V = V_{in} - V_{out} \qquad (3)$$

across the cell membrane. The membrane is composed of a lipid bilayer and behaves as a poor conductor, a dielectric. For this reason it acquires a capacitance, C_m, as shown in Figure 1. Further, K^+ are in greater concentration inside the cell than outside, and the opposite is true for Na^+ ions: this gives rise to the voltage sources denoted E_K and E_{Na} in the figure. The channels that allow ions to pass across the membrane also offer resistance to the current flow. The special insight that the Hodgkin-Huxley model uncovers is that these ion channels are under a special type of *voltage control*:

$$I_{ion} = g_{ion}(w)V_{ion}, \qquad (4)$$

where the conductance $g_{ion}(w)$ depends on some state variable(s)[9] w – called *gating variable(s)* – whose dynamics is dependent on voltage:

$$\frac{dw}{dt} = \frac{w_\infty(V) - w}{\tau_w(V)} \tag{5}$$

for some functions $w_\infty(V)$ and $\tau_w(V)$. The exact forms of these functions and the conductances are given in the Appendix. Equations (4) and (5) are exactly the equations of memristive devices, Equations (1) and (2). Together with each ionic current equal to product of the conductance and the difference between the membrane potential and the voltage source corresponding to that ion – called the *Nernst reversal potential*, V_{ion}^R, each takes the form

$$I_{ion} = g_{ion}(w)(V - V_{ion}^R). \tag{6}$$

Additionally, a "leak channel" accounts for a passive ion flux across the membrane. Using Kirchhoff's Law, the circuit of Figure 1 gives the equation of the membrane potential,

$$\frac{dV}{dt} = (-g_{Na}m^3h(V - E_{Na}) - g_Kn^4(V - E_K) - g_L(V - E_L) + I_{app} + I_{stim}(t))/C_m, \tag{7}$$

where I_{app} and $I_{stim}(t)$ represent terms corresponding to either current injected by the experimenter, or due to presynaptic activity.

Together with three (differential) equations for the gating variables of the type Equation (5), two of which correspond to the sodium channel, m and h, and one to the potassium channel, n, these four equations constitute the Hodgkin and Huxley model of the neuron. The complete details are as stated in the Appendix.

2.4 Excitability

Neurons have an interesting electrical property not found in Ohmic wires. If a neuron is "stimulated", i.e., a brief current pulse is injected into it, one of two possible behaviors can occur depending on the strength and duration of the stimulus: if the stimulus is weak the neurons membrane voltage responds a little and quickly returns to its original state; on the other hand if the stimulus is strong, the membrane potential results in a

---

[9]For simplicity the equations here are written with a single w, but conductance depends, in general, on several w_i.

dramatic change, the action potential. Figure 3 shows this. Further, if a tonic stimulus – a constant current – is applied, for some current values a series of action potentials can result repeatedly one after the other.

2.5 A geometric explanation of excitability

2.5.1 Model reduction

A four dimensional system is difficult to analyze directly. We will therefore use two approximations in order to reduce the number of variables in the system: these approximations will allow us to retain the essential dynamics, while reducing the system to a planar one that can be examined more effectively.

1. The first of the two approximations to treat the m dynamics as driven by its steady-state equation, i.e., we take $m = m_\infty$. Physically this corresponds to assuming that the m dynamic is much faster than any other variable; this is plausible since τ_m is typically an order of magnitude smaller than the corresponding τ's of the other gating variables.

2. The other is based on a trick introduced by Richard Fitzhugh in the 1960s. We observe that a graph of n versus h for the model very closely hugs the straight line $h = 0.9 - 1.05\,n$: Figure 2

   shows a trajectory obtained from the full system together with the approximate relationship between the two variables.

With these two approximations, the dm/dt and dh/dt equations are no longer necessary; the essential dynamics of the system is contained in V and n.

2.5.2 The phase plane

Figures 4(a) and 4(b) introduce the major geometric structures in the n versus V plane that can be used to understand the generation of an action potential due to a current stimulus of a large enough strength. Two major curves are plotted: these are the V- and n-nullclines as shown in Figure 4(a), i.e., the curves for which $\frac{dV}{dt} = 0$ and $\frac{dn}{dt} = 0$. Nullclines possess the useful property that their intersections give the points of equilibria of the system. Figure 4(a) shows the nullclines as well as the fixed point

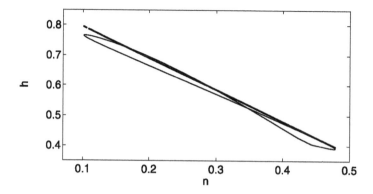

Figure 2: *The line* $h = 0.9 - 1.05\ n$ *conveniently approximates a h versus n trajectory (light curve) of the full four-dimensional system (computed with* $I_{app} = 10$).

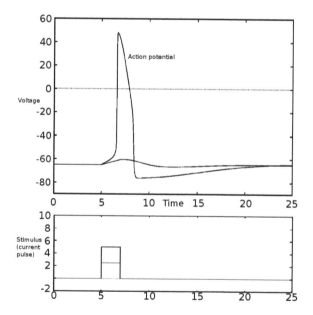

Figure 3: *An action potential. A "gentle" current stimulus causes the membrane potential to deflect but it rapidly returns to rest. A somewhat stronger pulse, say one of the same width but a larger amplitude greater than some threshold level, cause the membrane potential to rise steeply, become positive, then fall to a larger negative value than rest, before finally recovering to steady state.*

for $I_{app} = 0$, and Figure 4(b) that for $I_{app} = 10$, the amplitude of a strong stimulus pulse. It turns out that this fixed point is stable when $I_{app} = 0$: the same figure also plots several trajectories that start at different initial conditions in the plane and are seen to converge to this fixed point, i.e., if the system is started in a state away from rest, it returns to rest. The Figure 4(b) shows the phase plane at $I_{app} = 10$: notice that a significant difference is in the position of the V-nullcline; it is raised relative to that in 4(a). In this new position, it turns out that the fixed point is no longer stable; instead, however, a periodic orbit appears, and is stable as well, as evidenced by the various trajectories that lead towards it in 4(b).

In order to make sense of the action potential of Figure 3 we reason as follows: at $I_{app} = 0$ there is a rest state of the neurons. Once current is switched on to $I_{app} = 10$, the fixed point loses stability and the system approaches a large amplitude oscillation. As I_{app} is returned to 0 again, the oscillation dies out towards a rest state once again. This therefore results in the characteristic action potential of Figure 3.

2.5.3 The firing rate curve

When there is no current applied the model predicts a stationary resting state of the neuron. Figure 5(a) shows that if a constant current is injected, for certain values of I_{app} a repeated sequence of action potentials can fire continuously in the neuron[10]. A plot of the frequency of action potential firing as a function of the current is shown in Figure 6. What is perhaps somewhat surprising is that for even higher values of current (not shown) the neuron again assumes a stationary state, albeit at a more elevated voltage level.

2.6 Predictions

It is often contended that for a model to be of value, it ought to be able to explain some observation or situation beyond its original scope. As an example of this principle, we will point out several such *predictions* of this model. Figure 6 shows a frequency-response curve computed from the model, i.e., the frequency of oscillations as a function of applied (tonic) current. The model indicates that the neuron is at a steady state (in *n* as

---

[10]Note that certain neurons, such as those that are involved in involuntary rhythms such as breathing or heartbeats, can fire spontaneously even at $I_{app} = 0$.

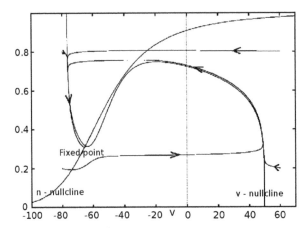

(a) The phase plane at $I_{app} = 0$: The nullclines intersect in a stable fixed point that attracts various trajectories towards it.

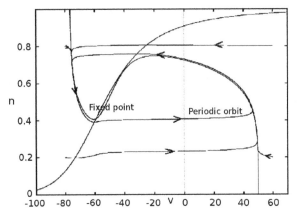

(b) The phase plane at $I_{app} = 10$: The fixed point at the intersection of the nullclines is now unstable; a periodic orbit shows up in the system instead, to which various trajectories can be seen to be attracted.

Figure 4: *As I_{app} switches between 0 (left panel) – in which state there is a stable fixed point – and 10 (right panel) – in which state there is a stable periodic orbit – briefly, the trajectory escapes into an action potential of Figure 3 before returning to rest (left panel again).*

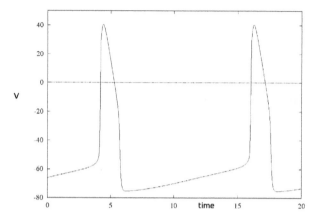

(a) Periodically repeating action potentials at a value of applied (constant) current, $I_{app} = 10$, in the reduced system.

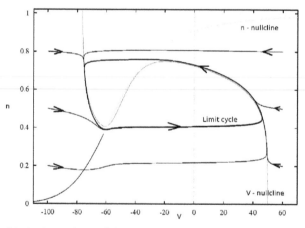

(b) A phase plane of the system for this value of the current shows several trajectories that converge towards a periodic solution (a limit cycle) in time. The fact that there is an attracting periodic orbit in the system translates into a train of action potentials that repeat incessantly, two pulses of which are shown in (a).

Figure 5: *For certain values of constant current, for example* $I_{app} = 10$, *the voltage can oscillate continuously.*

Color image of this figure appears in the color plate section at the end of the book.

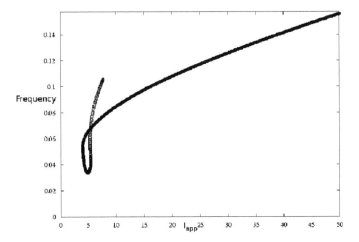

Figure 6: *The frequency response of the system – when it has oscillations – as a function of I_{app}. Note that some of the oscillations are actually unstable (open circles), and as such they are not observed in practice.*

well, even though only V is plotted in the diagram) for low values of I_{app}. The transition to oscillations is abrupt, around a critical value of I_{app} approximately 7.75. Further, the oscillations are seen to emerge with a finite frequency, i.e., *oscillations of arbitrarily small frequency are not found in this model.* Similarly a second, similarly abrupt, transition from oscillations to steady state occurs at $I_{app} = 270.3$ (not shown)[11].

We hasten to point out that this situation is but one of the many different dynamic properties characteristic of neurons. In fact, a classification of neuronal dynamics is not yet complete. Hodgkin and Huxley had originally identified two broad classes of neuronal dynamics: the above model is a Class II neuron; in a Class I neuron oscillations can have an arbitrary low frequency for example. Ermentrout and Rinzel provided a geometric description of Hodgkin and Huxley's Classes I and II (see (John Rinzel and Bard Ermentrout, 1998) for a detailed account of the ideas introduced here); many more types of neuronal dynamics are now known (Izhikevich, 2007).

---

[11]These "transitions" are examples of what are known as *Hopf bifurcations*. Indeed, Figure 6 is an example of a *bifurcation diagram*, a map of the asymptotic states of the system for a range of values of some parameter (here the current I_{app}).

2.7 The role of memristances in neural dynamics

We have seen that the memristances of sodium and potassium channels are fundamental to the generation of that unambiguouly charateristic electrical signal of nerve cells, the action potential. The presence of memristances corresponding to the voltage gating of these channels introduces essential *nonlinearities* into the system; these make the system rich in behavior but also that much more difficult to anlayze.

Memristive devices are passive, that is they cannot generate power, and they cannot store energy either (Chua and Kang, 1976). Nevertheless, the memristance of an element depends on an internal state and dynamical history: it can therefore *store information*. Circuit elements that can store information – which is analog moreover, and not just digital – without the need for the power source mark a powerful change in electronics: indeed it has been suggested that such concepts may be at the very heart of the function of the brain, learning and behavior (Ventra *et al.*, 2009) (see also (Pershin *et al.*, 2009) for a memristive model of adaptive learning in the amoeba). It is also interesting to note that memristances seem to arise naturally at nanoscales, which suggests memristance may be precisely the property ion channels have evolved to exploit.

3 Epilogue

To the author's knowledge no fabricated device has yet been announced that implements the Hodgkin-Huxley neural equations with memristances. However, with an increasing number of physical realizations of memristive devices, it is very likely that a new era of synthetic circuitry with realistic neuronal architectures is just around the corner. So far, an "in silico" Hodgkin-Huxley neuron has stood for a computational code of the model; the only other "implementation" has been the neuron itself. The next generation of conductance-based neurons in silico may very well be true blue silicon hardware: It is probable that we may soon begin to expect little off-the-shelf IC chips containing "truly" neural circuitry, possibly even miniaturized together with nanodevices (Snider, 2008). When that happens, an entirely new tool will become available to us to explore not only artificial neural networks, but also biological neural organization. The technological and engineering impact of this development will be outstanding, and it will likely lead farther still towards finally cracking the neural codes that underlie behavior and computation and cognition.

In all that follows, the mathematical insights of the past decades will be crucial to the way forward. Perhaps more exciting even, the promise of *new* mathematics, yet unexplored, awaits us.

4 Pointers and further reading

Mathematical biology has existed for many decades, but recent years have seen it flourish like never before. Indeed, it is common to hear it said that "math biology is the physics of the next millennium". I have used a classical example from mathematical neuroscience to convey but a flavor of the subject (see also (Gakkhar, 2011) for a detailed discussion of mathematical modeling in biology in general); contemporary research in math biology spans a very wide spectrum of topics: mathematical modeling of cellular processes such as growth and division and signal transduction, genomics, proteomics and bioinformatics, ecology and evolution, physiology of systems such as the cardiac and auditory and numerous others, bioengineering and biomechanics, developmental biology, network and synthetic biology, circadian clocks and synchronization in biological rhythms, just to name a few[12].

5 Acknowledgements

I thank Aurnab Ghose and Neelima Sharma for a careful reading of the manuscript and their suggestions.

6 Appendix

The Hodgkin-Huxley(Hodkin and Huxley, 1952) equations, in modern notation, are the following:

$$\frac{dV}{dt} = (-g_{Na}m^3h(V - V_{Na}) - g_Kn^4(V - V_K) - g_L(V - V_L)$$
$$+ I_0 + I_{stim}(t))/C_m$$

---

[12]Many of these topics in this list are based on the numerous Workshops held at the Mathematical Biosciences Institute, The Ohio State University, USA (see www.mbi.org). A good resource to learn more about research not only in neuroscience but also computational and mathematical neuroscience are the web pages of the Society for Neuroscience, www.sfn.org.

$$\frac{dm}{dt} = a_m(V)(1 - m) - b_m(V)m$$

$$\frac{dh}{dt} = a_h(V)(1 - h) - b_h(V)h$$

$$\frac{dn}{dt} = a_n(V)(1 - n) - b_n(V)n$$

where the gating variables obey the following functions:

$$a_m(V) = \phi\, 0.1(V + 40)/(1 - \exp(-(V + 40)/10))$$
$$b_m(V) = \phi\, 4\exp(-(V + 65)/18)$$
$$a_h(V) = \phi\, 0.07\exp(-(V + 65)/20)$$
$$b_h(V) = \phi\, 1/(1 + \exp(-(V + 35)/10))$$
$$a_n(V) = \phi\, 0.01(V + 55)/(1 - \exp(-(V + 55)/10))$$
$$b_n(V) = \phi\, 0.125\exp(-(V + 65)/80)$$

with parameters $I_0 = 10$, $V_{Na} = 50$, $V_K = -77$, $V_L = -54.4$, $g_{Na} = 120$, $g_K = 36$, $g_L = 0.3$, $C_m = 1$, $\phi = 1$, $I_p = 0$, $p_{on} = 5$, $p_{off} = 10$. Units: voltage is in mV, conductances g_{Na}, g_K and g_L are in mS/cm^2, capacitance is in μF/cm^2, current is in μA/cm^2 and time is in ms.

References

A New Biology for the 21st Century, National Academies Press, USA (2009), available online at http://www.nap.edu/catalog/12764.html.

Beauchamp, J.L.J. 2004. Countering Terrorism: The Role of Science and Technology, *Engineering & Science*. **4**:26–35.

Chua, L.O. and Kang, S.M. 1976. Memristive devices and systems, *Proc. IEEE*. **64**:209–223.

Chua, L.O. 1971. Memristor - the missing circuit element, *IEEE Trans. Circuit Theory*. **18**:507–519.

Chua, L.O. 1980. Device modeling via non-linear circuit elements, *IEEE Trans. Circuits and Systems*. **27**:1014–1044.

Di Ventra, M., Pershin, Y.V. and Chua, L.O. 2009. Circuit elements with memory: memristors, memcapacitors and meminductors, *Proc. of the IEEE*. **97(10)**:1717–1724.

Gakkhar, S. 2011. *Mathematical biology: a modelling approach*, this volume.

Gardner, J.W. and Bartlett, P.N. 2000. Electronic Noses. Principles and Applications, *Meas. Sci. Technol.* **11(7)**:1087.

Hodkin, A. and Huxley, A. 1952. A quantitative description of membrane current and its application to conduction and excitation in nerve. *J. Physiol.* **117**: 500–544.

http://en.wikipedia.org/wiki/Dog.

http://en.wikipedia.org/wiki/Neuron.

http://science.nasa.gov/science-news/science-at-nasa/2004/06oct_enose/

Izhikevich, Eugene M. 2007. *Dynamical Systems in Neuroscience: The Geometry of Excitability and Bursting* The MIT Press.

Johnston, Daniel and Samuel Miao-Sin Wu. 1994. *Foundations of Cellular Neurophysiology*, The MIT Press.

Pershin, Y.V., La Fontaine, S. and Di Ventra, M. 2009. Memristive model of amoebaâs learning, *Phys. Rev. E.* **80(2)**:021926-1–021926-6.

Reed, Michael C. 2004. Why is mathematical biology so hard?, *Notices of the AMS.* **51(3)**:338–342.

Rinzel, John and Bard Ermentrout. 1998. Analysis of neural excitability and oscillations, Chapter 7 in *Methods in neuronal modeling: from ions to networks*, eds. Christ of Koch and Idan Segev. MIT Press.

Snider, Greg S. 2008. *Cortical computing with memristive nanodevices*, SciDAC Review (Winter 2008), pp. 58–65.

Stevens, Charles F. 2000. Models are common; good theories are scarce, *Nature Neuroscience.* 1177(3).

Strukov, D.B., Snider, G.S., Stewart, D.R. and Williams, R.S. 2008. The missing memristor found, *Nature.* **453**:80–83.

CHAPTER 21

Mathematics and Biology: The Growing Synergy

Dhiraj Kumar and Samrat Chatterjee

Immunology group, International Centre for Genetic Engineering and
Biotechnology, Aruna Asaf Ali Marg, New Delhi-110067, India.
e-mail: samrat_ct@rediffmail.com

The exquisiteness of Science lies in postulating testable hypotheses.
Practically therefore, it is predictive in nature - driven by a clear cause
and affect relationship. The cause and effect relationships of events are
manifested along most of the day-to-day proceedings that we witness. Scientists can accurately predict the trajectory of a spacecraft even before it
is launched! Similarly precise cause and effect relationship allows a mechanical engineer to break open a car into its tiniest of the components,
reassemble them and make it work again. This precise nature of cause and
effect relationship however mostly fails to get manifested in biological sciences. Not convincing? Let us consider a living organism- anything from
bacteria to humans and put the question forward- can we break open the
living organism into its tiniest of components and reassemble them into a
living organism again? The answer so far remains undisputedly negative.
Clearly we do not understand the rules of the games in biological system
as precisely as we know them in other fields of Science like Mathematics,
Physics or Chemistry.

What makes biological system so difficult to understand? Especially
as we know now that most of the fundamental laws of nature including
laws of thermodynamics, mass action etc are indeed strictly followed in
Biology. Most likely the answer lies in one specific feature that stands out

in biology, the feature of self-organization, the intelligent design. None of the mechanical objects display properties of self-organization- a hallmark of living organism. It however highlights the prevalence of natural laws and cross talk among them that are yet to be precisely understood. Curiously whatever scientific and engineering marvels mankind has achieved in the past two centuries were achieved effortlessly millions of years ago by nature. The simplest life form that evolved on earth was able to utilize energy more efficiently than any of the man made contemporary advanced designs could achieve. It was possible because, as we must reemphasize, there are some fundamental laws of nature that are yet to be discovered. A clear understanding of biology, can therefore also allow us to unveil some universal laws that nature has adopted in its pursuit for intelligent design.

1 Biology: an observational science

Biology has so far stayed largely as an observational science. Except for the tremendous success in surgical medicine, which to an extent resembles engineering, our understanding of even the simplest of event in Biology is still not unequivocal and in majority of cases we are restricted to just the superficial and isolated understanding of the process. Moreover, the amount of information that is now available in biology is tremendous and integrating them together to get a comprehensive unified understanding remains major challenge to biologist. The numbers involved are mind-numbing. For example, a human genome-that is contained within each of its 10^{13} cells contains nearly 3 billions base pairs in its DNA. Out of this total DNA complement, only a small fraction of them could be characterized in terms of their functions as a result of their transcription and translation into a protein. What are the functions of the rest of the DNA complement is only vaguely understood. The unresolved question in biology lingers at every level of our understanding like- at molecular level, how functions of proteins are regulated, effect of small ions, lipids and carbohydrates and their metabolism, at cellular level how cells divide, migrate from one place to another and responds to changes in environmental conditions, at tissue and organ level how various biological functions are executed and at individual level how overall growth and homeostasis is maintained. As we go along discussing difficulties in various aspects that needs to be explored for better understanding, we realize one key aspect of biological system. That is biological systems are organized into a modular hierarchy from molecular interactions at the very base to cells to tissues to organ and organ system and then to organism at the top. The hierarchy in organization does not end here as various organisms are then organized into society, ecosystems and eventually into the biosphere. Going back to our argument of biology as observational science, over past one century tremendous knowledge has been gathered in the field of biology, almost revolutionizing the field of bio-medical sciences and medicine. However majority of the knowledge gathered stay as independent piece of information, largely because of shear complexity of various biological processes and our limitations in monitoring biological systems in totality. Implicit to these limitations are the obvious difficulty in integrating the accumulated knowledge in a fashion that not only captures all the features of the biological system but also makes it amenable to predictions that can be tested experimentally. The above mentioned facts also highlight the difficulty in discovering a comprehensive, universally applicable law in Biology at present. Therefore taking cues from the organizational principles of complex system, researchers have now started integrating biological systems at different hierarchy level separately that could then be integrated together in future to get universal models of biological systems. As we will see in this chapter how mathematics in its various forms as tools or guiding principle have become crucial in achieving these objectives. In this context, it is imperative to mention that the scope for the topic in general about role of mathematics in biology is immense, however to remain focused on the overall goal to get a feeling of this fact, we will address two crucial aspects of biology where in recent past Mathematics have acquired significant role. These are the organizational principles and the dynamics. As we shall see in this chapter, these are the two most fundamental aspects that are executed at every level of biology and are critical to overall understanding of biology.

2 Organization of Biological systems

As we witnessed in previous sections biological systems are organized in a particular fashion, and mostly represent self-organization. Self organization leads to another phenomenon called emergence, a property typically displayed by complex systems, where features of a system are significantly diverse and versatile as compared to the collective features of the components of the system individually. The property of emergence can be easily comprehended by yet another powerful aspect of self-organization -scale. What appears to be a dynamic, ever changing organizational set-up at the scale of the interacting components that comprise it, looks to be a single, functional entity from a higher scale. We consider the human body to explore this property- human body as such a complete functional entity is comprised of various organs and organ system (Figure 1). Organs alone are incapable of independent existence, let alone their ability to perform functions. Organs in turn are comprised of tissues and cells, while cells comprise myriads of molecular components intricately connected with each other. At each scale, the constituent components function without realizing their contribution to the overall system. However, put together through self-organization, they not only retain their individual features but also impart certain novel properties to the higher scale. To exemplify it, let us consider an intracellular event of signaling at molecular level and its implications in defining biological output. A eukaryotic cell, that encounters certain cue from its environment, either biochemical entity or physical stress, initiates a series of events, including biochemical modifications and molecular re-localization. A family of proteins called the MAP kinase family of protein kinases that are evolutionarily conserved are an interesting example. The MAP kinase pathway has been implicated in regulating cellular processes as diverse as growth, differentiation, movement and cell death. Owing to the diversity of processes regulated by this pathway, it has been extensively modeled in order to understand how responses at the cellular levels are determined by the dynamics at the molecular level. We now therefore have a very good understanding of MAP kinase pathway and multiple regulatory features associated with it. It is now known that various kinetic parameters and magnitude of activation of terminal MAP kinase pathway can distinguish between various environmental contexts and hence determine the output at the cellular level. The crucial thing therefore is how kinetic and magnitude parameters of MAPK are regulated. Through iterative improvisation of the MAPK models, now we know that there are various organizational principles that impart characteristic features to this pathway.

Thus presence of a negative feedback loop incorporates oscillation, while two step phosphorylation-activation process makes it a bi-stable system. Presence of phosphatases in the kinase cascade provides flexibility in

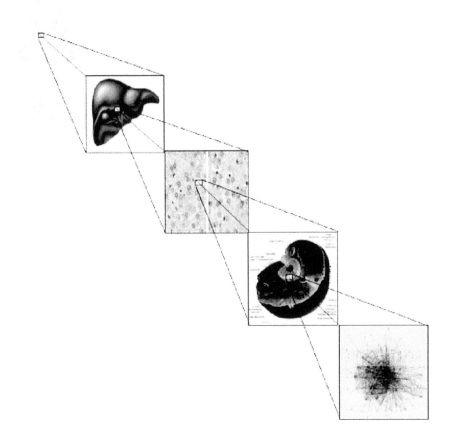

Figure 1: *Hierarchical organization of Biological system An organism is reduced to its organ, tissue, cellular and sub-cellular level. At each level of organization, complex interplay among the components regulates the dynamic property of the system. Also there are cross talks across-the-scale, for example small molecules (cytokines, hormones) affecting cells, tissues or the whole organ. At every scale bio-molecules are rendered with the responsibility of communicating among and across the scales. (Adapted from: Systems biology and medicine, Kumar, D and Rao, KVS (2009), Current trends in Science, Indian Academy of Sciences Platinum Jubilee Special).*

Color image of this figure appears in the color plate section at the end of the book.

the response and integration of multiple phosphatases along the MAP kinase axis into organized micro-structures provides wide diversity in the overall stability and activation status. A universal observation from studies on MAP kinase pathway - that also finds resonance with most similar studies on diverse biological systems is that - biological systems are organized in a modular fashion and topological features of these modules play significant role in determining overall functioning of the larger system. In fact hierarchical and modular organization is hallmark of most of the complex systems including biological networks (eg. Signaling network, Protein-protein interaction network, transcriptional regulatory network) and social networks (eg. Internet, Tweeter network, Authorship network etc). Interestingly though, biology as such does not has the requisite framework to study the modular organization, network properties and dynamics ensued thereof. Equally interesting is the fact that these properties admirably falls within the purview of Graph Theory, an established branch in mathematics and physics. Arguably then therefore studying the key organizational principles and their implications thereof on biological behavior and systems regulation becomes an exciting field of application for graph theory. Readers are suggested to consult the book Hand Book of Graphs and Networks, Edited by Bornholdt and Schuster for a specialized understanding of this topic.

In this chapter, we continue our focus on the micro-structures that are apparent in Figure 2. A closer look at the smaller graph in Figure 2 shows that they could be broken into smaller structures of three nodes, and could be re-assembled to the depicted structure. Extending this observation, even the larger network in the Figure 2 could be easily broken down into such smaller structures. These structures of two to four interacting nodes in a particular fashion are called motifs. Often motifs are considered as the building blocks of networks mainly because motifs display unique dynamics, and are dependent on the nature of interactions among the components in the motif. Consequently, motifs can be classified into various classes based on the nature of relationship among the components like Feedback loops, Feed-forward loops, bifans and scaffolds, each of them further classified based on activating or inhibiting influence on the interacting partners within the motif. The motifs are capable of exhibiting specific properties, like oscillations, delay, amplification, adaptability, robustness and sensitivity. Interestingly these motifs were identified as recurring units of structure organized in larger networks like transcriptional regulatory network.

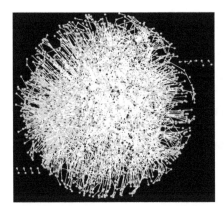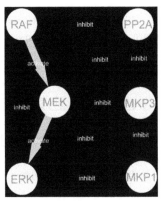

Figure 2: *Complex Biological Network and regulatory relationship among the components at modular level. The figure shows a complex human signaling network (left panel) that incorporates events in regulating cancer. A small module, the MAP kinase module (right panel) has been showed separately to highlight how individual components are organized into the network and influence functioning of their target components and the system as a whole (see text for details). The MAPK module is adapted from Chaudhri et al, JBC 2009.*

Color image of this figure appears in the color plate section at the end of the book.

As a logical extension to it, we are then bound to ask two questions- how these motifs are organized into larger network of interacting cellular components and second whether there are certain organizational principle that define the overall property of a network on the basis of how motifs are organized in the network. These questions can best be addressed through computational and mathematical modeling, where properties of each of the components at the level of motifs could be defined through analytical and/or numerical means and then integrated together to get the global understanding of the system. In the following sections, we will look at some of the mathematical tools or models that can be used to study the dynamics of micro-structures. Interestingly, if we remember from our discussions in the previous sections, the organization at different scale in the system follow similar set of rules, some of the examples discussed here are from very diverse backgrounds. Nonetheless these examples can easily be extended to any interaction/relationship structure at any scale like protein-protein interaction level, transcriptional level, metabolic level, cellular level or at organism level in ecosystems.

3 Mathematical models: Historical perspective and applicability in biology

In the last section we have seen that how mathematics has surrounded biology and in today's world biology can not be separated from mathematics. But this did not happen suddenly. The use of mathematics in biology has a long history and it takes a while for the researcher to accept the importance of mathematics in biology. Here we will discuss in brief, how in bits and pieces the mathematics is introduced in different fields. The history of biomathematics will help us to understand the fundamental theories behind the development of this field.

3.1 In ecology

Ecology is one of the prominent biological field where the mathematics is applied. The first definitive mathematical treatment of population dynamics was Thomas Malthus' (1798) *Essay on the Principle of population*, where he proposed the following growth model

$$\frac{dN}{dt} = rN, \tag{1}$$

where $N(t)$ denotes a population at any time t and r is the growth rate of the population which is the difference between the birth rate and the death rate.

If we observe the solution of the system (1), it grows exponentially. This is unrealistic since no population can grow indefinitely with time. To overcome the unrealistic unbounded exponential growth shown by the model (1), forty years later, Verhulst (1838) proposed a new model, known as logistic growth model, where the intraspecific competition within the population due to limited available resources, was considered. He proposed the following differential equation,

$$\frac{dN}{dt} = rN\left(1 - \frac{N}{K}\right), \tag{2}$$

where K denotes the carrying capacity of the environment.

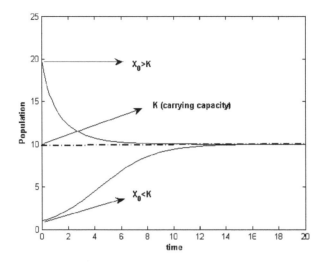

Figure 3: *Logistic growth curve. Figure showing the dynamic with two initial points (N_0), one greater than the carrying capacity K and the other one less than K.*

Putting $\frac{dN}{dt} = 0$ and solving the right hand side of the equation (2) as an algebraic equation we get $N = 0$, K. Thus $N = 0$ and $N = K$ are the two equilibrium points (i.e., the points where $\frac{dN}{dt} = 0$) for the equation (2). Among these two equilibrium points, $X = K$ is the stable point, i.e., solution starting with any initial point will ultimately goes to the point K, see Figure 3.

The logistic growth model can be easily extended to other biological systems like growth pattern of bacteria, yeast or cells of multicellular organism.

The above models are simple but till date they are central theoretical construct for single-species population dynamics.

The use of mathematical tools in multi-species ecological problem begins with the famous Lotka-Volterra equations, also known as the predator-prey equations proposed independently by Alfred J. Lotka in 1925 and Vito Volterra in 1926. This is a classical model which uses the equations from the population dynamics of the lynx and the snowshoe hare, and is popularized due to the extensive data collected on the relative populations of the two species by the Hudson Bay company during the 19th century. The

Lotka-Volterra equation is given by

$$\frac{dX}{dt} = \alpha X - \beta XY, \equiv F_1(X, Y), \tag{3}$$

$$\frac{dY}{dt} = -\delta Y + \gamma XY, \equiv F_2(X, Y),$$

where $X(t)$ and $Y(t)$ denote prey and predator densities, respectively, as functions of time. Furthermore, all constants are assumed to be positive with $X(0) \geq 0$ and $Y(0) \geq 0$.

Here the non zero equilibrium point (also known as interior equilibrium point) is given by $E^* = \left(\frac{\delta}{\gamma}, \frac{\alpha}{\beta}\right)$. To observe the stability property of E^* we use the most common method of Jacobian matrix. The Jacobian matrix for the system (3) around E^* is given by

$$J^* = \left(\begin{array}{cc} \frac{\partial F_1}{\partial X} & \frac{\partial F_1}{\partial Y} \\ \\ \frac{\partial F_2}{\partial X} & \frac{\partial F_2}{\partial Y} \end{array} \right)_{\left(\frac{\delta}{\gamma}, \frac{\alpha}{\beta}\right)} \tag{4}$$

The eigenvalues corresponding to the matrix J^* (given in (4)) is purely imaginary and so E^* is not a stable point rather the solutions form a close trajectories around E^*, see Figure 4.

After the pioneering work of Lotka and Volterra, lots of work have been done in this field taking into account all possible realistic features into the models. Todays' model not only contains ordinary differential equations (ODEs), but also other form of differential equations like delay differential equations (DDEs), stochastic differential equations (SDEs) and discrete difference equation. We will discuss them later in this chapter.

3.2 In epidemiology

Another important field of study that use lots of mathematical tools is epidemiology. Hamer (1906) and Ross (1908) were the first to formulate specific theories about the transmission of infectious disease in simple but precise mathematical statements and to investigate the properties of the resulting models. The ideas of Hamer and Ross were extended and explored in more detail by Soper (1929) who deduced the underlying mechanisms responsible for the often observed periodicity of epidemics, and by Kermack and McKendrick (1927) who established the threshold theory. This theory, according to which the introduction of a few infectious individuals

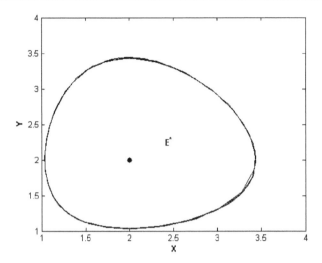

Figure 4: *Phase plane diagram. The figure showing phase plane for the system (3), where the solutions form a close trajectories around E^*.*

Color image of this figure appears in the color plate section at the end of the book.

into a community of susceptible will not give rise to an epidemic out break unless the density or number of susceptible is above a certain critical value. It is, in conjunction of the mass action principle, a corner stone of modern Epidemiology.

Here we will discuss the famous Kermack and McKendrick (1927) model, which was proposed in the same time when Lotka-Volterra model was proposed for the ecology. It was proposed to explain the rapid rise and fall in the number of infected patients observed in epidemics such as the plague (London 1665-1666, Bombay 1906) and cholera (London 1865). It assumes that the population size is fixed (i.e., no births, deaths due to disease, or deaths by natural causes), incubation period of the infectious agent is instantaneous, and duration of infectivity is same as length of the disease. It also assumes a completely homogeneous population with no age, spatial, or social structure. With these assumptions they wrote the following set of differential equations,

$$\frac{dS}{dt} = -\beta S I,$$
$$\frac{dI}{dt} = \beta S I - \gamma I, \tag{5}$$

$$\frac{dR}{dt} = \gamma I,$$

where $S(t)$ is the number of susceptible people, $I(t)$ is the number of people infected, and $R(t)$ is the number of people who have recovered and developed immunity to the infection, β is the infection rate, and γ is the recovery rate.

The key value governing the time evolution of the equation (5) is the so-called epidemiological threshold,

$$R_0 = \frac{\beta S}{\gamma}. \tag{6}$$

R_0 (given in (6)) is defined as the number of secondary infections caused by a single primary infection; in other words, it determines the number of people infected by contact with a single infected person before his death or recovery. When $R_0 < 1$, each person who contracts the disease will infect fewer than one person before dying or recovering, so the outbreak will peter out ($\frac{dI}{dt} < 0$). When $R_0 > 1$, each person who gets the disease will infect more than one person, so the epidemic will spread ($\frac{dI}{dt} > 0$).

4   Some other kinds of differential equations

Researchers mainly try to do their work based on the simple deterministic models made up of system of non-linear ordinary differential equations, but it is becoming clear that the simplest models cannot capture the rich variety of dynamics observed in natural systems. There are many possible approaches to deal with these complexities. One approach which is gaining prominence is the inclusion of time delay terms in the differential equations. The system of differential equations which incorporate the time delay factor are called delay differential equations (DDEs). The delays or lags can represent gestation times, incubation periods, transport delays, or can simply lump complicated biological processes together, accounting only for the time required for these processes to occur.

Other than time delay another approach to deal with such complexities is to include the effect of environmental fluctuation in the system. This is done either by taking parameters involved in the system as random parameters or by adding the randomly fluctuating driving force directly to the deterministic system and the newly form system of differential equations are called stochastic differential equation (SDEs). A stochastic model provides a more realistic picture of a natural system than its deterministic counterpart. Deterministic model will prove ecologically useful only if the dynamical patterns they reveal are still in evidence when stochastic effects are introduced.

Another kind of equation use to define a discrete system is known as difference equation. Since most of the biological process are time dependent and so the observed data is not continues rather discrete in nature. To define those process this kind of equations are used.

In the following subsections we gave examples showing different kinds of models. The mathematical tools use for the analysis is beyond the scope of this chapter and hence we are not giving the analysis of these models.

4.1 Delay differential equation (DDEs)

In biological models it is often more appropriate to allow the rates of change of the variables to depend on the previous history. An interesting example of such phenomenon is gene expression, translation and decay. This is always a time lag (delay) between transcription of genes (from DNA to RNA) and translation of RNA into proteins. a model that simultaneously includes dynamics of transcription, translation and post translational modifications (like phosphorylation, methylation and acetylation of proteins) can benefit significantly from the Delay Differential Equations. Thus prediction requires the specification not merely of initial values but of values over an extended time. Due to this we need time delays in simple dynamical models. Moreover, statistical analysis of ecological data has shown that there is evidence of delay effects in the population dynamics of many species. The models with time delay have the advantage of combining a simple, intuitive derivation with a wide variety of possible behaviour regimes for a single system The main aim of delay differential equation is to asses the qualitative or quantitative differences that arise due to the inclusion of the time delays.

Dynamical systems with delays were first studied in various disciplines during the years 1920-40. Systemic work with mathematical models in medicine and biology began with the epidemiological studies of Ross (1911). After some year Volterra (1926) also proposed a model incorporating a delay in the response of a population's death rate to changes in population density cause by an accumulation of pollutants in the past.

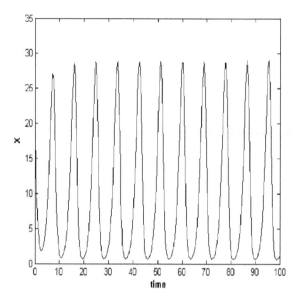

Figure 5: *Delay differential equation and periodicity. Figure showing nonconstant periodic solution oscillation in the system (7) for $\tau > \frac{\pi}{2r}$.*

Here we are presenting a very simple model to show how the time lags are included in the differential equation. The model is known as delay logistic equation with discrete delay (also sometimes called Wright's equation) and looks like,

$$\frac{dX(t)}{dt} = rX(t)\left(1 - \frac{X(t-\tau)}{K}\right), \tag{7}$$

where τ denotes the time taken by the population to modifies its environment. Here $X(0) > 0$. One could calculate that in equation (7) if $\tau < \frac{37}{24r}$ then $X(t) \rightarrow K$ as $t \rightarrow \infty$ and if $\tau > \frac{\pi}{2r}$ then system (7) has nonconstant periodic solution oscillating with respect to K, see Figure 5.

Delay models are now becoming more common and appearing in many branches of biological modelling, like primary infection, drug therapy, immune response, chemostat models, circadian rhythms, epidemiology, the respiratory system, tumor growth and neural networks.

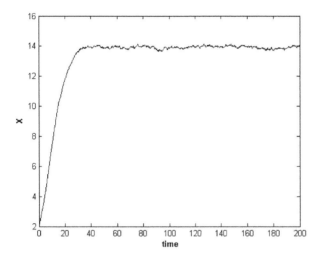

Figure 6: *Fluctuation in stochastic differential equations. Figure depicting the system (8) for small value of σ, where the outcome is not constant rather fluctuating and is away from the stable equilibrium point k = 10 (for the system (2)). Here r = 0.2 and σ = .05*

4.2   Stochastic differential equation (SDEs)

In real life it is practically impossible to explain an entire biological or ecological process through a system of deterministic differential equations as uncertainty is inevitable in such systems. To understand those uncertainty we need Stochastic differential equation (SDEs). The basic difference between the stability of a deterministic model and a stochastic model is that, in a deterministic model we seek for the constant equilibrium population and then investigate their stability. If the population approaches towards that constant equilibrium point as time increases, then the model system is said to be deterministically stable or simply stable. But in stochastic model the equilibrium population distribution fluctuates randomly around some average values and so in this situation we generally seek for a "probabilistic cloud", described by the equilibrium probability distribution. The model systems with this type of compact cloud of population distribution is called a stochastically stable system.

To illustrate this we present here a small example. Let us consider a simple SDE which is formed by adding randomness in the logistic equation (2),

$$dX = rX\left(1 - \frac{X}{K}\right)dt + \sigma(X - K)d\xi_t, \tag{8}$$

where σ is real constant and known as the intensity of fluctuations, ξ_t is the standard Wiener processes, independent of each other.

We consider (8) as an Itó stochastic differential system of type

$$dX_t = \chi(t, X_t)dt + g(t, X_t)d\xi_t, \quad X_{t0} = X_0, \tag{9}$$

where the solution $\{X_t, t > 0\}$ is a Itó process, χ is the slowly varying continuous component or drift coefficient and g is the rapidly varying continuous random component or diffusion coefficient. (A simulation result of equation (8) with a particular parameter value is given in Figure 6, showing the stability of the system under stochastic perturbations.)

4.3 Discrete difference equation

A difference equation also sometimes called as recurrence relation is an equation that recursively defines a sequence: each term of the sequence is defined as a function of the preceding terms.

An example of a recurrence relation is the logistic map,

$$X_{n+1} = rX_n(1 - X_n) \tag{10}$$

We have seen that in continuous case, logistic equation (also called as logistic flow) has two equilibrium point and one of them is stable in nature. There is no oscillation occurs for logistic flow. Surprisingly in case of logistic map, not only oscillation occurs but it leads to 'chaos', see the bifurcation diagram (Figure 7) showing the route to chaos for increase in r.

Here we shall take the opportunity for a very small discussion on chaos.

4.3.1 Chaos

Chaos theory is a field in mathematics where the behavior of dynamical systems that are highly sensitive to initial conditions are studied. This sensitivity is popularly referred to as the butterfly effect. Small differences in initial conditions (such as those due to rounding errors in numerical computation) yield widely diverging outcomes for chaotic systems, rendering long-term prediction impossible in general. This happens even though

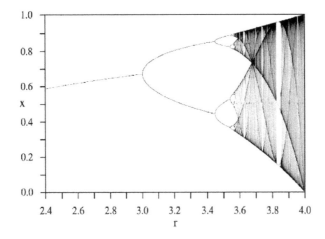

Figure 7: *Logistic map and Bifurcation diagram. Figure showing bifurcation for the system (10) with r as the bifurcation diagram. The diagram showing the route to chaos starting from stability to periodic oscillation to multi-periodic oscillation and finally to chaos.*

these systems are deterministic, meaning that their future behaviour is fully determined by their initial conditions, with no random elements involved. In other words, the deterministic nature of these systems does not make them predictable. One typical example of such system is the logistic map (discussed above). It could be seen from the Figure 7 that when *r* crosses 3.6, the system no more shows any predictive outcome rather the overlapping between the outcomes is so much that one cannot distinguish between two points.

Explanation of such behavior may be sought through analysis of a chaotic mathematical model with the help of bifurcation diagram (as shown above) and other analytical techniques such as recurrence plots, Poincaré maps and Lyapunov exponents.

5 Future scope

An interesting aspect that we noticed through out this chapter was challenges being posed by the biological systems and the ability of mathematics to provide solutions for them. It is always amusing to see how mathematics, which is also a product of human thought, can so admirably be close to reality. The power of Mathematics lies in its ability to define/explain almost every natural process. Biology and biological processes we understand so far are in isolated pieces and yet to be integrated to get a universal understanding. The integration of enormous amount of data along with defining the rules that govern various processes is almost impossible to achieve without securing substantial support from mathematics. We have seen examples of signaling motifs and their integration through mathematical and computational models. Similar studies at modular level needs to be performed at transcriptional, metabolic and other biochemical level. We have also seen how different modelling of the same system can yield very different results, like the logistic flow (continues) and logistic map (discreet). Even with the limited examples that we discussed here, two things are clear- a universal law that governs the organization of biological systems has to be formulated. Here it is important to highlight that self-organization remains one of the most active field of study for biologists, physicists and mathematicians across the globe. Secondly, we need to understand how dynamical relationships are determined and executed at different organizational scales. All dynamical models discussed here were fundamental in nature however can be specifically used for specific problems. Further, due to the large data generated today and the uncertainty and error involve in those data, the two most prominent field which needs special attention from bio-mathematician are- the stochastic differential equations (SDEs) and the chaos theory. Therefore evidently the scope for mathematics in biology is unlimited and the synergy between the two fields will undoubtedly grow rapidly with time.

6 Further reading

1. Bornholdt, S.,Heinz Georg Schuster, H. S.,(Eds.) Handbook of Graphs and Networks: From the Genome to the Internet. *Wiley-Vch*, (2003).

2. Chaudhri V. K., Kumar, D., Misra, M., Dua, R., Rao, K, V., Integration of a phosphatase cascade with the Mitogen Activated Protein Kinase pathway provides for a novel signal processing function. *J Biol Chem.*, (2009).

3. Kot, M., Elements of Mathematical Ecology. *Cambridge Univ. press*, (2001).

4. Kuang, Y., Delay Differential Equations with Application in Population Dynamics. *Academic Press. New York*, (1993).

5. Lakshmanan, M., Rajasekar, S., Nonlinear dynamics- integrability, chaos and patterns. *Springer-Verlag*, (2003).

6. Murray, J. D., Mathematical Biology: I. An Introduction (Interdisciplinary Applied Mathematics). *Springer- third edition*, (2001).

CHAPTER 22

Mathematical Biology: A Modelling Approach

Sunita Gakkhar

Department of Mathematics,
Indian Institute of Technology
Roorkee 247667, India.
e-mail: sungkfma@gmail.com

1 Mathematical biology

Leonardo da Vinci once suggested, "No human enquiry can be a science unless it pursues its path through mathematical exposition and demonstration". Mathematics provides inroads to almost any scientific discipline. Biology is not an exception. Mathematical biology or theoretical biology integrates mathematics with biology. It is an interdisciplinary research area with applicability in biology, medicine and biotechnology. In Mathematical Biology or biomathematics the stress is on mathematics. Mathematical concepts are applied to biological problems. On the other hand, in Theoretical Biology, more emphasis is on biological aspects of the problem. This interdisciplinary field needs practitioners that can speak each other's language. Biology departments at research universities and medical schools routinely carry out interdisciplinary projects that involve mathematicians, physicists, statisticians, and computational scientists. And mathematics departments frequently engage professors whose main expertise is in the analysis of biological problems. Mathematical biology is a fast growing area to become the science of the 21st century.

Applying mathematics to biological problems is not a recent development. The power of mathematics to synthesize models and analyze their solutions has important consequences. For instance, the Kermack-McKendrick threshold theorem identifies a dimensionless parameter, Basic Reproduction Number to determine when a population is at risk for propagation of an epidemic disease. Biology has enormous influence on development of mathematics. Brownian motion discovered by a botanist is central to probability theory. Early developments in probability theory motivated by studies of Mendel's and Darwin's theories of genetics and evolution led to development of statistical methods that now form a major part of design and interpretation of most experiments. The interaction of biology and mathematics has enriched both the disciplines. Irregular solutions of nonlinear difference equations arising in ecology models were eventually explained as chaos. One of the algorithms to solve complex problems comes from artificial intelligence. In evolutionary computation, a number of random solutions to a problem are created and subsequently allowed to "breed," thereby creating more solutions. The iterative process involving Darwinian "survival of the fittest" and "mutation" eventually converges to the solutions. After all, if it works in nature, why not here?

Mathematics is extensively employed in the following major biological research areas such as Cell organization, Ecology and ecosystems, Evolution and diversity, Genome organization and expression, Immune system, orgsanism function and disease, Molecular biology, Neurobiology, Plant biology and agriculture, Industrial biotechnology. Mathematical modelling is in the core of applications of mathematics in these areas. It provides a platform to translate a real life problem into the language of mathematics so that mathematical analysis and tools can be applied to provide its solution. Mathematical models make it possible to bring high level computer languages and digital computers to bear on biology problems, and shape the foundations of the emerging area of computational biology.

2 Mathematical Modeling

Models are abstractions / representations of the real world objects, situations, etc. Abstraction means generalization i.e. taking the most important components of real system and ignoring less important components. Importance is evaluated by the relative effect of system components on its

Real system   **Model**

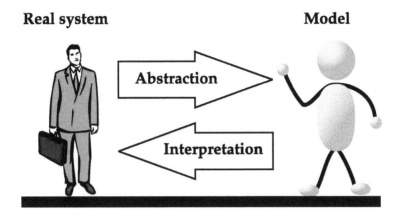

Figure 1: *Real system Vs model*

Color image of this figure appears in the color plate section at the end of the book.

dynamics. The relationship between the reality and a model is depicted in the following diagram.

Typically a model will refer only to some aspects of the phenomenon under consideration. The two models representing the same phenomenon may be different due to several reasons such as:

- Differing requirements of the model's end users

- Conceptual or esthetic differences amongst the modelers

- Decisions made during the modeling process.

For this reason, users of a model need to understand the model's original purpose and the validity of assumptions.

Mathematical Models are idealized abstractions that represent situations, activities, processes, etc. by means of mathematical techniques. It may not be possible to translate a real problem into a mathematical problem in its generality. The real world system is to be first idealized or simplified and then represented using mathematical notations. Sometimes the idealization or simplifications are so drastic that the real problem is approximated by another problem and then mathematical representation may be given.

A mathematical model usually describes a system by a set of variables and a set of equations that establish relationships between the variables.

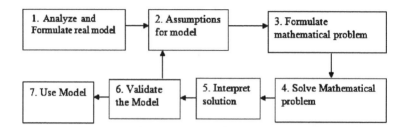

Figure 2: *Stages in mathematical modeling*

The values of the variables can be practically anything; real, integer, boolean values or strings. These can be grouped as independent variables, input variables, state variables, random variables, output variables and parameters. The state variables are dependent on some variables and parameters of the system as well as on input variables. Furthermore, the output variables are dependent on the state of the system represented by the state variables. The model is the set of functions that describe the relations between the different variables.

The mathematical model is used to describe the behavior of the system. It can provide qualitative and quantitative predictions about future events. We use models to gain insights into the mechanisms underlying observed phenomena, predict patterns of behaviour, design effective field experiments and evaluate strategies for management.

Mathematical Modeling refers to the process of generating or developing a model as a conceptual representation of some phenomenon.

3   Main Stages in the Modelling Process

The main stages in the modeling process are illustrated in the Figure 2

1. Analyze and Formulate real model: Analyze the problem in the problem domain itself. The real world problem may be very complex in nature. The first step is to identify the real world problem to be investigated. Understand its essential characteristics relevant to the situation. At the same time, find the irrelevant aspects of the problem, which can be ignored. Identify the physical, chemical, social or biological laws to be relevant in the situation. Data may be collected to get some insight into the problem.

2. Assumptions for model: Assumptions for model are made to idealize or simplify the real life problem.

3. Formulate mathematical problem: This means to identify the variables and then to postulate the mathematical relationship between them. The relationship is in terms of algebraic, transcendental, differential, difference, integro-differential equations or inequations. In some situations where the correct mathematical function may be difficult or uncertain tabular models can be used. The usual tables are one or two-dimensional. Tabular models can also incorporate fuzziness.

4. Solve Mathematical problem: The next step is to solve the equations involved in the model. Use of analytical methods is preferred as far as possible. Whenever analytical solutions are not possible, numerical methods or computer-oriented methods may be used. Many times simulations may be carried out to get some insight into the solution of the problem. Even the analysis of the model without solving it may also give rise to some interesting behavior of the system. Since the real life problems are more complex and challenging, existing tools for solving the problem may not be sufficient. New tools may need to be developed for solving the more challenging problems. If some changes in the initial assumptions will lead to further simplifications in the mathematical model that helps in finding the analytical solution of the problem, then the model is further simplified and then solved.

5. Interpret solution: The mathematical solutions so obtained are translated back into the problem domain. Interpretation means that model components (parameters, variables) and model behavior can be related to components, characteristics, and behavior of real systems. If model parameters have no interpretation, then they cannot be measured in real systems.

6. Validate the Model: The comparison is made between the predicted behavior of the mathematical model and the observed behavior of the real system. If the two are in good agreement then the model is validated. If the correlation is not adequate then one must return to the assumptions made in the modeling process. The cycle is

again iterated till the results are satisfactory. An important part of the modeling process is the evaluation of an acquired model. Usually the engineer has a set of measurements from the system, which are used in creating the model. Then, if the model is well developed then the model will adequately show the relations between system variables and the measurements at hand. Does the model describe well the properties of the system between the measured data (interpolation)? Does the model describe well outside the measured data (extrapolation)? A common approach is to split the measured data into two parts: training data and verification data. The training data is used to train the model, that is, to estimate the model parameters. The verification data is used to evaluate model performance. Assuming that the training data and verification data are not the same, we can assume that if the model describes the verification data well, then the model describes the real system well.

7. Use Model: The validated model can be used to meet the objectives.

4 Examples

4.1 Single species population model

The problem of estimating population variation with time is a matter of interest in many contexts. One may be interested to answer the following questions:

- How long will it take for a population to grow to a specific size?

- What will be the population size after n years (or generations)?

- How long the population can survive at non-favorable conditions?

Step 1: Let us identify the phenomena responsible for changes in a biological population. Self-reproduction is the main feature of all living organisms. This is what distinguishes them from non-living things. Any model of population dynamics must include reproduction. At the same time mortality of living organisms is equally important. We first collect some data about the birth and death processes of the population. For example how many births are taking place per unit time, what is the mortality rate of the population? What are the factors that affect these processes?

Step 2: The following simplifying assumptions can be made about the population:

- Continuous reproduction (e.g., no seasonality)

- All organisms are identical (e.g., no age structure)

- Environment is constant in space and time (e.g., resources are un-limited)

- Population is not migrating.

- Number of new births/deaths depends on the size of population. Bigger is the size of population more births/deaths will take place.

The following conservation law is assumed:
Change in population = births - deaths
Step 3: Let the size of population at any time t is $N(t)$. Let us assume that in a small interval of time δt both births and deaths are proportional to the population size, i.e.
Births = $bN(t)$ and Deaths = $dN(t)$ (b and d are positive constants)
Therefore $\delta N = [bN(t) - dN(t)]\delta t$
In the limiting case as δt tends to 0

$$\frac{dN}{dt} = rN; \quad r = b - d \tag{1}$$

The birth rate b, is the number of offspring / organisms produced per one existing individual in the population per unit time. Death rate d is the probability of dying per one individual. The parameter r in the model can be interpreted as a difference between the birth (reproduction) rate and the death rate. "Instantaneous rate of natural increase" and "Population growth rate" are generic terms used for r as they do not imply any relationship to population density. The parameter r is also called Malthusian parameter.

The mathematical model for population growth under the above assumptions is given by (1). This is a first order differential equations which can be easily solved together with the initial condition $N(0) = N_0$. This model is called exponential model. The behavior of the population depends upon the sign of the parameter r.

$r > 0$ population is increasing
$r = 0$ population is constant

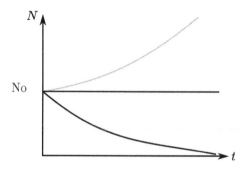

Figure 3: *Exponential growth model*

$r < 0$ population is decreasing. See Figure 3.

Step 4: The solution of Initial Value Problem is obtained as

$$N = N_0 exp(rt) \tag{2}$$

N_0 is the constant of integration.

Step 5: The population is growing exponentially for positive r. This is why the model is also known as exponential growth model. It tends to infinity as t tends to infinity. That is, population grows without any bounds.

Step 6: Let us verify the model for the data given in table 1.

Consider $t = (T - 1789)/10$ then the rate r is computed as $log(5.3/3.9)$. Figure 4 shows the population vs time plots. The upper curve is the solution curve (2) as obtained from the model. The lower curve is for the population data given in Table 1.

One may observe that the data is in agreement for some time initially, but later on the deviation becomes too large for model to be acceptable for large values of time. However, exponential model is robust; it gives reasonable precision even if these conditions do not met. Organisms may differ in their age, survival, and mortality. But the population consists of a large number of organisms, and thus their birth and death rates are averaged.

Therefore, if the model is to be applied for large t, it is to be modified. We must look back at the assumptions and identify the cause of the large deviation for large t. From this we will go back to step 2 and reiterate the process till satisfactory results are obtained.

Step 2. The assumption 3 is at fault, as the limited availability of resources will restrain the population to grow unboundedly. To accommodate this,

Table 1:

Time t	T Year	Observed size of population ($\times 10^6$)	Model 1 Prediction of population size ($\times 10^6$)
0	1789	3.9	3.9
1	1799	5.3	5.3
2	1809	7.2	7.202564
3	1819	9.6	9.7881
4	1829	12.9	13.30178
5	1839	17.1	18.07677
6	1849	23.2	24.56587
7	1859	31.4	33.38439
8	1869	38.6	45.36853
9	1879	50.2	61.65467
10	1889	62.9	83.78711
11	1899	76	113.8645
12	1909	92	154.739
13	1919	106.5	210.2863
14	1929	123.2	285.7737

the growth rate r should depend on population size.

Step 3: As population size increases the growth rate should decline. The growth rate is assumed to be linearly decreasing and reaches 0 when $N = K$:

$$r = r_0 \left(1 - \frac{N}{K}\right) \tag{3}$$

The term "Intrinsic rate of increase" is used for the parameter r in (3). It is equal to the population growth rate at very low density at which no environmental resistance is experienced. Parameter r_0 is the maximum possible rate of population growth, which is the net effect of reproduction and mortality excluding density-dependent mortality. The parameter K is the upper limit of population growth and it is called carrying capacity. It is interpreted as the amount of resources expressed in the number of organisms that can be supported by these resources. If population size

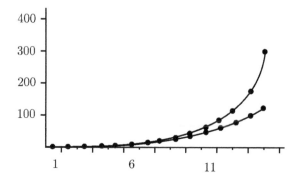

Figure 4: *Comparison with exponential growth model*

Color image of this figure appears in the color plate section at the end of the book.

exceeds K, then population growth rate becomes negative and population size declines.

The dynamics of the population is now described by the differential equation:

$$\frac{dN}{dt} = rN = r_0 N \left(1 - \frac{N}{K}\right) \tag{4}$$

Logistic model (4) combines two ecological processes: reproduction and competition. Both processes depend on population numbers (or density). The rate of both processes corresponds to the mass-action law with coefficients: r_0 for reproduction and r_0/K for competition.

Step 4. The differential equation has the following solution:

$$N_t = \frac{N_0 K}{N_0 + (K - N_0)exp(-r_0 t)} \tag{5}$$

Three possible model outcomes are shown in Figure 5:

1. Population increases and reaches a plateau ($N_0 < K$). This is the logistic curve given by equation (5).

2. Population decreases and reaches a plateau ($N_0 > K$)

3. Population does not change ($N_0 = K$ or $N_0 = 0$)

Logistic model has two equilibrium points: $N = 0$ and $N = K$.

The first equilibrium is unstable because any small deviation from this equilibrium will lead to population growth. The second equilibrium is

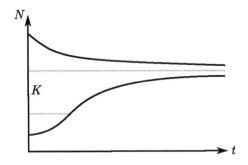

Figure 5: *Logistic growth model*

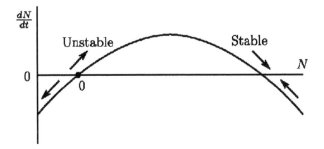

Figure 6: *Stability of logistic model*

stable because after small disturbance the population returns to this equilibrium state.

Step 5. As $t \to \infty$, $N \to K$ therefore the growth is no longer unlimited.

An equilibrium is considered stable (for simplicity we will consider asymptotic stability only) if the system always returns to it after small disturbances. If the system moves away from the equilibrium after small disturbances, then the equilibrium is unstable.

In Figure 6, population growth rate, dN/dt, versus population density, N, is plotted. This is often called a phase-plot of population dynamics. If $0 < N < K$, then $dN/dt > 0$ and thus, population grows and the point in the graph moves to the right. If $N < 0$ or $N > K$ (of course, $N < 0$ has no biological sense), then population declines and the point in the graph moves to the left. The arrows show that the equilibrium $N = 0$ is unstable, whereas the equilibrium N=K is stable. From the biological point of view, this means that after small deviation of population numbers from $N = 0$, the population never returns back to this equilibrium. Instead, population numbers increase until they reach the stable equilibrium $N = K$. After any deviation from N=K the population returns back to this stable equilibrium.

The difference between stable and unstable equilibrium point is in the slope of the line on the phase plot near the equilibrium point. A negative slope characterizes stable equilibrium points whereas a positive slope characterizes unstable equilibrium points.

4.2 Modelling of Two Interacting Species

A general dynamical system for the evolution of two interacting species is given as:

$$\frac{dX}{dt} = g_1(X) + f_1(X, Y)Y$$
$$\frac{dY}{dt} = g_2(Y) + f_2(X, Y)X \tag{6}$$

where g_i, known as intrinsic growth function, represents the growth of biological species in the absence of other species. It incorporates birth, death, immigration and emigration. The following two forms for the growth function are used extensively when no immigration or emigration is taking place:

$$g(X) = X \qquad \text{Exponential growth}$$
$$g(X) = X\left(1 - \frac{X}{K}\right) \qquad \text{Logistic growth} \tag{7}$$

The functions $f_i(X, Y)$, $i = 1, 2$ capture the interactions of two species. The interaction between biological species is classified into the following three categories:

- Prey Predator

- Competition

- Mutualism

A Simple Model for predator-prey interactions is given by the following system of equations

$$\frac{dx}{dt} = a(x - by), \quad \frac{dy}{dt} = -y(c - dx) \tag{8}$$

These are known as Lotka-Volterra equations. Here, species x is they prey and y is the predator.

The critical point of interest is $(x_c, y_c) = (c/d, a/b)$ obtained by setting $\dot{x} = \dot{y} = 0$ in (8). We define new variables X and Y as

$$X = x - \frac{c}{d}, \quad Y = y - \frac{a}{b}$$

Substituting into the predator-prey equations and linearizing by neglecting all degree 2 terms, we obtain

$$\dot{X} = \frac{c}{d}(-bY), \quad \dot{Y} = -\frac{a}{b}(-dX)$$

or

$$\ddot{X} = -\frac{b}{d}c\dot{Y} = \left(-\frac{b}{d}c\right)\left(\frac{d}{c}a\right)X = -acX$$

Its solution is given as

$$X = A \cos \sqrt{ac}t + B \sin \sqrt{ac}t$$

By substitution we find that

$$Y = -\frac{d}{bc}\dot{X} = \frac{d}{bc} \sqrt{ac}\left[A \sin \sqrt{ac}t - B \cos \sqrt{ac}t\right]$$

We can easily see that the solutions oscillate about the critical point $(c/d, a/b)$ and the fundamental natural frequency of the system is $\sqrt{ac}$. The solutions for X and Y may not at first glance appear to be ellipses. However, by introducing some phase angle δ such that $\tan \delta = B/A$ we can write

$$X = \sqrt{A^2 + B^2} \cos\left(\sqrt{ac}t - \delta\right),$$

$$Y = \sqrt{A^2 + B^2}\frac{d}{b}\sqrt{\frac{a}{c}} \sin\left(\sqrt{ac}t - \delta\right)$$

Where $\cos \delta = A/\sqrt{A^2 + B^2}$ and $\sin \delta = B/\sqrt{A^2 + B^2}$. Hence we can obtain, by setting $\alpha = \sqrt{A^2 + B^2}$ and $\beta = \sqrt{A^2 + B^2}(d/b)\sqrt{a/c}$, the familiar equation of ellipse

$$\frac{X^2}{\alpha^2} + \frac{Y^2}{\beta^2} = 1$$

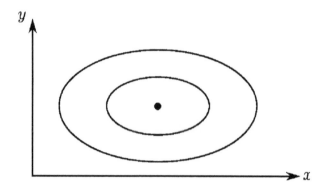

Figure 7: *Phase plane trajectories*

Figure 7 shows trajectories in the phase plane. These two closed trajectories were plotted using the same model parameters but different initial prey densities. Trajectories are closed curves. Since the major and minor axes α and β depend upon the initial conditions A and B, it does not forget the initial conditions. This model of Lotka and Volterra has no asymptotic stability, as it does not converge to an attractor. It is not very realistic. The model does not consider any competition among prey or predators. As a result, prey population may grow infinitely without any resource limits. Predators have no saturation; their consumption rate is unlimited. The rate of prey consumption is proportional to prey density. Thus, it is not surprising that model behavior is unnatural showing no asymptotic stability. The general Lotka Volterra dynamical model for interaction of two species is given as

$$\frac{dx}{dt} = x(a_1 - b_1 x - c_1 y) \tag{9}$$

$$\frac{dy}{dt} = y(a_2 - b_2 y - c_2 x) \tag{10}$$

In this model a_i, b_i and c_i are model parameters. The signs of these parameters in equations (9) and (10) depend upon the type of interactions between species. In prey predator interactions where x is the prey density, a_1 is positive and a_2 is negative. Further, b_1 is positive and b_2 is negative. Also c_1 and c_2 are both negative representing the intra species competition.

In the following section, a general mathematical analysis is presented which is applicable to system of differential equations for predicting the dynamical behavior of underlying system.

Equilibrium is a state of a system, which does not change with time. If the dynamics of a system is described by a differential equation or a system of differential equations, then equilibrium can be estimated by setting the time derivative(s) of state variable(s) to zero. Detection of stability in multi species models is not that simple as in one-variable single species models. Let us consider a predator-prey model with two variables: X is the density of prey and Y is the density of predators. Dynamics of the model is described by the system of two differential equations:

$$\frac{dX}{dt} = F(X, Y), \quad \frac{dY}{dt} = G(X, Y) \tag{11}$$

This is the 2-variable model in a general form. The first step to discuss the stability of a system is to find the equilibrium densities of prey (X^*) and predator (Y^*). For this, we need to solve the system of equations:

$$F(X^*, Y^*) = 0, \quad G(X^*, Y^*) = 0 \tag{12}$$

For a linear model, only one equilibrium point is obtained. For nonlinear models, more equilibrium points may be obtained. The stability of the system is to be discussed for each of these equilibrium points.

The second step is to linearize the model at the equilibrium point $(X = X^*, Y = Y^*)$ by writing $X = x + X^*$, $Y = y + Y^*$, x and y are small perturbations about the equilibrium point. The following linearized system about the equilibrium point is obtained by neglecting the higher degree terms of (11):

$$\frac{dx}{dt} = \frac{\partial F}{\partial X}x + \frac{\partial F}{\partial Y}y$$
$$\frac{dy}{dt} = \frac{\partial G}{\partial X}x + \frac{\partial G}{\partial Y}y$$

(Here all the derivatives are evaluated at equilibrium values.) This can be written as

$$\begin{bmatrix} \frac{dx}{dt} \\ \frac{dy}{dt} \end{bmatrix} = J \begin{bmatrix} x \\ y \end{bmatrix}$$

where the Jacobian Matrix J is given by

$$J = \begin{bmatrix} \partial F/\partial X & \partial F/\partial Y \\ \partial G/\partial X & \partial G/\partial Y \end{bmatrix}$$

The eigenvalues of matrix J determines the stability of the system. The number of eigenvalues is equal to the number of state variables. In our case there will be two eigenvalues. Eigenvalues are generally complex numbers. For stability, the two eigenvalues should have negative real parts. If at least one eigenvalue has a positive real part, then the equilibrium is unstable.

The necessary and sufficient conditions for stability are:

$$trace(J) < 0; \qquad det(J) > 0 \qquad (13)$$

Eigenvalues are used here to reduce a two-dimensional problem to a couple of one-dimensional problems. Eigenvalues have the same meaning as the slope of a line in phase plots. Negative real parts of eigenvalues indicate a negative feedback. It is important that all eigenvalues have negative real parts. If one eigenvalue has a positive real part then there is a direction in a two-dimensional space in which the system will not tend to return back to the equilibrium point. For n dimensional problem with n state variables, the conditions (13) are not sufficient for stability. The more generalized Routh-Hurwitz criterion specifies the necessary and sufficient conditions for all eigenvalues to have negative real parts.

There are two types of stable equilibrium points in a two-dimensional space: Node and Focus. If the eigenvalues are real then we have node otherwise focus. For the node and focus, the trajectories are shown in the Figure 8. While Figure 8(a) and (b) shows node and focus respectively, Figure 8 (c) shows the neutral stability of the point $(0,0)$. We say that the system has a center at $(0,0)$.

There are 3 types of unstable equilibrium points in a two-dimensional space: node, focus, and saddle. Figure 9 (a) shows $(0,0)$ as unstable node and 9 (b) shows saddle point. For details see the references given at the end of the chapter.

Since we are discussing the eigenvalues of the linearized system, therefore, the stability behavior of the complete system (11) will not be the same as has been discussed. This behavior of the system is the local behavior of the system applicable only for the small variations about the equilibrium point or in the small neighborhood of the equilibrium point. The global behavior of the nonlinear system may be quite different than this.

The Poincare Bendixton theorem states that the two species non-linear systems allow only the following two types of dynamical behaviors:

1. Approach to equilibrium or limit point. The approach may be exponential or oscillatory

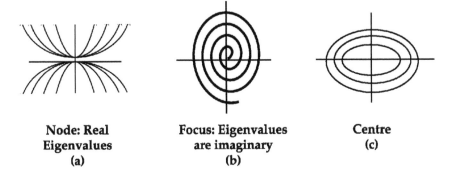

Node: Real
Eigenvalues
(a)

Focus: Eigenvalues
are imaginary
(b)

Centre
(c)

Figure 8: *Trajectories in the phase plane*

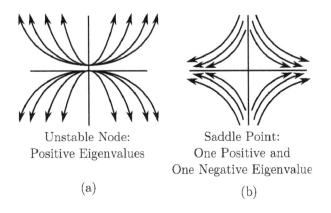

Unstable Node:
Positive Eigenvalues

(a)

Saddle Point:
One Positive and
One Negative Eigenvalue

(b)

Figure 9: *Unstable trajectories in the phase plane*

2. Existence of limit cycle. It is a closed curve in a phase plane to which all trajectories approach

When the dynamics of a population model is traced over a large number of generations, then the model exhibits an asymptotic behavior, which is almost independent from initial conditions. Asymptotic trajectory is called "attractor" because model trajectories converge to this asymptotic trajectory. For a nonlinear model, there may be several attractors. Each attractor has its domain of attraction. Model trajectory converges to that attractor in which domain initial conditions were located.

In Figure 10, there are two attractors: a limit cycle and a stable equilibrium. The domain of attraction of the limit cycle is shown in light color

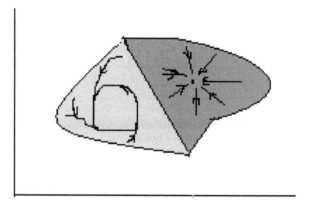

Figure 10: *Region of attraction*

at the left while that of stable equilibrium point is on the right in the dark color. Different regions of attractions never overlap. For different initial conditions, trajectories converge to different attractors.

The behavior of nonlinear systems for three or more interacting species system is much more complex than this. The system may have quasi-periodic solutions or chaos.

5 Some Remarks

It may be noted that

- Many system properties are not represented in the model. For example, age structure is not represented in both the exponential and logistic models.

- Some model properties cannot be found in real systems. For example, solutions of differential equations are always smooth, while trajectories of real systems are always noisy.

- The model may have an equilibrium density, but real populations may not have it because of many reasons like:

  - Population density cannot be measured with infinite accuracy.
  - Weather fluctuations always add noise to system's dynamics.
  - Time series are never long enough to talk about limits and convergence.

6 Classification of Mathematical Models

There are number of criterion available to classify mathematical models:

1. Subject wise: The model may be classified according to the subject matter of the model or the discipline to which a model is applicable. For example we may have mathematical models in physics, economics, medicine, biology or engineering. The subject that deals with mathematical modeling of problems in physics is known as Mathematical physics. Mathematical biology deals with mathematical models in biology. Mathematical Economics or Econometrics is the study of mathematical models in economics and so on. Sometimes the subject is not as vast as physics, chemistry or biology; it is very specialized as mathematical modeling of transportation problems or mathematical modeling in blood flows or optimal utilization of renewable resources. Even these specialized areas are so vast that one can spend a lifetime working on the mathematical modeling of specialized problems.

2. Techniques being used: This classification is based on the type of mathematical techniques being utilized for solving the mathematical representation of the real life problems. We can have mathematical modeling through algebra, modeling using differential equations, modeling through difference equations, mathematical modeling through graphs, through calculus of variation and so on. In such cases the effort is mainly on the development of the mathematical tools and models are used only as illustrations. The mathematical technique is applied to variety of problems.

3. Mathematical models may be classified according to its use. For example we may have mathematical models for prediction or forecasting, for description, for optimization or for control. A model suitable for prediction may not be suitable for control.

4. Mathematical models can also be classified according to their nature. The model may be classified as:

   - Linear or nonlinear: if the model equations, objective functions and constraints are linear then the model is linear. If one or more of them are nonlinear then the model is nonlinear. Linear models are easy to solve and interpret as compared to nonlinear models.

- Static or Dynamic: A static model does not account for time variations. Dynamical models are typically represented with difference equations or differential equations.

- Deterministic or stochastic: No variable or parameter of the deterministic model behaves randomly. In a deterministic system, every action produces a reaction, and every reaction, becomes the cause of subsequent reactions. The totality of these cascading events can theoretically predict the state of the system that will exist at any moment in time. Deterministic models behave the same way for a given set of conditions. In stochastic models randomness is present.

- Discrete or continuous: All the variables are continuous in Continuous models in contrast to discrete models.

- Lumped parameter or distributed parameter: If the model is homogeneous, the parameters are lumped. If the model is heterogeneous then the parameters are distributed. Models with distributed parameters are typically represented by partial differential equations.

Sometimes a deterministic linear and static model may abstract a real problem. Such models are generally simpler than the corresponding stochastic, nonlinear and dynamical models. Essentially most realistic models are nonlinear, dynamic and stochastic. However, linear and static models are used as simpler models that can be easily handled and give good approximation to the real problem. Similarly continuous models are preferred over discrete ones. Analytical solutions may be obtainable for continuous models in contrast to discrete models.

A mathematical model should be as close to the reality as possible. Models are constantly improved to make them more realistic. We may have hierarchy of models. Taking more variables and incorporating more details, a model may be closer to the real problem. More complex and realistic models succeed over simpler and less realistic models. However, a very realistic model may be too complex to be tractable. Therefore a tradeoff is made between tractability of the model and reality. For given problem, many different mathematical models are available. They differ in their precision and agreement with the observation. The model to be selected depends upon the use of the model. The trade off is also possible between oversimplification and over ambitious models. Over simplified models may not represent reality and over ambitious model may be too complex to analyze. A model may explain only one aspect of the real problem. Different models are desirable to deal with different aspects of the problem. Normally models give predictions as expected qualitatively. It may be the quantitative descriptions that are more important. The unexpected predictions may often lead to breakthroughs or lead to better understanding of the system.

Given a mathematical model for a real problem, the following questions may be asked: Is it realistic enough? How can it be made more realistic? Is it oversimplified or over ambitious? Does it suggest new ideas or concepts? Is it robust, fragile or sensitive? Is it consistent? What aspects can it explain? What aspects the model cannot explain? What will be the use of the model? How the parameters will be estimated? How the validation will be carried out? Can it be simplified?

Mathematical modeling problems are classified as:

- A white box model is a system where all necessary information is available. It is preferable to use as much a priori information as possible to make the model more accurate. A priori information may be available in the form of some mathematical relationship between different variable. For example, drug administration problem, it is known that drug is exponentially decaying in the blood. However, the initial amount in the blood and the rate at which the medicine is decaying is still not known. These have to be estimated before the model is to be used. The problem will be increasingly complex if the model requires large number of parameters to be estimated. The example at hand is not a perfect white box.

- A black box model is a system of which there is no a priori information available. We try to estimate both the functional form of relation between variables and the parameters. Neural networks, usually, is a method to tackle such situations.

When a model is sufficiently robust and general it may be termed as a law, where the independent and dependent variables may be interchangeable and therefore Laws are non-directional. Laws must have strong theoretical basis. Interchangeability of dependent and independent variables is possible in model but it is impossible in nature.

References

Jones, D.S. and Sleeman, B.D. 2003. *Diffential Equations and Mathematical Biology*, Chapmaan and Hall.

Hirsch, M.W., Smale, S. and Devaney, R.L. 2004. *Differential Equations, Dynamical systems and An Intoduction to Chaos*, Academic Press.

CHAPTER 23

A Mathematical Approach to Brain and Cognition

Shripad Kondra and Prasun Kumar Roy

National Brain Research Centre,
Manesar, Gurgaon 122050, India.
e-mail: kondra@nbrc.ac.in,pkroy@nbrc.ac.in

1 Introduction

Recent technological and experimental advances in the capabilities to record signals from neural systems have led to an increase in the types and volume of data collected in neuroscience experiments and hence, is the need for appropriate techniques to analyze those using combinations of likelihood, Bayesian, state-space, time-series and point process approaches. This chapter will cover mathematical and statistical techniques used in neuronal modeling, quantitave EEG analysis and functional neuroimaging studies.

2 Neuronal Modeling

Computational neuroscience is somewhat distinct from psychological connectionism and theories of learning from disciplines such as machine learning, neural networks and statistical learning theory. Computational neuroscience emphasizes descriptions of functional and biologically realistic neurons (and neural systems) and their physiology and dynamics. These

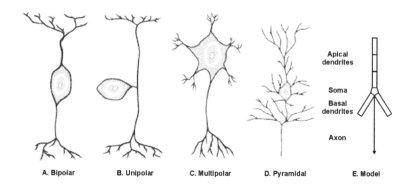

Figure 1: *A-D: Schematic representation of different types of neurons, E: Simplified discrete compartmental model of the pyramidal neuron.*

approaches capture the essential features of the biological system at multiple spatial-temporal scales, from membrane currents, protein and chemical coupling to network oscillations, columnar and topographic architecture as well as learning, memory, imagination and consciousness. These computational models are used to frame hypotheses that can be directly tested by current or future biological and/or psychological experiments. Currently, the field is undergoing a rapid expansion. There are many software packages, such as GENESIS and NEURON, that allow rapid and systematic *in silico* modeling of realistic neurons.

2.1   Neurons

Neurons form the basic circuitry of the brain. Therefore, a sufficient in-depth knowledge about neurons is necessary for the study of the brain. A typical human brain consists of approximately 100 billion neurons, each neuron having at least 10,000 connections with other neurons. Electro-chemical signals flow in neurons, originating at the dendrites or cell body in response to stimulation from other neurons and propagating along axon branches which terminate on the dendrites or cell bodies of perhaps thousands of other neurons. Figure 1 shows a typical pyramidal cell. The connections between the ends of axons and the dendrites or cell bodies of other neurons are specialized structures called synapses (Dayan and Abbott, 2001).

A great deal is known about the biophysical mechanisms responsible for generating neuronal electrochemical activity which provides the basis for various mathematical models for biological processes. As it is generally believed that neurons communicate with each other via action potentials, these models basically represent neuronal behavior in terms of membrane potential and action potential. The neuron like other cells has a lipid bilayer cell membrane which acts like a capacitor separating the charges inside and outside the cell. The cell membrane consists of many channels most of which are selective to only single ion and allow it to pass. There are also ion pumps which spend energy to maintain the balance of concentration inside and outside of the membrane. Action potentials in neurons are also known as nerve impulses or spikes, and the temporal sequence of action potentials generated by a neuron is called its spike train. A neuron that emits an action potential is often said to fire. Action potentials result from special types of voltage-gated ion channels. As the membrane potential is increased, sodium ion channels open, allowing the entry of sodium ions into the cell. This is followed by the opening of potassium ion channels that permit the exit of potassium ions from the cell. The inward flow of sodium ions increases the concentration of positively-charged cations in the cell and causes depolarization, where the potential of the cell is higher than the cell's resting potential. The sodium channels close at the peak of the action potential, while potassium continues to leave the cell. The efflux of potassium ions decreases the membrane potential or hyperpolarizes the cell. For small voltage increases from rest, the potassium current exceeds the sodium current and the voltage returns to its normal resting value, typically $-70\,\text{mV}$. Figure 2 shows the generation of action potential and the ion channels involved.

2.2   Modeling

In the case of modeling a biological neuron, physical analogues such as weight and transfer function are used as parameters. The input to a neuron is often described by an ion current through the cell membrane that occurs when neurotransmitters cause an activation of ion channels in the cell. We describe this by a physical time-dependent current I(t). The cell itself is bound by an insulating cell membrane with a concentration of charged ions on either side that determines a capacitance Cm. Finally, a neuron responds to such a signal with a change in voltage, or an electrical po-

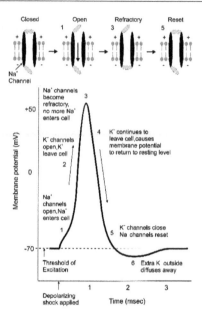

Figure 2: *Action Potential*

Color image of this figure appears in the color plate section at the end of the book.

tential energy difference between the cell and its surroundings, which is observed to sometimes result in a voltage spike called an action potential. This quantity, then, is the quantity of interest and is given by Vm. A review of different biological neuronal models along with simulink models are explored in (Mishra *et al.*, 2006)

2.2.1   Cable Theory

The flow of currents within an axon can be described quantitatively by the cable theory (Rall, 1989) and its elaborations, such as the compartmental model (Segev *et al.*, 1989). Figure 3 shows the equivalent electrical circuit of a basic neural compartment. Cable theory was developed in 1855 by Lord Kelvin to model the transatlantic telegraph cable (Kelvin, 1885) and was shown to be relevant to neurons by Hodgkin and Rushton in 1946 (Hodgkin and Rushton, 1946). In simple cable theory, the neuron is treated as an electrically passive, perfectly cylindrical transmission cable, which can be described by a partial differential equation (Rall, 1989):

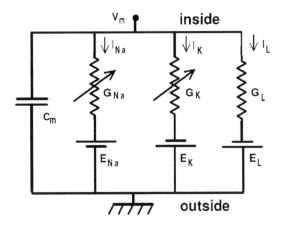

Figure 3: *Electrical equivalent circuit, the variable resistances represent voltage-dependent conductances (Bower and Beeman, 1998)*

$$\tau\frac{\partial V}{\partial t} = \lambda^2\frac{\partial^2 V}{\partial x^2} - V \qquad (1)$$

where $V(x, t)$ is the voltage across the membrane at a time t and a position x along the length of the neuron, and where λ and τ are the characteristic length and time scales on which those voltages decay in response to a stimulus. These scales can be determined from the resistances and capacitances per unit length (Purves *et al.*, 2001) as shown in eq.(2)

$$\tau = r_m c_m$$
$$\lambda = \sqrt{\frac{r_m}{r_l}} \qquad (2)$$

2.2.2 Integrate and Fire Model

One of the earliest models of a neuron was first investigated in 1907 by Lapicque (Abbott, 1999). A neuron is represented in time by eq.(3)

$$I(t) = C_m\frac{dV_m}{dt} \qquad (3)$$

which is just the time derivative of the law of capacitance, Q = CV. When an input current is applied, the membrane voltage increases with time until it reaches a constant threshold V_{th}, at which point a delta function spike occurs and the voltage is reset to its resting potential, after which the model continues to run. The firing frequency of the model thus increases linearly without bound as input current increases. The model can be made more accurate by introducing a refractory period t_{ref} that limits the firing frequency of a neuron by preventing it from firing during that period. Through some calculus involving a Fourier transform, the firing frequency as a function of a constant input current thus looks like eq. (4)

$$f(I) = \frac{I}{C_m V_{th} + t_{ref} I} \tag{4}$$

A remaining shortcoming of this model is that it implements no time-dependent memory. If the model receives a below-threshold signal at some time, it will retain that voltage boost forever until it fires again. This characteristic is clearly not in line with observed neuronal behavior.

2.2.3   Hodgkin-Huxley Model

The most successful and widely-used models of neurons have been based on the Markov kinetic model developed from Hodgkin and Huxley's 1952 work based on data from the squid giant axon, There axons are so large in diameter (roughly 1 mm, or 100-fold larger than a typical neuron) that they can be seen with the naked eye, making them easy to extract and manipulate.

We note as before our voltage-current relationship, this time generalized to include multiple voltage-dependent currents: as in eq. (5)

$$C_m \frac{dV_m}{dt} = -\sum_i I_i(t, V) \tag{5}$$

Each current is given by Ohm's Law as in eq. (6)

$$I(t, V) = g(t, V) \cdot (V - V_{eq}) \tag{6}$$

where $g(t, V)$ is the conductance, or inverse resistance, which can be expanded in terms of its constant average and the activation and inactivation fractions m and h, respectively, that determine how many ions can flow through available membrane channels. This expansion is given by eq. (7)

$$g(t, V) = \bar{g} \cdot m(t, V)^p \cdot h(t, V)^q \tag{7}$$

and our fractions follow the first-order kinetics as in eq. (8)

$$\frac{dm(t, V)}{dt} = \frac{m_\infty(V) - m(t, V)}{\tau_m(V)} \tag{8}$$

with similar dynamics for h, where we can use either τ and m_∞ or α and β to define our gate fractions. With such a form, all that remains is to individually investigate each current one wants to include. Typically, these include inward Ca^{2+} and Na^+ input currents and several varieties of K^+ outward currents, including a leak current. The end result can be at the small end 20 parameters which one must estimate or measure for an accurate model, and for complex systems of neurons not easily tractable by computer. Careful simplifications of the Hodgkin-Huxley model are therefore needed.

2.2.4   Numerical Methods for Neuronal Modeling

A neural simulation program solves a set of coupled equations by replacing the differential equation with a difference equation that is solved at discrete time intervals. Typically, smaller time intervals lead to greater accuracy but slower execution time, as more time steps are required for the solution over a given time period. The difference between Ordinary differential equations (ODEs) and Partial differential equations (PDEs) is that ODEs are equations in which the rate of change of an unknown function of a single variable is prescribed, usually the derivative with respect to time. In contrast, PDEs involve the rates of change of the solution with respect to two or more independent variables, such as time and space. The numerical methods for both ODEs and PDEs involve replacing the derivatives in the differential equations with finite difference approximations to these derivatives. This reduces the differential equations to algebraic equations. The two major classes of finite difference methods are characterized by whether the resulting algebraic equations explicitly or implicitly define the solution at the new time value.

The Hodgkin-Huxley equations can be used as illustrative examples for the numerical methods. If one clamps a section of a squid giant axon, the membrane potential will no longer depend on the spatial location within the clamped region. This reduces the original PDE to a system of ODEs, and leads us to model the membrane potential with the following system of ODEs as in eq. (9)

$$C\frac{dV}{dt} = -\bar{g}_{Na}m^3h(V - E_{Na}) - \bar{g}_K n^4(V - E_K) - \bar{g}_{leak}m^3h(V - E_{leak}), where$$

$$\frac{dm}{dt} = (1 - m)\alpha_m(V) - m\beta_m(V),$$

$$\frac{dh}{dt} = (1 - h)\alpha_h(V) - h\beta_h(V),$$

$$\frac{dn}{dt} = (1 - n)\alpha_n(V) - n\beta_n(V), \tag{9}$$

If instead of space clamping the Loligo giant axon, one allows the voltage across the membrane of the axon also to vary with longitudinal distance along the axon x, then the membrane potential satisfies a PDE. This PDE is similar to the space clamped ODE case except that equation is replaced with eq. (10)

$$C\frac{\partial V}{\partial t} = \frac{\alpha}{2R}\frac{\partial^2 V}{\partial x^2} - \bar{g}_{Na}m^3h(V - E_{Na}) - \bar{g}_K n^4(V - E_K) - \bar{g}_{leak}m^3h(V - E_{leak}), \tag{10}$$

The Hodgkin-Huxley ODE model in eqn 9 is a system of four first-order ODEs. We can define the four dimensional vector of functions to obtain the single vector differential equation as in eq. (11)

$$\frac{dy_i}{dt} = f(t, y_1, y_2, \ldots y_N), i = 1, \ldots N \tag{11}$$

The following section gives a discussion of the use of numerical methods for solving the equations that arise in neural simulations. for a detailed review see (Bower and Beeman, 1998; Mascagni and Sherman, 1989)

2.2.5   Explicit Methods

Explicit methods are so called because the new values are given explicitly in terms of functions of the old values.

- **Forward Euler Method:** For a time increment Δt we approximate $y(t + \Delta t)$ by eq. (12)

$$y(t + \Delta t) = y(t) + f(t)\Delta t \tag{12}$$

This approximation is equivalent to keeping only the first derivative in a Taylor series expansion as in eq. (13)

$$y(t + \Delta t) = y(t) + \frac{dy}{dt}\Delta t + \frac{1}{2}\frac{d^2 y}{dt^2}(\Delta t)^2 + \frac{1}{6}\frac{d^3 y}{dt^3}(\Delta t)^3 + \cdots \quad (13)$$

- **Adams-Bashforth Methods:** The Adams-Bashforth methods approximate these missing higher derivatives by making use of past values of f(t) in the approximation for $y(t + \Delta t)$. The general form is shown in eq. (14)

$$y(t+\Delta t) = y(t)+\Delta t(a_0 f(t)+a_1 f(t-\Delta t)+a_2 f(t-2\Delta t)+\cdots+a_n f(t-n\Delta t)) \quad (14)$$

where the coefficients a_n may be found by expanding $f(t - n\Delta t)$ in a Taylor's series. If we evaluate f at n previous times in Eq. 14, we say that this is an $(n + 1)^{th}$ order Adams-Bashforth method, because it corresponds to keeping terms through the $(n + 1)^{th}$ derivative in the Taylor's series expansion. These methods are computationally very efficient, as they achieve higher accuracy by making use of free information that has already been calculated at previous time steps.

- **Exponential Euler Method:**

$$\frac{dy}{dt} = A - By$$
$$y(t + \Delta t) = y(t)D + (A/B)(1 - D), \quad (15)$$
$$D = e^{-B\Delta t},$$

Although it is difficult to rigorously analyze the error introduced by this approximation as shown in eq.(15), simulation results show that it is highly accurate for most models that contain active channels and only a few compartments. In these cases, it allows much larger integration steps than other methods.

2.2.6   Implicit Methods

For implicit methods, the right-hand side of the equation involves a function of the new value of y, which has yet to be determined.

- **Backward Euler Method:**

$$y(t + \Delta t) = y(t) + f(t + \Delta t)\Delta t. \tag{16}$$

Thus, Eq. 16 gives an implicit definition of $y(t + Dt)$, rather than an explicit expression that can be evaluated. This means that we need some additional method to solve the equations that arise at each step.

- **Crank-Nicholson Method:** This method is based upon the trapezoidal rule of numerical integration. It uses an average of the forward and backward Euler methods in order to achieve a partial cancelation of errors. This occurs because the neglected second derivative terms are equal and opposite in the two approximations. The approximation is then

$$y(t + \Delta t) = y(t) + f(t + \Delta t)\Delta t \tag{17}$$

The right-hand sides of Eqs. 16 and 17 involve functions that depend on the unknown quantity on the left-hand side. In general, we would have to use iterative methods, such as predictor-corrector or Newton's methods (Acton, 1990) in order to find the solution.

3   Electroencephalogram (EEG)

The EEG is generated when currents flow either into (a current sink) or out of (a current source) a cell across charged neuronal membranes. The EEG does not derive from summated action potentials. Instead, the EEG recorded at the scalp is largely attributable to graded postsynaptic potentials (PSPs) of the cell body and large dendrites of vertically oriented pyramidal cells in cortical layers 3 to 5 (Lopes da Silva, 1991). The extracellular EEG recorded from a distance represents the passive conduction of currents produced by summating activity over large neuronal aggregates To measure the EEG extracranially, electrodes are attached to the scalp with a conducting paste or liquid; each electrode is connected with an electrically "neutral" lead attached to the chest, chin, nose, or ear lobes (a reference montage) or with an active lead located over a different scalp area (a bipolar montage). Differential amplifiers are used to record voltage changes over time at each electrode; these signals are then digitized with 12 or more bits of precision and are sampled at a rate high enough to prevent aliasing of the signals of interest. EEGs are conventionally described as patterns of activity in four frequency ranges: δ (less than 4 Hz), θ (4–7 Hz), α (8-12 Hz), and β activity (13–35 Hz). Complex mental activity and sustained attention result in increased signal power in the lower frequency ranges (below α) and decreased signal power at the higher ranges (α and above). (Toga and Mazziotta, 2002) Derivatives of the EEG technique include evoked potentials (EP), which involves averaging the EEG activity time-locked to the presentation of a stimulus of some sort (visual, somatosensory, or auditory). Event-related potentials refer to averaged EEG responses that are time-locked to more complex processing of stimuli; this technique is used in cognitive science, cognitive psychology, and psychophysiological research.

Advances in EEG acquisition technology have led to chronic recording from multiple channels and resulted in an incentive to use computer technology, automate detection and analysis, and use more objective quantitative approaches. This has provided the impetus to the field of quantitative EEG (qEEG) analysis (Tong and Thakor, 2009). Statistical techniques like Principal component analysis (PCA), independent component analysis (ICA) and dipole modeling which are most commonly used in analysis of EEG patterns, are discussed below.

3.1   Principal Component Analysis (PCA)

Multivariate measures take more than one channel of EEG into account simultaneously. This is used to consider the interactions between the channels and how they correlate rather than looking at channels individually. This is useful if there is some interaction (e.g., synchronization) between different regions of the brain leading up to a seizure. PCA takes a dataset in a multidimensional space and linearly transforms the original dataset to a lower dimensional space using the most prominent dimensions from the original dataset. PCA is used as a seizure detection technique itself (Milton and Jung, 2003). This linear transform has been widely used in data analysis and compression. The following presentation is adapted from (Gonzalez and Woods, 1992). Some of the texts on the subject also include (Oja, 1983, 1989). Principal component analysis is based on the statistical representation of a random variable. Suppose we have a random vector population x, where

$$x = (x_1, \ldots, x_n)^T$$

and the mean of that population is denoted by

$$\mu_x = E\{x\}$$

and the covariance matrix of the same data set is

$$C_x = E\left\{(x - \mu_x)(x - \mu_x)^T\right\}.$$

The components of C_x, denoted by c_{ij}, represent the covariances between the random variable components x_i and x_j. The component c_{ii} is the variance of the component x_i. The variance of a component indicates the spread of the component values around its mean value. If two components x_i and x_j of the data are uncorrelated, their covariance is zero. The covariance matrix is, by definition, always symmetric.

From a symmetric matrix such as the covariance matrix, we can calculate an orthogonal basis by finding its eigenvalues and eigenvectors. The eigenvectors e_i and the corresponding eigenvalues λ_i are the solutions of the equation

$$C_x e_i = \lambda_i e_i, i = 1, \ldots, n$$

For simplicity we assume that the λ_i are distinct. These values can be found, for example, by finding the solutions of the characteristic equation

$$|C_x - \lambda I| = 0$$

where I is the identity matrix having the same order as C_x and $|\cdot|$ denotes the determinant of the matrix. If the data vector has n components, the characteristic equation becomes of order n. This is easy to solve only if n is small. Solving eigenvalues and corresponding eigenvectors is a non-trivial task, and many methods exist. One way to solve the eigenvalue problem is to use a neural solution to the problem (Oja, 1983). The data is fed as the input, and the network converges to the wanted solution. By ordering the eigenvectors in the order of descending eigenvalues (largest first), one can create an ordered orthogonal basis with the first eigenvector having the direction of largest variance of the data. In this way, we can find directions in which the data set has the most significant amounts of energy. Suppose one has a data set of which the sample mean and the covariance matrix have been calculated. Let A be a matrix consisting of eigenvectors of the covariance matrix as the row vectors. By transforming a data vector x, we get

$$y = A(x - \mu_x)$$

which is a point in the orthogonal coordinate system defined by the eigen-vectors. Components of y can be seen as the coordinates in the orthogonal base. We can reconstruct the original data vector x from y by

$$x = A^T y + \mu_x$$

using the property of an orthogonal matrix $A^{-1} = A^T$. The A^T is the trans-pose of a matrix A. The original vector was projected on the coordinate axes defined by the orthogonal basis. The original vector was then recon-structed by a linear combination of the orthogonal basis vectors. Instead of using all the eigenvectors of the covariance matrix, we may represent the data in terms of only a few basis vectors of the orthogonal basis. If we denote the matrix having the K first eigenvectors as rows by A_K, we can create a similar transformation as seen above

$$y = A_K(x - \mu_x), \quad \text{and}$$
$$x = A_K^T y + \mu_x.$$

This means that we project the original data vector on the coordinate axes having the dimension K and transforming the vector back by a linear com-bination of the basis vectors. This minimizes the mean-square error be-tween the data and this representation with given number of eigenvectors. If the data is concentrated in a linear subspace, this provides a way to compress data without losing much information and simplifying the rep-resentation. By picking the eigenvectors having the largest eigenvalues we lose as little information as possible in the mean-square sense. One can e.g. choose a fixed number of eigenvectors and their respective eigenvalues and get a consistent representation, or abstraction of the data. This preserves a varying amount of energy of the original data. Alternatively, we can choose approximately the same amount of energy and a varying amount of eigenvectors and their respective eigenvalues. This would in turn give approximately consistent amount of information in the expense of varying representations with regard to the dimension of the subspace. We are here faced with contradictory goals: On one hand, we should simplify the prob-lem by reducing the dimension of the representation. On the other hand we want to preserve as much as possible of the original information con-tent. PCA offers a convenient way to control the trade-off between loosing information and simplifying the problem at hand. As it will be noted later, it may be possible to create piecewise linear models by dividing the input data to smaller regions and fitting linear models locally to the data. Sample mean and sample covariance matrix can easily be calculated from the data. Eigenvectors and eigenvalues can be calculated from the covariance matrix. Consider an example where the first eigenvalue corresponding to the first eigenvector is $\lambda_1 = 0.18$ while the other eigenvalue is $\lambda_2 = 0.0002$. By comparing the values of eigenvalues to the total sum of eigenvalues, we can get an idea how much of the energy is concentrated along the particular eigenvector. In this case, the first eigenvector contains almost all the energy. The data could be well approximated with a one-dimensional representation.

3.2   Independent Component Analysis (ICA)

To rigorously define ICA (Jutten and Herault, 1991; Comon, 1994), we can use a statistical latent variables model. Assume that we observe n linear mixtures $x_1, \ldots, x_n$ of n independent components as in eq. (18)

$$x_j = a_{j1}s_1 + a_{j2}s_2 + \cdots + a_{jn}s_n, \text{for all j} \tag{18}$$

We have now dropped the time index t; in the ICA model, we assume that each mixture x_j as well as each independent component s_k is a random variable, instead of a proper time signal. The observed values $x_j(t)$, e.g., the microphone signals in the cocktail party problem, are then a sample of this random variable. Without loss of generality, we can assume that both the mixture variables and the independent components have zero mean: If this is not true, then the observable variables x_i can always be centered by subtracting the sample mean, which makes the model zero-mean. It is convenient to use vector-matrix notation instead of the sums like in the previous equation. Let us denote by the random vector whose elements are the mixtures $x_1, \ldots, x_n$, and likewise by **s** the random vector with elements $s_1, \ldots, s_n$. Let us denote by **A** the matrix with elements a_{ij}. Generally, bold lower case letters indicate vectors and bold upper-case letters denote matrices. All vectors are understood as column vectors; thus x^T, or the transpose of x, is a row vector. Using this vector-matrix notation, the above mixing model is written as shown in eq. (19)

$$x = As \tag{19}$$

Sometimes we need the columns of matrix **A** ; denoting them by a_j the model can also be written as shown in eq.(20)

$$x = \sum_{i=1}^{n} a_i s_i \tag{20}$$

The statistical model in Eq. 21 is called independent component analysis, or ICA model. The ICA model is a generative model, which means that it describes how the observed data are generated by a process of mixing the components s_i. The independent components are latent variables, meaning that they cannot be directly observed. Also the mixing matrix is assumed to be unknown. All we observe is the random vector x, and we must estimate both A and s using it. This must be done under as general assumptions as possible. For simplicity, we are also assuming that the unknown mixing matrix is square, but this assumption can be sometimes relaxed. Then, after estimating the matrix, we can compute its inverse, say W, and obtain the independent component simply by eq.(21)

$$s = Wx \tag{21}$$

The Figure 4 shows an example of application of ICA for obtaining artifact free EEG, First the independent components for the multi-channel EEG recordings are found and components which correspond to eye blinks, temporal muscle activity etc. are canceled and remaining components are mixed back to get the artifact corrected EEG.

3.3   Dipole Modeling

Dipole modeling provides another method for generating hypotheses concerning the neuroanatomical loci responsible for generating neuroelectric events measured at the scalp (Scherg and Von Cramon, 1985; Fender, 1987). Dipole modeling uses iterative numerical methods to fit a mathematical representation of a focal, dipolar current source, or collection of such sources, to an observed scalp-recorded EEG or MEG field. Source modeling does not, in general, provide a unique or necessarily physically correct answer about where in the brain activity recorded at the scalp is generated. This is so because solving for the source of an EEG or MEG distribution recorded at the scalp is a mathematically ill-conditioned **inverse**

Source channel decomposition using ICA

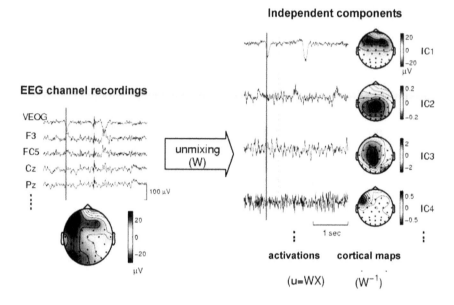

Summation of selected ICA components

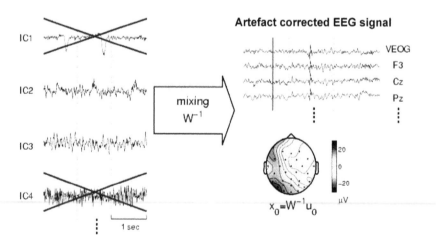

Figure 4: *Independent component analysis for EEG artifact removal (Jung et al., 2000)*

Color image of this figure appears in the color plate section at the end of the book.

problem which has no unique solution; additional information and/or assumptions are required in order to choose among candidate source models. Some of this a priori information is obvious-for example that the potentials must arise from the space occupied by the brain. Thus, anatomical data from structural MR images can usefully constrain inverse solutions (Spinelli *et al.*, 2000). In contrast, other assumptions border on presupposing unknown information (i.e., that the potentials arise only from the cortex).

Single-Dipole Fitting: Single-dipole fitting can be seen as the localization of the electrically active neurons as a single **hot spot** (Tong and Thakor, 2009). Consider the case of a single current dipole located at position r_v eq.(22) with dipole moment j_v eq.(23), where

$$r_v = (x_v \quad y_v \quad z_v)^T \tag{22}$$

$$j_v = (j_x \quad j_y \quad j_z)^T \tag{23}$$

denotes the position vector, with the superscript T denoting vector/matrix transposition, and To introduce the basic form of the equations, consider the nonrealistic, simple case of the current dipole in an infinite homogenous medium with conductivity σ. Then the electric potential at location $r_e \in R^{3\times1}$ for $r_e \neq rv$ is as in eq.(24)

$$\phi(r_e, r_v, j_v) = k_{e,v}^T j_v + c \tag{24}$$

where

$$k_{e,v} = \frac{1}{4\pi\sigma} \frac{(r_e - r_v)}{\|r_e - r_v\|^3} \tag{25}$$

denotes what is commonly known as the lead field as defined in eq.(25). c is a scalar accounting for the physics nature of electric potentials, which are determined up to an arbitrary constant. A slightly more realistic head model corresponds to a spherical homogeneous conductor in air. The lead field in this case is shown in eq.(26)

$$k_{e,v} = \frac{1}{4\pi\sigma} \left[2\frac{(r_e - r_v)}{\|r_e - r_v\|^3} + \frac{r_e\|r_e - r_v\| + (r_e - r_v)\|r_e\|}{\|r_e\|\|r_e - r_v\| \left[\|r_e\|\|r_e - r_v\| + r_e^T(r_e - r_v) \right]} \right] \tag{26}$$

in which this notation eq.(27) is used:

$$\|X\|^2 = tr(X^T X) = tr(XX^T) \tag{27}$$

and where tr denotes the trace, and X is any matrix or vector. If X is a vector, then this is the squared Euclidean L_2 norm; if X is a matrix, then this is the squared Frobenius norm. The equation for the lead field in a totally realistic head model (taking into account geometry and full conductivity profile) is not available in closed form. Numerical methods for computing the lead field can be found in (Fuchs *et al.*, 2007). Formally, we can now state the single-dipole fitting problem. Let ϕ_e (*for* $e = 1, \ldots, N_E$) denote the scalp electric potential measurement at electrode e, where N_E is the total number of cephalic electrodes. All measurements are made using the same reference. Let $\phi_e(rv, jv)$ (*for* $e = 1, \ldots, N_E$) denote the theoretical potential at electrode e, due to a current dipole located at r_v with moment j_v. Then the problem consists of finding the unknown dipole position r_v and moment j_v that best explain the actual measurements. The simplest way to achieve this is to minimize the distance between theoretical and experimental potentials. Consider the functional: shown in eq.(28)

$$F = \sum_{e=1}^{N_E} \left[\phi_e(r_v, j_v) - \hat{\phi}_e \right]^2 \tag{28}$$

This expresses the distance between measurements and model, as a function of the two main dipole parameters: its location and its moment. The aim is to find the values of the parameters that minimize the functional, that is, the least squares solution. Many algorithms are available for finding the parameters, as reviewed in (Scherg and Von Cramon, 1985; Sukov and Barth, 1998).

Multiple-Dipole Fitting: A straightforward generalization of the previous case consists of attempting to explain the measured EEG as being due to a small number of active brain spots. Based on the principle of superposition, the theoretical potential due to N_V dipoles is simply the sum of potentials due to each individual dipole. Therefore, the functional in eq.(28) generalizes to eq.(29)

$$F = \sum_{e=1}^{N_E} \left[\sum_{v=1}^{N_V} \phi_e(r_v, j_v) - \hat{\phi}_e \right]^2 \tag{29}$$

and the least squares problem for this multiple-dipole fitting case consists of finding all dipole positions r_v and moments j_v, for $v = 1 \ldots N_V$ that minimize F. Figure 5 shows example of multiple dipoles fitted of an epileptic spike using EEGLab toolbox.

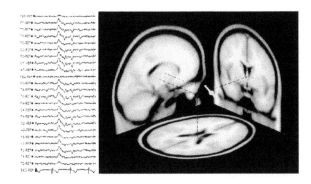

Figure 5: *Dipoles fitted using EEGlab toolbox*

Color image of this figure appears in the color plate section at the end of the book.

4   Functional Magnetic Resonance Imaging

Magnetic Resonance Imaging (MRI) of the brain is well-recognized for its excellent spatial resolution, allowing neuroanatomic structures to be viewed in sharp detail. Functional Magnetic Resonance Imaging (fMRI) is a type of specialized MRI scan. It measures the hemodynamic response (change in blood flow) related to neural activity in the brain or spinal cord of humans or other animals. It is one of the most recently developed forms of neuroimaging. Since the early 1990s, fMRI has come to dominate the brain mapping field due to its relatively low invasiveness, absence of radiation exposure, and relatively wide availability.

4.1   fMRI Paradigm design

There are two types of designing a paradigm in fMRI such as **block design** and **event-related design**. Figure 6 shows the experimental designs for task activation studies. The predicted fMRI response is obtained by convolving the administered task with the hemodynamic response function, which may be assumed or measured. Because the hemodynamic response is prolonged and delayed, blocks of sequential stimuli produce a response with a sustained peak, whereas the responses to widely spaced individual stimuli can still be segregated. Data analysis procedures typically use pixel-by-pixel regression analysis to identify pixels in which signal changes correlate significantly with the predicted responses. Each block

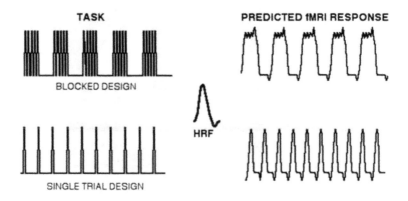

Figure 6: *fMRI paradigms*

will have duration of a certain number of fMRI scans and within each block only one condition is presented. By making the conditions differ in only the cognitive process of interest, the fMRI signal that differentiates the conditions should represent this cognitive process of interest. The fMRI activation can be found using a technique called statistical parametric mapping.

4.2 Statistical Parametric Mapping (SPM)

The analysis of functional neuroimaging data involves many steps (see Figure 7) that can be broadly divided into
(i) spatial processing,
(ii) estimating the parameters of a statistical model, and
(iii) making inferences about those parameter estimates with their associated statistics (Toga and Mazziotta, 2002). Statistical parametric maps (SPMs) are image processes with voxel values that are, under the null hypothesis, distributed according to a known probability density function, usually the Student t or F distributions. These are known colloquially as t or F maps. The success of statistical parametric mapping is due largely to the simplicity of the idea. Namely, one analyzes each and every voxel using any standard (univariate) statistical test. The resulting statistical parameters are assembled into an image-the SPM. SPMs are interpreted as spatially extended statistical processes by referring to the probabilistic behavior of Gaussian fields (Friston *et al.*, 1994). The general linear model

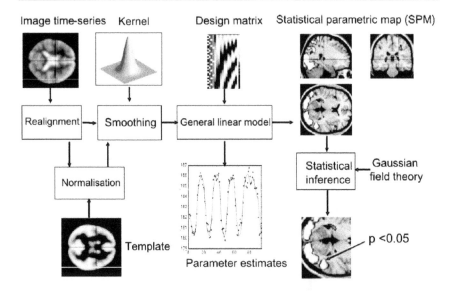

Figure 7: *Steps involved in SPM (Toga and Mazziotta, 2002)*

Color image of this figure appears in the color plate section at the end of the book.

is an equation, $Y = X + \epsilon$, that expresses the observed response variable Y in terms of a linear combination of explanatory variables X plus a well behaved error term Figure 8 shows the general linear model in details. The general linear model is variously known as **analysis of covariance** or **multiple regression analysis** and subsumes simpler variants, like the **t test** for a difference in means, to more elaborate linear convolution models such as finite impulse response (FIR) models. The matrix X that contains the explanatory variables (e.g., designed effects or confounds) is called the design matrix. Each column of the design matrix corresponds to some effect one has built into the experiment or that may confound the results. These are referred to as regressors. Parameter estimation is done using ordinary least square method as in eq.(30)

$$\|y - X\beta\|^2 = \epsilon^T \epsilon. \tag{30}$$

Ordinary least estimates are shown in (31)

$$\hat{\beta} = (X^T X)^{-1} X^T y. \tag{31}$$

To test an hypothesis, we construct **test statistics**. The Null Hypothesis H_0 is typically what we want to disprove (no effect). The test statistic sum-

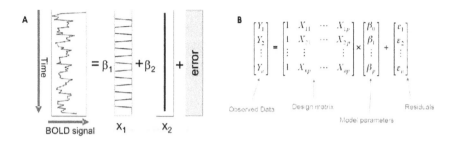

Figure 8: *General linear model regression (A. Single voxel, B. Multi Voxel)*

Color image of this figure appears in the color plate section at the end of the book.

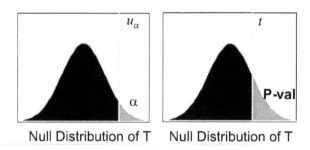

Null Distribution of T   Null Distribution of T

Figure 9: *test statistics (Null distribution)*

marizes evidence about H_0. Typically, test statistic is small in magnitude when the hypothesis H_0 is true and large when false. We need to know the distribution of T under the null hypothesis. Threshold u_α controls the false positive rate. We reject the null hypothesis in favor of the alternative hypothesis if t > u_α. A p-value summarizes evidence against H_0. This is the chance of observing value more extreme than t under the null hypothesis as shown in Figure 9. We are usually not interested in the whole β vector. A contrast selects a specific effect of interest as in (33): a contrast c is a vector of length p as in (32). $c^T \beta$ is a linear combination of regression coefficients β.

$$c^T = [1\ 0\ 0\ 0\ 0\ \cdots] \tag{32}$$

$$c^T \beta = 1 \times \beta_1 + 0 \times \beta_2 + 0 \times \beta_3 + 0 \times \beta_4 + 0 \times \beta_5 + \cdots \tag{33}$$

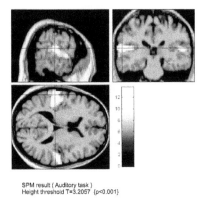

SPM result (Auditory task)
Height threshold T=3.2057 {p<0.001}

Figure 10: *Final fMRI activation (Auditory task)*

Color image of this figure appears in the color plate section at the end of the book.

The question to test is whether the box-car amplitude is greater than zero. Therefore the null hypothesis will be $c^T\beta = 0$.

$$T = \frac{Contrast\ of\ estimated\ parameters}{\sqrt{Variance\ estimate}} \quad (34)$$

$$T = \frac{\hat{\beta}}{var(c^T\hat{\beta})} \quad (35)$$

Inferences using SPMs can be of two sorts depending on whether one knows where to look in advance. With an anatomically constrained hypothesis, about effects in a particular brain region, the uncorrected P value associated with the height T (34) and (35) or extent of that region in the SPM can be used to test the hypothesis. With an anatomically open hypothesis (i.e., a null hypothesis that there is no effect anywhere in a specified volume of the brain) a correction for multiple dependent comparisons is necessary. The theory of Gaussian random fields provides a way of correcting the P value that takes into account the fact that neighboring voxels are not independent by virtue of continuity in the original data. Figure 10 shows the example of fMRI activation during an auditory task for $p < 0.001$.

5 Conclusions

The chapter has given a brief exposure to the popular mathematical tools used for modeling, mapping and extracting useful information from the brain. Every great leap in neuroscience has been preceded by the development of instrumentation or methodology. The rate of methodological progress has been steadily increasing, resulting in more comprehensive and useful maps of brain structure and function. There also has been an ever-closer relationship between neuroscience and those disciplines (especially mathematics and statistics) that can help accelerate the development of tools to map the brain.(Toga and Mazziotta, 2002)

6 Further reading

Mathematical techniques used in structural and functional neuroimaging studies, like

- diffusion tensor imaging,
- state-of-the-art algorithms for computational anatomy and algorithms for meta-analysis of functional images,
- methods to relate imaging information to genetics,
- cutting-edge methods for image registration and segmentation indispensable steps in all brain image analysis,
- sophisticated modeling of cortical anatomy and function,
- statistics of anatomical variation in development and disease,
- creation of anatomical templates and atlases to represent populations, and
- automatic labeling of brain structures

are discussed in the special issue of NeuroImage, entitled Mathematics in Brain Imaging (Thompson *et al.*, 2009).

We hope that the contents of this chapter will generate the interest of mathematicians or anyone interested in the mathematical developments in the field.

References

Abbott, L.F. 1999. Lapique's introduction of the integrate-and-fire model neuron (1907), *Brain Research Bulletin* 50 **(5/6)**:303–304.

Acton, F.S. 1990. *Numerical methods that work*, The Mathematical Association of America.

Bower, J.M. and Beeman, D. 1998. *The book of GENESIS*, Springer New York.

Comon, P. 1994. Independent component analysis, a new concept?, *Signal processing* **36**:287–314.

Dayan, P. and Abbott, L.F. 2001. *Theoretical neuroscience: Computational and mathematical modeling of neural systems*, MIT press.

Fender, D.H. 1987. Source localization of brain electrical activity, *Methods of analysis of brain electrical and magnetic signals* **1**:355–403.

Friston, K.J., Worsley, K.J., Frackowiak, R.S.J., Mazziotta, J.C. and Evans, A.C. 1994. Assessing the significance of focal activations using their spatial extent, *Human Brain Mapping* **1**:214–220.

Fuchs, M., Wagner, M. and Kastner, J. 2007. Development of volume conductor and source models to localize epileptic foci, *Journal of Clinical Neurophysiology*. **24**:101–119.

Gonzalez, R.C. and Woods, R.E. 1992. *Digital image processing*, Addison Wisley.

Hodgkin, A.L. and Rushton, W.A.H. 1946. The electrical constants of a crustacean nerve fibre, *Proceedings of the Royal Society B* **133**:444–479.

Jung, T.P., Makeig, S., Humphries, C., Lee, T.W., Mckeown, M.J., Iragui, V. and Sejnowski, T.J. 2000. Removing electroencephalographic artifacts by blind source separation, *Psychophysiology*.

Jutten, C. and Herault, J. 1991. Blind separation of sources, part I: An adaptive algorithm based on neuromimetic architecture, *Signal processing* **24**:1–10.

Kelvin, W.T. 1885. On the theory of the electric telegraph, *Proceedings of the Royal Society* **7**:382–399.

Lopes da Silva, F. 1991. Neural mechanisms underlying brain waves: from neural membranes to networks, *Electroencephalography and Clinical Neurophysiology*. pp 81–93.

Mascagni, M.V. and Sherman, A.S. 1989. Numerical methods for neuronal modeling, *Methods in Neuronal Modeling: From Synapses to Networks*, eds. Koch C and Segev I; Cambridge MA: Bradford Books, MIT Press.

Milton, J. and Jung, P. 2003. *Epilepsy as a dynamic disease*, Springer.

Mishra, D., Yadav, A., Ray, S. and Kalra, P.K. 2006. Exploring Biological Neuron Models, *Directions*, The Research Magazine of IIT Kanpur **7**:13–22.

Oja, E. 1983. *Subspace methods of pattern recognition*, Research Studies Press Hertfordshire, UK.

Oja, E. 1989. Neural networks, principal components, and subspaces, in *International Journal of Neural Systems*. **1/1:**pp 61–68.

Purves, D., Augustine, G.J., Fitzpatrick, D., Katz, L.C., LaMantia, A.S., McNamara, J.O. and Williams, S.M. 2001. *Neuroscience*, Sinauer Associates 52–53.

Rall, W. 1989. Cable Theory for Dendritic Neurons, *Methods in Neuronal Modeling: From Synapses to Networks*, eds. Koch C. and Segev I.; Cambridge MA: Bradford Books, MIT Press. p. 962

Scherg, M. and Von Cramon, D. 1985. A new interpretation of the generators of BAEP waves IV: results of a spatio-temporal dipole model, in *Electroencephalography and Clinical Neurophysiology*. **62:**290–299.

Scherg, M. and Von Cramon, D. 1985. Two bilateral sources of the late AEP as identified by a spatio-temporal dipole model, in *Electroencephalography and Clinical Neurophysiology*. **62:**32–44.

Segev, I., Fleshman, J.W., Burke, R.E. 1989. Compartmental Models of Complex Neurons, *Methods in Neuronal Modeling: From Synapses to Networks*, eds. Koch C and Segev I; Cambridge MA: Bradford Books, MIT Press. p. 6396.

Spinelli, L., Andino, S.G., Lantz, G., Seeck, M. and Michel, C.M. 2000. Electromagnetic inverse solutions in anatomically constrained spherical head models, *Brain topography* **13:**115–125.

Sukov, W. and Barth, D.S. 1998. Three-dimensional analysis of spontaneous and thalamically evoked gamma oscillations in auditory cortex, in *Journal of neurophysiology*. **79:**2875–2884.

Thompson, P.M., Miller, M.I., Poldrack, R.A., Nichols, T.E., Taylor, J.E., Worsley, K.J. and Ratnanather, J.T. 2009. Special issue on mathematics in brain imaging, in *Neuro Image* p. 45.

Toga, A.W. and Mazziotta, J.C. 2002. *Brain mapping: The methods*, Academic Press.

Tong, S. and Thakor, N.V. 2009. *Quantitative EEG Analysis Methods and Clinical Applications*, Artech House Publishers.

Part VI
Mathematics over the Millenia

CHAPTER 24

Rational Quadrilaterals from Brahmagupta to Kummer

R. Sridharan[1],*

*Chennai Mathematical Institute,
Chennai 603 103, India.
e-mail: rsridhar@cmi.ac.in

1 Introduction

From very early times in India, the emphasis on geometry has been from the numerical point of view and this has possibly been due to cultural reasons, since in India, most abstract concepts have had their origin in religion, or were intended for practical applications. The simplest geometrical figure, the right angled triangle(along with the so called theorem of Pythagoras) has already been mentioned in the Sulva Sutras[2] (the earliest of which can be dated at least to the eighth century B.C.), was studied from the point of view of constructing Vedic altars. One of the commonest type of questions treated by the ancient Indians was the construction of right angled triangles all of whose sides are rational(integral), such as triangles with sides (3,4,5), (5,12,13), etc. which of course must have been probably known to other ancient civilisations too. The study of such triangles must

---

[1]To Bharat, the mainstay of my life, with my love.

[2]The Sulva Sutras are part of the Vedic literature, which includes mathematical texts containing detailed instruction for the building of sacrificial altars.

have continued to evolve in India and it is reasonable to believe that triangles with rational sides and rational areas were also studied here in India from very early times. However, the first documentary evidence of their study comes from the work of that great Indian astronomer-mathematician Brahmagupta(born 598 AD). He infact used such triangles for the construction of cyclic quadrilaterals with rational sides and rational diagonals. The aim of this article is to show how the study of such quadrilaterals which began in this "remote" part of the world, as far as the westerners were concerned, waited for thirteen centuries , before it came to be known in England due to the English translation of the work of Brahmagupta and Bhaskara II by (Colebrooke, 1817), crossed the Channel to move to France and eventually came to the attention of the nineteenth century German mathematician Kummer, who not only wrote a critical paper on this work of Brahmagupta, but also generalised the problem to construct(not necessarily cyclic) quadrilaterals with rational sides and diagonals by new methods, belonging to that part of mathematics now called 'Arithmetic Geometry'(which consists, in particular, of methods of solving indeterminate polynomial equations).

Kummer, in this work (Kummer, 1848) used some earlier techniques developed by Euler and this work of Euler in turn had its origin in the work of the Italian mathematician Fagnano. We therefore have here the story of the migration of a mathematical idea from India to Europe leading to development of mathematics in an apparently unrelated area and shows the basic unity of mathematical thought(and if one wishes to speculate further, one could perhaps say that the course of mathematics is in some sense pre-determined).

2   Brahmagupta's Construction

The essential idea of Brahmagupta to construct rational cyclic quadrilaterals seems to have been based on the idea of juxtaposing rational right angled triangles. One method consists in juxtaposing two rational right angled triangles after making the hypotenuses equal. As we know that the sides of these right angled rational triangles can be parametrised(upto rational factors) by $(c^2 - d^2, 2cd, c^2 + d^2)$ and $(a^2 - b^2, 2ab, a^2 + b^2)$ and then the quadrilateral $ABCD$ Figure 1 has as its sides $AB = (a^2 + b^2)(c^2 - d^2)$, $BC = 2cd(a^2 + b^2)$, $CD = (c^2 + d^2)(a^2 - b^2)$ and $DA = 2ab(c^2 + d^2)$.

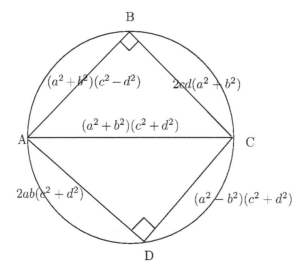

Figure 1: *First construction*

The diameter AC has length $(a^2 + b^2)(c^2 + d^2)$, and since the quadrilateral is cyclic, the other diagonal is also rational(due to a theorem of Ptolemy which says that the quadrilateral is cyclic if and only if the sum of the products of its opposite sides is equal to the product of its diagonals).

The second construction is obtained by scaling two rational right angled triangles with sides $(a^2 - b^2, 2ab, a^2 + b^2)$ and $(c^2 - d^2, 2cd, c^2 + d^2)$ and joining them along BE and ED to get a quadrilateral (Figure 2 with sides $AB = (a^2 + b^2)(c^2 - d^2)$, $BC = 2ab(c^2 + d^2)$, $CD = 2cd(a^2 + b^2)$, $DA = (a^2 - b^2)(c^2 + d^2)$.

One checks that the above quadrilateral is cyclic, for instance, by noting that $BE.ED = AE.EC$; inspecting the diagonals, it is clear that this construction is different from the first.

Remark 1. Prithudakaswami, (850 AD), the famous commentator of Brahmagupta, in his commentary (See Colebrooke (1817)) of Brahmagupta's work gives a numerical example (cf. Figure 3) by taking, for instance, the triangles with sides $(3, 4, 5)$ and $(5, 12, 13)$.

We shall now prove a proposition showing how to construct a rational quadrilateral with rational diagonals by adjoining any two rational right angled triangles which is very similar to the spirit of the second construction of Brahmagupta. We begin by proving a lemma which shows how

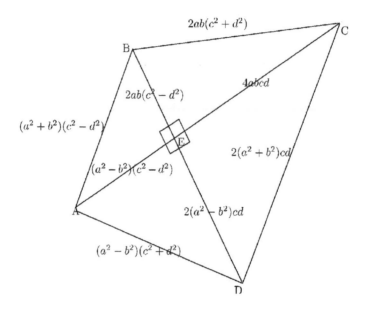

Figure 2: *Second construction*

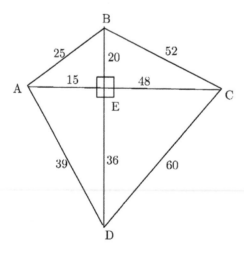

Figure 3: *Prithudaka's Example*

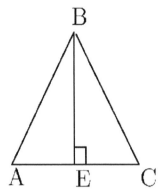

Figure 4: *Rational Triangle*

to construct rational triangles with rational area by adjunction of any two rational right angled triangles.

Lemma 1. *Let ABE and EFC be two right angled triangles with rational sides. Then, by replacing EFC by a similar triangle, if necessary, we may assume that EF = EB. Then the triangle ABC obtained by the juxtaposition of the above right angled triangles along BE is a rational triangle with rational area.*

Proof. First of all, we remark that in any rational triangle ABC, the cosines of all the angles are rational, which follows from the cosine formula.

Hence $AE = AB \cos A$ is rational and $EC = BC \cos C$ is also rational.We also know by our choice that BE is rational. This proves the lemma. $\square$

Remark 2. There is an easily proved converse to the above lemma which says that any triangle with rational sides and rational area can be obtained by juxtaposing two right angled triangles with rational sides. In fact, we need only to drop an altitude of the triangle.

Proposition 1 (Brahmagupta). *Let ABC be a rational triangle which is obtained by adjoining the right angled triangles ABE, BEC. Let BE produced meet the circumcircle of the triangle ABC at D. Then ABCD is a quadrilateral with rational sides and rational diagonals.*

Proof. The triangles AED and BEC are obviously similar so that $\frac{AD}{BC} = \frac{AE}{BE} = \frac{DE}{EC}$.

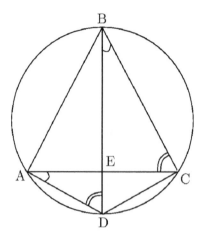

Figure 5: *Brahmagupta's Cyclic Quadrilateral*

However, BC, BE, EC are all rational so that AD, DE are rational. Similarly, $\frac{DC}{AB} = \frac{DE}{AE}$, so that CD is also rational, and the proposition is proved.

□

H T Colebrooke's English translation of the work of Brahmagupta and Bhaskara II was published in London (Colebrooke, 1817) in 1817 and it drew the attention of the French geometer Chasles(1793–1880). He was very much impressed by the work of the Indians on geometry and wrote an appreciative note (Chasles, 1837). Kummer(1810–1893) became aware through this of the work of Brahmagupta and Bhaskara II and wrote a critical paper, as we remarked, where, apart from commenting on the work of Brahmagupta, gave a construction of an *infinity* of **not necessarily** cyclic quadrilaterals with rational sides and rational diagonals(we note that Brahmagupta's construction always gave rational quadrilaterals with rational areas). The rest of this article is devoted not merely to reproduce the proofs of Kummer, but also to give a broad indication of his main ideas. We shall end this note by showing that a particular construction of Kummer of such quadrilaterals breaks down in general if **one of the diagonals bisects the other**. This result arises from a proposition, which we prove which says that this happens in special cases where the absolute value of the cosine of the angles between the diagonals is for example, $1/3$. We in fact show this by proving that there cannot exist a rational triangle one of whose medians makes an angle with the base whose cosine has absolute value $1/3$. Very

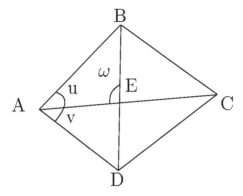

Figure 6: *Rational Quadrilateral*

interestingly, this elementary geometric fact is equivalent to the celebrated theorem of Fermat(with a rather incomplete proof of Euler, and a complete proof for the first time in (Itard, 1973)) that there do not exist four rational squares in a non-trivial arithmetic progression.

3   Kummer's Construction

We begin with the following beautiful and elementary fact established by Kummer.

Lemma 2. *Let ABCD be a quadrilateral such that all its sides and diagonals are rational. Then the diagonal intercepts BE,ED,AE,EC are all rational.*

Proof. We begin by remarking the fact, which we noted already, that in a triangle with all its sides rational, all the cosines of its angles are rational. In triangle ABC, $\cos u$ is therefore rational. Similarly, considering the triangle ACD, it follows that $\cos v$ is rational. On the other hand, in $\triangle BAD$, $\cos(u+v)$ is rational, since BD is rational. Hence $\cos u \cos v - \sin u \sin v = \cos(u+v)$ is rational. Therefore, $\frac{\sin u}{\sin v}$ is rational, since, in any case, $\sin^2 v = 1 - \cos^2 v$ is rational. If ω denotes angle BEA then, $\frac{AB}{\sin \omega} = \frac{BE}{\sin u}$. Similarly, in $\triangle AED$ we have $\frac{AD}{\sin(\pi - \omega)} = \frac{ED}{\sin v}$. We also have $\frac{BE}{ED} = \frac{BE}{ED} \frac{\sin u \sin v}{\sin u \sin v}$, but $\frac{BE}{\sin u} = \frac{AB}{\sin \omega}$ and $\frac{ED}{\sin v} = \frac{AD}{\sin \omega}$, so that $\frac{BE}{ED} = \frac{AB}{\sin \omega} \frac{\sin u \sin \omega}{\sin v \, AD}$, which is rational since each factor on the right hand side is rational. However,

$BD = BE + ED$ is given to be rational. Therefore, BE and ED are both rational. Similarly, AE and EC are rational. Hence the lemma is proved. □

The next proposition of Kummer gives a parametrisation of any rational triangle with an angle whose cosine is a given rational. The original proof of Kummer is quite complicated. We present a simpler proof.

Proposition 2. *Let ABE be a rational triangle. Let angle $BEA = \omega$ and $\cos \omega = c$. Let $a = AB, \beta = BE, \alpha = AE$. Then, for any arbitrary positive rational ξ the ratios $\frac{a}{\beta}$ and $\frac{\alpha}{\beta}$ are given by $\frac{a}{\beta} = \frac{\xi^2 + k^2}{2\xi}$ and $\frac{\alpha}{\beta} = \frac{(\xi+c)^2 - 1}{2\xi}$. And this indeed parametrises all rational triangles with a fixed c: here $k^2 = 1 - c^2$.*

Proof. From the cosine formula, we have $a^2 = \alpha^2 + \beta^2 - 2c\alpha\beta$. Dividing by β^2, we get $(\frac{a}{\beta})^2 = (\frac{\alpha}{\beta})^2 + 1 - 2c\frac{\alpha}{\beta}$, so that $(\frac{a}{\beta})^2 = 1 + (\frac{\alpha}{\beta} - c)^2 - c^2 = (\frac{\alpha}{\beta} - c)^2 + k^2$. i.e., $(\frac{a}{\beta})^2 - (\frac{\alpha}{\beta} - c)^2 = k^2$, which incidentally shows that we can parametrise by putting $\frac{a}{\beta} + \frac{\alpha}{\beta} - c = \xi$ and $\frac{a}{\beta} - \frac{\alpha}{\beta} + c = \frac{k^2}{\xi}$ for a rational number ξ. Solving this pair of equations for $\frac{a}{\beta}$ and $\frac{\alpha}{\beta}$ proves the proposition. □

Remark 3. Dickson (1921) in fact shows that $\xi = \frac{\sin \omega}{\sin u}(1 + \cos u)$, where u is the angle at A. This can be seen easily from the expressions for $\frac{a}{\beta}$ and $\frac{\alpha}{\beta}$.

We shall now show how the above proposition was used by Kummer to write down a polynomial indeterminate equation whose solution leads to rational quadrilaterals. Kummer realised that to construct such quadrilaterals it is enough to construct four different triangles AEB, BEC, CED, DEA with a common vertex E and with a single compatibility condition which we write below.

Let $AB = a$, $BC = b$, $CD = d$, $DA = e$ with a common vertex E we shall force such a compatibility condition. Suppose for the moment that such a quadrilateral exists. Let us call the diagonal intercepts $AE = \alpha$, $BE = \beta$, $CE = \gamma$, $DE = \delta$.

Repeating the parametrisation described in Proposition 2 for triangle ABE, for the triangles CEB, CED and DEA, we have rational parameters (η, y, x) such that: $\frac{\gamma}{\beta} = \frac{(\eta-c)^2 - 1}{2\eta}$, $\frac{b}{\beta} = \frac{\eta^2 + k^2}{2\eta}$, $\frac{\delta}{\gamma} = \frac{(y+c)^2 - 1}{2y}$, $\frac{d}{\gamma} = \frac{y^2 + k^2}{2y}$ and $\frac{\delta}{\alpha} = \frac{(x-c)^2 - 1}{2x}$, $\frac{e}{\alpha} = \frac{x^2 + k^2}{2x}$. We note that this choice implies a single condition

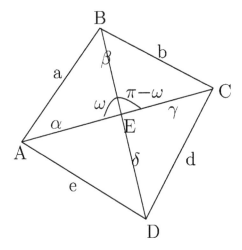

Figure 7: *Kummer's Construction*

that $\frac{\alpha}{\beta}\frac{\delta}{\alpha} = \frac{\delta}{\beta} = \frac{\delta}{\gamma}\frac{\gamma}{\beta}$; using the explicit values for the ratios we have the condition

$$\left(\frac{(\xi+c)^2-1}{2\xi}\right)\left(\frac{(x-c)^2-1}{2x}\right) = \left(\frac{(y+c)^2-1}{2y}\right)\left(\frac{(\eta-c)^2-1}{2\eta}\right). \qquad (1)$$

Conversely, for any set of rational parameters, ξ, x, y, η and for a fixed rational number c with $|c| < 1$, we do obtain a quadrilateral all of whose sides and diagonals are rational, provided (1) is satisfied, since the equations involving the ratios of the diagonal intercepts are then consistent and give rational values for $\alpha, \beta, \gamma, \delta$ and the ratios of the sides $\frac{a}{\beta} = \frac{\xi^2+k^2}{2\xi}$, $\frac{b}{\beta} = \frac{\eta^2+k^2}{2\eta}, \frac{d}{\gamma} = \frac{y^2+k^2}{2y}$ and $\frac{e}{\alpha} = \frac{x^2+k^2}{2x}$. We may take one of the values, say β, to be 1. We then assign arbitrary rational values to the parameters ξ, η, c(with $|c| < 1$) and look for rational solutions for x, y.

If we do this, what we end up with is a polynomial equation of the form $F(x, y) = 0$,where F has rational coefficients and we look for rational solutions for x, y. Let us look at the form of the polynomial F(having made, as we said, definite choices for ξ, η and c). Then the equation is

$$\alpha\left(\frac{(x-c)^2-1)}{2x}\right) = \gamma\left(\frac{(y+c)^2-1}{2y}\right) \qquad (2)$$

which written out in long hand takes either of the two equivalent forms

$$F : xy^2\gamma - (\alpha x^2 - 2cx(\alpha+\gamma) - k^2\alpha)y - k^2x\gamma = 0 \qquad (3)$$

$$F : yx^2\alpha - (y^2\gamma + 2cy(\alpha + \gamma) - k^2\gamma)x - k^2y\alpha = 0 \qquad (4)$$

In other words, we need to show that (x, y) should satisfy the polynomial equation $F(X, Y) = 0$ which is of degree two in each of the variables X and Y. This is a problem that goes back to Euler. In modern terminology, $F(X, Y) = 0$ represents a curve of genus 1 over $\mathbb{Q}$, and what we are looking for are 'rational points' of this curve over $\mathbb{Q}$.

Perhaps at this stage, we should also remark that Kummer also considered the problem of construction of quadrilaterals with rational sides and diagonals whose area is also rational. Since the area in question is the sum of the areas of the four constituent triangles that we had constructed, it is obviously $1/2(\alpha\beta + \beta\gamma + \gamma\delta + \delta\alpha)\sin\omega$. Thus, for the area of the quadrilateral also to be rational, we simply require that apart from $\cos\omega$, $\sin\omega$ should also be rational. The way that this can be achieved is by demanding further for rational solutions of $\sin^2 t + \cos^2 t = 1$ and therefore choosing $\cos\omega = \frac{r^2-1}{r^2+1}$, we have, $\sin\omega = \frac{2r}{r^2+1}$ for r rational.

4   Fagnano's Doubling of the Lemniscate and Euler's Ascent Method

The so called Lemniscate curve which is an 'algebraic curve' defined by the equation $r^2 = \cos 2\theta$ in polar coordinates in the plane, or equivalently, by its cartesian equation $(x^2 + y^2)^2 = x^2 - y^2$ is a curve which looks like a figure of reclining eight, was a very great favourite for the mathematicians of the eighteenth and nineteenth centuries. A very interesting thing about the curve is that, though it is algebraic, the arc-length of this curve from $(0,0)$ to an arbitrary point (x, y) of the curve is given by an elliptic integral, $\int_0^r \frac{dr}{\sqrt{1-r^4}}$, r denoting the radius vector. In other words, the infinitesimal length of the curve is given by the differential $\frac{dt}{\sqrt{1-r^4}}$. Fagnano studied the curve in extenso and showed in particular that he could halve the arc of the Lemniscate by ruler and compass. (It goes to the credit of Gauss(unpublished) and Abel much later, that they both gave a necessary and sufficient condition, like for a circle for a Lemniscate to be divided into n equal parts by ruler and compass). This work of Fagnano reached the Berlin Academy on 23rd December 1751 and was handed over to Euler(Jacobi remarks that this date should be fixed as the birthdate of elliptic functions). Euler got very much interested in the work of

Fagnano, and began working on the differential, $\omega = \frac{dt}{\sqrt{1-t^4}}$. To make matters short, Euler realised that Fagnano's work showed how to construct an algebraic solution apart from the trivial one of the differential equation $\frac{dx}{\sqrt{1-x^4}} = \frac{dy}{\sqrt{1-y^4}}$. Fagnano's work amounted to showing that the equation $x^2 y^2 + x^2 + y^2 - 1 = 0$ is indeed another algebraic solution(this fact can be easily verified by simple computation; the "general solution" is $c^2 x^2 y^2 + x^2 + y^2 = c^2 + 2xy \sqrt{1 - c^2}$ for a constant c). Fagnano must have been motivated to solve the above differential equation by analogy with the differential equation $\frac{dx}{\sqrt{1-x^2}} = \frac{dy}{\sqrt{1-y^2}}$, which has, apart from the trigonometric solution $\sin^{-1} x - \sin^{-1} y = constant$, an algebraic solution $x^2 + y^2 = 1$. This really should be thought of as the beginning of the theory of elliptic functions, which of course Euler did not realise during his lifetime and it needed the genius of Jacobi and Abel(independently) to invert the integral $\int_0^x \frac{dt}{\sqrt{1-t^4}}$ to define the so-called Lemniscate functions, which were indeed the first examples of elliptic functions. But we shall not go into this beautiful piece of history but merely state that equations of type $F(x, y) = 0$ such as the one given by Fagnano, where the degrees of x and y are at most two, began to be of great interest to Euler for the rest of his life, and he worked on them extensively.

Being a number-theorist, Euler was always interested in constructing rational and integral solutions for polynomial equations with rational coefficients, and he found a beautiful way of producing solutions of polynomial equations of the type $F(x, y) = 0$ (with degrees of x, y at most 2) provided one could start with one solution (x_1, y_1) of $F(x, y) = 0$. This is now called the *ascent method* of Euler (see Weil (1987)). We shall explain this a bit and show how Kummer used this method of Euler to produce solutions of equations (3), (4) mentioned earlier, in order to construct quadrilaterals with rational sides and diagonals. We shall first briefly discuss Euler's ascent method.

Let $F(x, y)$ be a polynomial over $\mathbb{Q}$. Suppose $F(x_1, y_1) = 0$ for some x_1, y_1 in $\mathbb{Q}$. The polynomial in y, given by $F(x_1, y)$ has $y = y_1$ as a root and since $F(x_1, y)$ is quadratic in y, there must be(in general) another root y_2, so that $F(x_1, y_2) = 0$. Thus starting with a solution (x_1, y_1) for $F(x, y) = 0$, we have found another, (x_1, y_2). Now fix y_2 and consider the quadratic equation $F(x, y_2) = 0$. This has $x = x_1$ as a root, and therefore(generally) has another root, $x = x_2 \neq x_1$ i.e., $F(x_2, y_2) = 0$. Thus we have produced a third solution (x_2, y_2) for $F(x, y)$, and we can continue this procedure provided it does not close up which could happen. There is a classical thoerem of the French geometer Poncelet, which asserts that, if there is one polygon of n-sides circumscribing a conic and inscribed in another, then any polygon constructed by choosing a point on the outer conic and drawing tangents to the inner one will also be another polygon with n sides. This can be proved for example by using Euler's method, and the process does close up in this case.

5   Back to Kummer's Construction

In order to construct an infinite number of rational solutions to equation (2) of section 3, Kummer wanted to use Euler's method. So Kummer first notes that there is at least one solution that leads to a non-trivial quadrilateral. The value, $(x, y) = (\eta, \xi)$ gives a genuine quadrilateral. We then take $\beta = 1$, have $\alpha = \frac{(\xi+c)^2-1}{2\xi}, \gamma = \frac{(\eta-c)^2-1}{2\eta}, \delta = \alpha\gamma$ as the diagonal intercepts and as for the sides we have $AB = \frac{\xi^2+k^2}{2\xi}, BC = \frac{\eta^2+k^2}{2\eta}, CD = \left((\eta - c)^2 - 1\right)\left(\frac{\xi^2+k^2}{4\xi\eta}\right), DA = \left((\xi + c)^2 - 1\right)\left(\frac{(\eta^2+k^2)}{4\xi\eta}\right)$. This formula also contains the most general expression for all cyclic quadrilaterals with rational sides and diagonals, since in order that a quadrilateral be cyclic, we must have $\delta = \alpha\gamma$. Further, if we need the area to be rational, we can choose as we said earlier, $c = \frac{r^2-1}{r^2+1}$ for some $r \in \mathbb{Q}$.

One must mention that this construction subsumes the one given by Brahmagupta of juxtaposing rational triangles.

As an example Kummer takes $c = 1/2, \xi = 3/4, \eta = 3, \alpha = 3/8, \beta = 1, \gamma = 7/8, \delta = 21/64$ or in natural numbers, take $\alpha = 24, \beta = 64, \gamma = 56, \delta = 21$ and the sides $AB = 56, BC = 104, CD = 49$ and $DA = 39$ and the diagonals $AC = 80, BD = 85$. We note that this quadrilateral does not have rational area. (Kummer's paper has $\xi = 2$, which is a typographical error.)

6   Kummer's Construction of an Infinite Number of Rational Quadrilaterals

As we remarked, Kummer constructed an infinity of such quadrilaterals by using the method of Euler, which we briefly discussed eariler. To do this,

Kummer first notes that equation (3) or (4) of section 3 need not work, since the Euler method would lead from value x to the value $\frac{-k^2}{x}$ and lead back to x the second time. And similarly for y. Thus he suitably modifies the equation (3), by writing $xy = z$ and treating the equivalent equations,

$$z^2 \gamma - (x^2 \alpha - 2c(\alpha + \gamma)x - k^2 \alpha)z - k^2 x^2 \gamma = 0 \tag{5}$$

$$(z\alpha + k^2 \gamma)x^2 - 2c(\alpha + \gamma)zx - z(\gamma z + k^2 \alpha) = 0 \tag{6}$$

in x and z.

If (z, z') are roots of (5) and (x, x') are two roots of (6), we obviously have $zz' = -k^2 x^2, z+z' = \frac{\alpha x^2 - 2c(\alpha+\gamma)x - k^2\alpha}{\gamma}, xx' = \frac{-z(\gamma z + k^2 \alpha)}{z\alpha+k^2\gamma}, x+x' = \frac{2c(\alpha+\gamma)z}{z\alpha+k^2\gamma}$. If we take $x = 0$ then we find the corresponding values for z are $z = 0$ and $z = -\frac{k^2 \alpha}{\gamma}$. The first substitution, $z = 0$ leads nowhere, and the second leads to the pair $(x, z) = (0, -\frac{k^2\alpha}{\gamma})$. Using this value of z, we get $x' = \frac{2c\alpha}{\alpha-\gamma}$ using (6), which leads to the new value $z = \frac{-k^2\alpha}{\gamma}$ and $x = \frac{2c\alpha}{\alpha-\gamma}$ which are new rational values for x and z and therefore for y also. We also find for the second value z' to which $x' = \frac{2c\alpha}{\alpha-\gamma}$ corresponds to be $z' = \frac{4c^2\alpha\gamma}{(\alpha-\gamma)^2}$, and we get

$$x'' = -\frac{2c\alpha(4c^2\gamma^2 + k^2(\alpha-\gamma)^2)}{(\alpha-\gamma)(4c^2\alpha^2 + k^2(\alpha-\gamma)^2)},$$

and a new value z'',

$$z'' = -\frac{k^2\alpha(4c^2\gamma^2 + k^2(\alpha-\gamma)^2)^2}{(4c^2\alpha^2 + k^2(\alpha-\gamma)^2)^2},$$

and continuing this process leads Kummer to determine an infinity of solutions for (2), thereby leading to a proof of the existence of an infinity of rational quadrilaterals(the details of which we do not go into). We simply give as illustrations, two examples due to Kummer.

Example 1. Take $c = 1/2, \xi = 3/2, \eta = 3$. Then $\alpha = 1, \beta = 1, \gamma = 7/8$. Further, $x' = 8$, then $\delta = 221/64$ or in whole numbers, $\alpha = 64, \beta = 64, \gamma = 56, \delta = 221$; further $AB = 64, BC = 104, CD = 199, DA = 259$ and the diagonals are $AC = 120, BD = 285$.

Example 2. Let $c = 4/5, \xi = 1, \eta = 3, \alpha = 28/25, \beta = 1, \gamma = 16/25$, thus $x' = 56/15, \delta = 1711/1500$ or in whole numbers $\alpha = 1680, \beta = 1500, \gamma = 960, \delta = 1711$; further the sides are $AB = 1020, BC = 2340, CD = 1105, DA = 3217$ and the diagonals $AC = 2640, BD = 3211$. It is a quadrilateral with area 2543112.

Everything looks fine, *except* that there is one snag in the argument. The crucial point to note is that $\alpha - \gamma$ occurs in the denominator and therefore these formulae do not make sense in the case where $\alpha = \gamma$. In fact, it is proved in (Nagarajan and Sridharan, 2006) that when $\gamma = \alpha$, i.e., one of the diagonals bisects the other, such quadrilaterals may not exist in general. There, we prove the following curious proposition in elementary geometry which leads to the proof of the non-existence of such quadrilaterals. It must be remarked that Kummer concludes with a different method of construction of infinitely many rational quadrilaterals with rational diagonals towards the end of his paper(the details of which we shall not go into).

Proposition 3. *Let ABC be a triangle and BE the median through B meeting AC at E. Let $\angle BEA = \omega$, where $\cos \omega = 1/3$. Then if $AB = c, AE = EC = \alpha, BE = \beta$ and $BC = a$, then the numbers $(\alpha - \beta)^2, c^2, a^2, (\alpha + \beta)^2$ are in an arithmetic progression.*

Proof. We have by the cosine formulae,

$$c^2 = \alpha^2 + \beta^2 - 2/3\alpha\beta$$
$$a^2 = \alpha^2 + \beta^2 + 2/3\alpha\beta$$

Adding these equations and subtracting the first from the second, we get the following equations,

$$a^2 + c^2 = (\alpha + \beta)^2 + (\alpha - \beta)^2$$
$$3a^2 - 3c^2 = (\alpha + \beta)^2 - (\alpha - \beta)^2$$

Now, adding the above equations and subtracting the second from the first and cancelling factors of 2 in both, we get,

$$2a^2 - c^2 = (\alpha + \beta)^2$$
$$2c^2 - a^2 = (\alpha - \beta)^2$$

But the last two equations imply that $(\alpha - \beta)^2, c^2, a^2, (\alpha + \beta)^2$ are in an arithmetic progression. $\square$

Corollary 1. *There cannot be a triangle ABC with rational sides with a rational median which makes an angle ω with the opposite side of cosine whose absolute value is $1/3$.*

Proof. Since with the notation of the Proposition 3, c, a, α, β are all rational numbers, Proposition 3 gives a contradiction to a well known claim of

Fermat(proved by Euler and more rigorously by Itard (1973)) that there cannot exist four rational squares in an arithmetic progression. Hence the corollary is proved. □

Acknowledgements

I thank Sujatha for her suggestion that I contribute an article to this volume. To M.D. Srinivas and Raja, I am for ever grateful for their kindness, help and encouragement. K.R. Nagarajan is as responsible as I am (cf. Nagarajan and Sridharan (2006)) for the existence of this article, what ever be its merits. I am grateful to him for allowing we to use some of the results of the paper. Finally, I must confess that I cannot find adequate words to express my indebtedness to Vinay, first for his generous offer to make this tex file out of my total scribble and what is more, his ever willing help at every stage, which he gave with such a sweet and gracious smile.

References

Chasles, M. 1837. Aperçu Historique, Bruxelles, Note 12, p. 440.

Colebrooke, H.T. 1817. *Algebra, Arithmetic and Mensuration from the Sanskrit of Brahmagupta and Bhaskara.*

Dickson, L.E. 1921. Rational triangles and quadrilaterals. *Amer. Math. Monthly* **28**:244–250.

Itard, J. 1973. Arithmétique et Théorie des Nombres. *PUF*, Paris.

Kummer, E.E. 1848. Über die Vierecke, deren Seiten und Diagonalen rational sind. *J. Reine Angew. Math.* **37**:1–20.

Nagarajan, K.R. and R. Sridharan. 2006. On Brahmagupta's and Kummer's quadrilaterals, *Elem. Math.* **61**:45–57.

Weil, A. 1987. Number Theory. An Approach through History, *Birkhäuser Verlag.*

CHAPTER 25

Magic Squares

Raja Sridharan* and M. D. Srinivas**

*School of Mathematics,
Tata Institute of Fundamental Research,
Mumbai 400005, India.
e-mail: sraja@math.tifr.res.in

**Centre for Policy Studies, Mylapore,
Chennai 600004, India.
e-mail: mdsrinivas50@gmail.com

1 Introduction

Let us begin by recalling that an $n \times n$ magic square is a square array consisting of n^2 numbers whose rows columns and principal diagonals add up to the same number. In this article, we discuss some aspects of the history of magic squares in India and in particular, some of the work of the renowned mathematician Nārāyaṇa Paṇḍita on magic squares. Nārāyaṇa Paṇḍita gave a systematic account of magic squares in the fourteenth chapter of his work *Gaṇitakaumudī* (c.1356 CE), entitled *Bhadra-gaṇita* or "auspicious mathematics" (See (Padmākara Dvivedi, 1936; Kusuba, 1993; Singh, 2002)).

The study of magic squares in India has a long history, going back to the very ancient times. For example, the work of the ancient seer Garga is supposed to contain several 3×3 magic squares. Later, a general class of 4×4 magic squares has been attributed to the Buddhist philosopher Nāgārjuna (c. 2nd Century CE). In the great compendium *Bṛhatsaṃhitā* of Varāhamihira (c. 550 CE), we find a mention of the following magic square (referred to as *sarvatobhadra* or "auspicious all around" (See (Datta and Singh, 1992))):

$$
\begin{array}{cccc}
2 & 3 & 5 & 8 \\
5 & 8 & 2 & 3 \\
4 & 1 & 7 & 6 \\
7 & 6 & 4 & 1
\end{array}
$$

In this connection, it is interesting to note the following. To the above magic square, let us add the following magic square:

$$
\begin{array}{cccc}
0 & 8 & 0 & 8 \\
8 & 0 & 8 & 0 \\
8 & 0 & 8 & 0 \\
0 & 8 & 0 & 8
\end{array}
$$

We then get the magic square

$$
\begin{array}{cccc}
2 & 11 & 5 & 16 \\
13 & 8 & 10 & 3 \\
12 & 1 & 15 & 6 \\
7 & 14 & 4 & 9
\end{array}
$$

which is one of the magic squares considered by Nārāyaṇa Paṇḍita. This magic square incidentally belongs to the class of so called pan-diagonal magic squares, which have the property that apart from the sum of the entries of each row, column and the principal diagonals, the sum of each of the entries of all "broken diagonals" (for example, 12,8,5,9 in the above square) add up to the same number (namely 34 in the above square).

From very ancient times, the Jains[1] were also interested in the study of magic squares. In the entrance of a Jaina temple at Khajuraho (c. 12th century), we find inscribed the following pan-diagonal magic square:

$$
\begin{array}{cccc}
7 & 12 & 1 & 14 \\
2 & 13 & 8 & 11 \\
16 & 3 & 10 & 5 \\
9 & 6 & 15 & 4
\end{array}
$$

---

[1]Jainism is a very ancient Indian religion that prescribes a path of non-violence towards all living beings.

The Prākṛta work, *Gaṇitasārakaumudī* of Ṭhakkura Pheru (c.1300) (Sarma *et al.*, 2009), discusses the classification of $n \times n$ magic squares in to following three types (a classification which is employed later by Nārāyaṇa Paṇḍita also): (1) Samagarbha, where n is of the form $4m$, where m is a positive integer, (2) Viṣamagarbha, where n is of the form $4m + 2$, where m is a positive integer, and (3) Viṣama where n is odd. Ṭhakkura Pheru also indicates methods of constructing samagarbha and viṣama magic squares.

2   An example of a 3×3 magic square

A very famous example of a 3×3 magic square is the so called Lo-shu[2] magic square. The story is told that around 2200 BCE, the emperor Yu observed a divine tortoise crawling out of the Yellow River. On the back of the turtle was inscribed the following 3x3 array of numbers (which is now called as the Lo-shu magic square):

$$
\begin{array}{ccc}
4 & 9 & 2 \\
3 & 5 & 7 \\
8 & 1 & 6
\end{array}
$$

It is a well known fact that in a 3×3 magic square with entries $1, 2, 3, \ldots, 9$, the middle cell should always be 5. To see this, consider magic square

$$
\begin{array}{ccc}
a & b & c \\
d & e & f \\
g & h & i
\end{array}
$$

whose entries are given by $1, 2, 3, \ldots, 9$. Now, the sum of entries in any row or column or diagonal should be 15. Hence we have,

$$(a + e + i) + (c + e + g) + (b + e + h) + (d + f + e) = 60$$

Hence,

$$(a + b + c + d + e + f + g + h + i) + 3e = 60$$

Since $a, b, \ldots i$, are in fact the numbers $1, 2, 3, \ldots, 9$, in some order, we have $45 + 3e = 60$ or $e = 5$.

---

[2]Also known as the Nine Halls Diagram, it is the unique normal magic square of order three.

3 Nārāyaṇa Paṇḍita's 4×4 pan-diagonal magic squares

We now consider the set of all 4×4 pan-diagonal magic squares with entries given by the first sixteen natural numbers $1, 2, \ldots, 16$. We note first that in this case, the sum of the entries in any row, column, diagonal or broken diagonal is 34. In his *Gaṇitakaumudī* (c.1356 CE), Nārāyaṇa Paṇḍita displayed 24 pan-diagonal 4×4 magic squares, with different cells being filled by different numbers from the arithmetic sequence $1, 2, \ldots, 16$, the top left entry being 1. Nārāyaṇa also remarked that (by permuting the rows and columns cyclically) we can construct 384 pan-diagonal 4×4 magic squares with entries $1, 2, \ldots, 16$. In 1938, Rosser and Walker (1938), proved that this is in fact the exact number of 4×4 pan-diagonal magic squares with entries $1, 2, \ldots, 16$. Vijayaraghavan (1941), gave a much simpler proof of this result, which we shall briefly outline here. We begin with the following Lemmas due to Rosser and Walker:

Lemma 1. *Let M be a pan-diagonal 4×4 magic square with entries $1, 2, \ldots, 16$, which is mapped on to the torus by identifying opposite edges of the square. Then the entries of any 2×2 sub-square formed by consecutive rows and columns on the torus add up to 34.*

Proof. Let M be of the form

$$
\begin{matrix}
a & b & c & d \\
e & f & g & h \\
i & j & k & l \\
m & n & o & p
\end{matrix}
$$

Consider the following sub-squares M_1, M_2, M_3 given below:

$$
M_1 = \begin{matrix} a & b \\ e & f \end{matrix}, \quad M_2 = \begin{matrix} c & d \\ g & h \end{matrix} \quad \text{and } M_3 = \begin{matrix} i & j \\ m & n \end{matrix}
$$

Let s_1, s_2, s_3 be the sums of the entries of M_1, M_2, M_3 respectively. Since M is a magic square, we have $s_1 + s_2 = 68$, and using the fact that M is pan-diagonal, we have $s_2 + s_3 = 68$ and $s_3 + s_1 = 68$.

From these relations we obtain $s_1 = s_2 = s_3 = 34$, proving the lemma for these particular cases. The proof for any other sub-square is similar. □

From the above Lemma, we also obtain relations such as

$$a + b + m + n = 34$$
$$a + d + m + p = 34$$

and so on. We now proceed to the following:

Lemma 2. *Let M be a 4 × 4 pan-diagonal magic square with entries* $1, 2, \ldots, 16$, *which is mapped on to the torus. Then, the sum of an entry of M with another which is two squares away from it along a diagonal (in the torus) is always* 17.

Proof. Let M be as above. We shall for instance show that

$$b + l = j + d = 17$$

For this consider the following sums

$$r_1 = a + c + i + k$$
$$r_2 = b + d + j + l$$
$$r_3 = e + g + m + o$$

Now, since M is pan-diagonal, we have $r_1 + r_2 = 68$, and similarly, $r_2 + r_3 = 68$ and also $r_3 + r_1 = 68$. Therefore $r_1 = r_2 = r_3 = 34$. In particular,

$$r_2 = b + l + j + d = 34$$

Again from the pan-diagonal character of M it follows that

$$b + g + l + m = d + g + j + m$$

and hence, using $r_2 = 34$,

$$b + l = j + d = 17$$

The rest of the cases can be handled similarly. □

Before we state the next Lemma, we need to define the notion of the "neighbours" of an element. The "neighbours" of an element of a 4 × 4 pan-diagonal magic square (which is mapped on to the torus as before) are the elements which are next to it along any row or column. For instance, in the magic-square M written above, the neighbours of f are b, g, j and e; the neighbours of n are j, o, b and m. Now we can state the following:

Lemma 3. *Let M be a 4 × 4 pan-diagonal magic square with entries* $1, 2, \ldots, 16$*, which is mapped on to the torus. Then the neighbours of the entry* 16 *have to be the entries* $2, 3, 5$ *and* 9 *in some order.*

Proof. Let M be as given above. Since a pan-diagonal magic square can be mapped on to a torus, it follows that shifting the rows and columns cyclically does not change the pan-diagonal nature of the square or the neighbours of any element. Hence, we may assume that, we have, say, $a = 1$.

From Lemma 2, it follows that $k = 16$. Hence, c and d can at most be 14 and 15. Since $a + b + c + d = 34$, we have $a + b \geq 5$, forcing $b \geq 4$. Similarly, $c \geq 4$ and $d \geq 4$.

Now from Lemma 1, we have

$$a+b+e+f = a+b+m+n = a+e+i+m = a+e+d+h = a+f+k+p = 34.$$

In view of this we have that the elements e, f, i, k, m, n are all greater than or equal to 4. It follows that two of the entries g, j, l, and o will have to be 2 and 3, thus proving part of the Lemma.

Suppose we take $g = 2$ and $o = 3$. (The other cases can be handled in a similar manner) Then in view of Lemmas 1 and 2, our magic square M has the following form.

$$
\begin{array}{cccc}
1 & b & 13 & d \\
14 & f & 2 & h \\
4 & j & 16 & l \\
15 & n & 3 & p
\end{array}
$$

It now follows by a simple case by case analysis and Lemmas 1 and 2 that either $j = 5, l = 9$ or $j = 9, l = 5$, and the magic square has either of the following two forms:

$$
\begin{array}{cccc}
1 & 8 & 13 & 12 \\
14 & 11 & 2 & 7 \\
4 & 5 & 16 & 9 \\
15 & 10 & 3 & 6
\end{array}
$$

$$
\begin{array}{cccc}
1 & 12 & 13 & 8 \\
14 & 7 & 2 & 11 \\
4 & 9 & 16 & 5 \\
15 & 6 & 3 & 10
\end{array}
$$

This proves the Lemma. □

We can now prove the main result that we mentioned earlier:

Proposition. *There are precisely* 384 *pan-diagonal* 4 × 4 *magic squares with entries* 1, 2, . . . , 16.

Proof. (Vijayaraghavan, 1941) We first observe that there is an action of S_4, the symmetric group of permutations of four symbols, on the set S of pan-diagonal 4×4 magic squares given as follows. Let M be a pan-diagonal magic square as given above. We consider the neighbours b, j, g and e of a fixed element say f. Then, we claim that any element of S_4 gives rise to a permutation of the neighbours of f and with suitable changes in the other entries, to another pan-diagonal magic square. Since S_4 is generated by transpositions, it is enough to check this where the permutation is a transposition. For example, if we fix g and j and permute b and e, we obtain the pan-diagonal magic square:

$$
\begin{array}{cccc}
a & e & d & h \\
b & f & g & c \\
n & j & k & o \\
m & i & l & p
\end{array}
$$

Now, we have already observed that a pan-diagonal magic square retains its pan-diagonal nature if we cyclically permute its rows or columns. Hence the group $\mathbb{Z}/4 \times \mathbb{Z}/4$ of cyclic permutations of rows and columns acts on the set S of all pan-diagonal 4 × 4 magic squares. The cardinality of each orbit of this action is 16. Further each orbit contains a magic square with the first entry $a = 1$ and $k = 16$. From what we have proved in Lemma 3 already, it follows that there are 24 magic squares with $a = 1$ and $k = 16$, which are obtained by the action of S_4 on S. Hence there are 24×16 = 384 pan-diagonal 4 × 4 magic squares with entries 1, 2, . . . , 16. □

4 Ancient Indian method for constructing magic squares of odd order

In this section, we shall briefly discuss the ancient Indian method for constructing magic squares of odd orders. The earliest knowledge that the

West acquired (in the modern period) of this Indian construction is through La Loubère [1], the French Ambassador to Siam, who wrote a letter to the French Secretary of State about it. He apparently learnt about this method from a French Doctor, M Vincent - who had lived in Surat in India and had learnt about this method from the Indians. We shall exhibit two magic squares which are constructed using this method, one of them a 3×3 square and the other a 5 × 5 square, which serve to illustrate the working of the method.

For the 3×3 magic square, we consider a square with elements 1, 2, . . ., 9 which is also assumed to be embedded in a torus. We place 1 in the middle of the top row. The next element 2 is to be placed in the cell obtained by going diagonally one step upwards towards the right. Since we are now on a torus, this cell is precisely the last cell in the right in the third row. The next number 3 has to be, for similar reasons, placed in the first cell in the left in the second row. This procedure is to be followed as long as the cell which is diagonally one step upwards towards the right of the given cell is vacant. Now the cell which the next number 4 has to occupy is not vacant, since it is already occupied by 1. In such a case, the rule to be followed is that we place this number in the cell immediately below, in the same column. Continuing this way we obtain the following 3 × 3 magic square:

$$
\begin{array}{ccc}
8 & 1 & 6 \\
3 & 5 & 7 \\
4 & 9 & 2
\end{array}
$$

The same procedure can be followed in order to construct a magic square of any odd order. For instance, we have the following 5 × 5 magic square obtained by the same procedure:

$$
\begin{array}{ccccc}
17 & 24 & 1 & 8 & 15 \\
23 & 5 & 7 & 14 & 16 \\
4 & 6 & 13 & 20 & 22 \\
10 & 12 & 19 & 21 & 3 \\
11 & 18 & 25 & 2 & 9
\end{array}
$$

We may note that Nārāyaṇa Paṇḍita has also discribed this simple method of constructing a magic square of odd order and also given examples of variants of this method.

5   Nārāyaṇa Paṇḍita's "folding" method for odd order magic squares

In this final section we discuss the so called "folding" method given by Nārāyaṇa Paṇḍita for constructing magic squares of odd order. Incidentally Nārāyaṇa Paṇḍita has presented a similar method for the construction of magic squares of doubly-even orders also.

We shall illustrate Nārāyaṇa's method, using a simple example, where we construct a 5×5 magic square having entries $1, 2, \ldots, 25$, and therefore having the magic sum (sum of the columns and rows) equal to 65, by using two 5×5 magic squares and "folding" them together. For this purpose, we choose a basic sequence $1, 2, 3, 4, 5$ and a second sequence $0, 1, 2, 3, 4$. We form a 5×5 square with $1, 2, 3, 4, 5$ in the central column and using cyclic permutations of these numbers to fill the succeeding columns. Once we reach the last column we go to the first column and proceed till we reach the column to the left of the central column that we started with. Thus our square looks as follows:

$$
\begin{array}{ccccc}
4 & 5 & 1 & 2 & 3 \\
5 & 1 & 2 & 3 & 4 \\
1 & 2 & 3 & 4 & 5 \\
2 & 3 & 4 & 5 & 1 \\
3 & 4 & 5 & 1 & 2
\end{array}
$$

We form a similar square with the sequence $0, 1, 2, 3, 4$:

$$
\begin{array}{ccccc}
3 & 4 & 0 & 1 & 2 \\
4 & 0 & 1 & 2 & 3 \\
0 & 1 & 2 & 3 & 4 \\
1 & 2 & 3 & 4 & 0 \\
2 & 3 & 4 & 0 & 1
\end{array}
$$

We multiply this square by 5 to get

$$
\begin{array}{ccccc}
15 & 20 & 0 & 5 & 10 \\
20 & 0 & 5 & 10 & 15 \\
0 & 5 & 10 & 15 & 20 \\
5 & 10 & 15 & 20 & 0 \\
10 & 15 & 20 & 0 & 5
\end{array}
$$

We "fold" the first square with this square (in the manner shown below) to obtain the desired 5×5 magic square having entries $1, 2, \ldots, 25$:

$$
\begin{array}{ccccc}
4+10 & 5+5 & 1+0 & 2+20 & 3+15 \\
5+15 & 1+10 & 2+5 & 3+0 & 4+20 \\
1+20 & 2+15 & 3+10 & 4+5 & 5+0 \\
2+0 & 3+20 & 4+15 & 5+10 & 1+5 \\
3+5 & 4+0 & 5+20 & 1+15 & 2+10
\end{array}
=
\begin{array}{ccccc}
14 & 10 & 1 & 22 & 18 \\
20 & 11 & 7 & 3 & 24 \\
21 & 17 & 13 & 9 & 5 \\
2 & 23 & 19 & 15 & 6 \\
8 & 4 & 25 & 16 & 12
\end{array}
$$

We shall not go into the very interesting mathematical basis of this method except to remark that it is in fact related to the construction of so called mutually orthogonal Latin Squares.

Acknowledgements

We thank Prof. Sujatha for kindly inviting us to contribute an article to this volume. We are grateful to Nivedita for her enthusiastic, timely and invaluable help in the preparation of this article.

References

Datta, B. and Singh, A.N. (Revised by K.S. Shukla). 1992. Magic Squares in India. *Ind. Jour. Hist. Sc.* **27**:51–120.

Dvivedi, Padmākara. 1936, 1942. *Gaṇitakaumudī of Nārāyaṇa Paṇḍita*, Ed. by Padmākara Dvivedi, 2 Vols, Varanasi.

Kusuba, T. 1993. Combinatorics and Magic-squares in India: A Study of Nārāyaṇa Paṇḍita's *Gaṇita-kaumudī*, PhD Dissertation, Brown University, Chapters 13–14.

Loubère, De La. 1693. Description Du Royaume De Siam, Paris, Tome II, pp. 235–288.

Rosser, B. and Walker, R.J. 1938. On the Transformation Groups of Diabolic Magic Squares of Order Four, *Bull. Amer. Math. Soc.* **44**:416–420.

Sarma, S.R., Kusuba, T., Hayashi, T. and Yano, M. 2009. *Gaṇitasārakaumudī* of Ṭhakkura Pheru, Ed. with English Translation and Notes by SaKHYa, Manohar, New Delhi (2009).

Singh, Paramanand. 2002. *The Gaṇita-kaumudī of Nārāyaṇa Paṇḍita*: Chapter XIV, English Translation with Notes, Gaṇita Bhāratī, **24**:34–98.

Vijayaraghavan, T. 1941. On Jaina Magic Squares, *The Math. Student.* **9**(3):97–102.

CHAPTER 26

Differences in Style, but not in Substance: Ancient Indian Study of Combinatorial Methods

Raja Sridharan*, R. Sridharan**
and M. D. Srinivas***

*School of Mathematics, Tata Institute of Fundamental Research,
Mumbai 400 005, India.
e-mail: sraja@math.tifr.res.in
**Chennai Mathematical Institute, Siruseri,
Chennai 603 103, India.
e-mail: rsridhar@cmi.ac.in
***Centre for Policy Studies, Mylapore,
Chennai 600 004, India.
e-mail: mdsrinivas50@gmail.com

Introduction

Hermann Weyl, the mathematician of great renown of the last century was invited to give two lectures in 1945 in a course on European History and Civilisation titled "Universities and Science in Germany", by the Princeton School of International Affairs. The text of these thoughtful lectures was published at the initiative of Prof. K Chandrasekharan, in India, in *The Mathematics Student* (Weyl, 1953), in 1953. It was indeed with great foresight, that Chandrasekharan thought of publishing the text in India, since these lectures of Weyl contained many valuable insights into how a mathematician thinks – in particular the universality of the language of mathematics.

The first part of the title of this article "Differences in style, but not in substance", is the phrase employed by Weyl in his lecture, while discussing the universal nature of mathematics; mathematics for the Chinese, Indians or Westerners is independent of the creed and nationality to which the people belong – though Weyl did realise and emphasises that there indeed could be individual differences in their approach to mathematics and this would depend on the cultural milieu of the people.

The aim of this article is to illustrate this "unity in diversity in Mathematics", through the development of Indian combinatorial methods, which have a very long history dating back to the Vedic times: Indeed the first work on Sanskrit Prosody (from which Indian combinatorics sprang) dates back to Piṅgala (around 300 BC). Towards the end of his Chandaḥśāstra, Piṅgala deals with combinatorics underlying prosody of Vedic chants - in the typical "sūtra style" which is cryptic to the point of being obscure and can be understood only through later commentaries like that of Halāyudha (c. 10th Century CE) and others. Combinatorics then evolved through time to apply not merely to Sanskrit Prosody, but to many other problems of enumeration, for instance to music (which we shall discuss here very briefly), medicine, architecture and so on. The aim of this article is indeed a very modest one, namely to give (hopefully) some idea of the problems that Indian combinatorics dealt with and their very unique and distinctive features. We refer to some of the articles listed in the references for a fuller and better understanding of the topic. If one thing should at all emerge from this article, it is the distinctive approach of the Indians – for instance a formula, in the Western sense, for an ancient Indian mathematician meant many times a 'table' which contains the formula implicitly!

1 A typical example of an Indian table which encodes many implicit formulae

Our point of departure is the following table of numbers which is an infinite descending 'staircase' arising from enumeration problems related to San-

Table 1: *Mātrā Meru*

							1	$s_1 = 1$
						1	1	$s_2 = 2$
						2	1	$s_3 = 3$
					1	3	1	$s_4 = 5$
					3	4	1	$s_5 = 8$
				1	6	5	1	$s_6 = 13$
				4	10	6	1	$s_7 = 21$
			1	10	15	7	1	$s_8 = 34$
			5	20	21	8	1	$s_9 = 55$
		1	15	35	28	9	1	$s_{10} = 89$
		6	35	56	36	10	1	$s_{11} = 144$
	1	21	70	84	45	11	1	$s_{12} = 233$
	7	56	126	120	55	12	1	$s_{13} = 377$

skrit Prosody, namely the prosody of the so called Mātrā Vṛttas, discussed in the next two sections.

We emphasise that what we have written above (Table 1) is only a finite part of an infinite staircase, but we shall give the rule by which it is constructed so that it can be continued and lead indeed to an unending staircase. We shall explain this procedure by a recursive rule to which we will come later. But note that if we sum the numbers in each row of this table, we get

$$s_1 = 1, s_2 = 2, s_3 = 3, s_4 = 5, s_5 = 8, \ldots$$

Those who have seen this sequence before would realise that what we have done is to generate the so called Fibonacci sequence.

The staircase is constructed as follows. On the topmost row which consists of a single square, one starts with the number 1. The rest of the steps have been so chosen that each pair of succeeding steps have the same number of squares. The first square from the left of the second step is filled with 1 and so also the next square. In the third step, the first square from the left is filled with the sum of the two numbers of the two slanting squares above it (here it is 2=1+1). The next square is filled with 1. This process is carried on inductively as follows. Suppose that we have filled in the squares up to the i-th row. Then the first number in the $(i + 1)$-st row will be 1 in case the i-th row starts with a number different from 1; otherwise it will be the sum of the numbers in the two slanting squares of the earlier two steps. The subsequent squares of the $(i + 1)$-st row are also filled with the sum of the numbers in the two slanting squares of the earlier two steps; except for the last square (on the right) which is always filled with 1. The figure above is thus explained.

As we noted, if we sum all the numbers in the same step, we obtain the numbers,

$$s_1 = 1, s_2 = 2, s_3 = 3, s_4 = 5, s_5 = 8, \ldots$$

We thus have generated by this "curious" procedure, some numbers which add up at the n-th level to the n-th Fibonacci number, which we know is defined by the recurrence relations $s_1 = 1, s_2 = 2, s_3 = 3$, and more generally $s_n = s_{n-1} + s_{n-2}$ for $n \geq 3$. These numbers occur in many different contexts. Fibonacci (the famous 12th century Italian mathematician) discovered them by looking at the breeding of rabbits! But in ancient India they were discussed very much earlier in connection with Prākṛt[1] Prosody and were explicitly written down in the 7th century A.D. by Virahāṅka (see (Velankar, 1962)). We will explain this in some detail in what follows.

We return to the table above to interpret it in other ways. We first note that the diagonals of the above table (read from top-right to bottom-left) are (1), (1,1), (1,2,1), (1,3,3,1), (1,4,6,4,1), ... We recognise immediately, that these are the binomial coefficients $\binom{n}{i}$.

These numbers were arranged by Pascal (in the 17th century) as a triangular array which is now called the Pascal triangle. Here, they occur (as a piece of miracle) in the above table too. There is a logical reason too for this which will be explained in Section 3. Note that in the n-th step of the figure, the numbers that occur in the squares are in order $\binom{n-\left[\frac{n}{2}\right]}{\left[\frac{n}{2}\right]}, \ldots \binom{n-1}{1}, \binom{n}{0}$. The fact that these add up (as was remarked earlier) to s_n gives the formula:

$$s_n = \sum_{i=0}^{\left[\frac{n}{2}\right]} \binom{n-i}{i}$$

It is indeed remarkable that the Indians came to this formula through the study of Prākṛt Prosody.

---

[1] Prākṛt is the name for a group of Middle Indic, Indo-Aryan languages, derived from old Indic dialects.

Table 2: *Pascal Triangle*

									1	1
								1	2	1
							1	3	3	1
						1	4	6	4	1
					1	5	10	10	5	1
				1	6	15	20	15	6	1
			1	7	21	35	35	21	7	1
		1	8	28	56	70	56	28	8	1
	1	9	36	84	126	126	84	36	9	1
1	10	45	120	210	252	210	120	45	10	1

We showed how the infinite staircase with which we started whose row sums give the Fibonacci numbers also gives rise to another infinite staircase (the so called Pascal triangle). We now show how from the Pascal triangle (Table 2) one can recover the infinite staircase that we started with.

The diagonals of this table (read from left to right) are $(1), (1, 1), (2, 1),$ $(1, 3, 1), \ldots,$ which are the rows of the first infinite staircase. The row sums of the above infinite staircase are $2, 4, 8, \ldots,$ in other words the powers of 2. The aim of the next section is to give some precise definitions in Prosody and explain these rather mysterious relations between these staircases.

2 Varṇa Vṛttas (studied by Piṅgala) and Mātrā Vṛttas (as a logical development of the former)

Classical Sanskrit Prosody consists in analysing metres[2] which one could convert into a mathematical language by a simple device. We recall that a syllable is the smallest unit of an utterance. There are two kinds of syllables in Sanskrit, short (laghu, denoted by *l*) and long or heavy (guru, denoted by *g*). For example क(= क् + अ) is a short syllable, where as का (= क् + आ) is a long one. The syllable ल preceding a conjunct consonant क्ष (= क् + ष) in लक्ष will also be counted as a long syllable. A *metre* is, for us, by definition, a finite sequence of syllables some long and some short,

---

[2]Metre is the basic rhythmic structure of a verse.

i.e., it is a sequence of the kind *lglgl*. The *length* of a metre is by definition the number of syllables in the metre. One of the things Piṅgala did
was to enumerate all metres of a given length in a specific manner and this
spreading is called *prastāra* - in this case *varṇa prastāra*. For example, the
prastāra of metres of length 1 is simply as shown (*g* always comes first by
convention) and has $2^1 = 2$ elements.

g
l

From this basic prastāra of metres of length 1, one can generate the *prastāra* of metres of length *n*, for any *n*, by an inductive process (as outlined
by Piṅgala in his book on metres). For instance, the prastāra of metres of
length 2 is obtained by a repetition of prastāra of metres of length 1 twice,
and adjoining *g* to the right of the first *prastāra* and adjoining *l* to the right
of the second prastāra. Thus the prastāra of metres of length 2 (shown
below) has $2^2 = 4$ elements.

g	g
l	g
g	l
l	l

An elementary induction (in retrospect) is enough to show how to write
down the prastāra of metres of length *n* which has obviously 2^n elements.
For instance, the prastāra of metres of length 3 can be constructed from the
prastāra of metres of length 2 and is:

g	g	g
l	g	g
g	l	g
l	l	g
g	g	l
l	g	l
g	l	l
l	l	l

This has $2^3 = 8$ elements. Incidentally, this number which is the number of rows in the prastāra is called *saṃkhyā*.

One can give a rule as to how to write the prastāra without the laborious method of induction, but we shall not go into it. In the context of a prastāra there are two questions that naturally arise which we call *naṣṭa* and *uddiṣṭa*. Suppose we would like to know a certain row of a prastāra, but it so happens that the rows of the prastāra have been erased by mistake. The problem of naṣṭa asks how one can exhibit the desired row without writing the prastāra once again. The other question uddiṣṭa is the converse of naṣṭa and asks how given the form of a metre (or, what is the same, a row of the prastāra) one can find its row number without writing down the whole prastāra. Both these questions were answered by Piṅgala beautifully in his own cryptic way showing that the sequence of long and short syllables that characterises a metre is nothing but a mnemonic for the dyadic expansion of the number $(i-1)$, where i is the associated row-number in the prastāra. Here, we take g, l to be the binary digits 0,1 and any metre, say *ggl* as a mnemonic for the dyadic expansion $0 \times 2^0 + 0 \times 2^1 + 1 \times 2^2 = 4$. Adding 1, we get the row number 5 associated with the metre *ggl* in the prastāra of metres of length 3.

Hidden in the above prastāra of metres of length n is the formula

$$1 + 2 + 2^2 + \cdots 2^n = 2^n - 1$$

This is a consequence of the fact that the last row of a prastāra is *lll...l* and the fact that the prastāra is a mnemonic for writing down the binary representation of numbers.

There is another very interesting question which can be asked referred to as *lagakriyā* which is: How many rows are there in a prastāra of metres of length n, which have a fixed number i ($i \leq n$) of gurus? The answer, if one thinks about it for a few minutes, is clearly the number of ways of choosing i gurus out of n syllables, which precisely is $\binom{n}{i}$ and this gives the reason for the discovery of binomial coefficients while dealing with prosody. This discovery was couched as a combinatorial diagram by Piṅgala called *Meru* and this is the discovery of the Pascal triangle (Table 2) ages before Pascal even thought of it.

The result that the sum of the numbers in the n-th row of the Pascal triangle is 2^n, or

$$\sum_{i=0}^{n} \binom{n}{i} = 2^n$$

is a re-interpretation of the statement that the total number of metres of length n is equal to the sum of the number of metres of length n having i gurus $(1 \leq i \leq n)$ which is equal to $\binom{n}{i}$ by lagakriyā.

We have seen in this section, how answering a "lagakriyā" question for metres of length n (for the so called *varṇa vṛttas*), leads to the fact that the sum of the numbers in the nth row of the second staircase (Pascal Triangle) is 2^n. We will show, in the next section, that answering a "lagakriyā" question for metres of *mātrā* or duration n (the so called *mātrā vṛttas*) leads to the fact that the sum of the numbers in the nth row of the first staircase is the nth Fibonacci number s_n.

3   Study of Mātrā Vṛttas

Our idea here is to interpret the nth Virahāṅka-Fibonacci number as the total number of metres of value n, where we shall define the value (*mātrā* or duration) of a metre as follows: In the evolution of Sanskrit Prosody it was recognised (especially with the advent of Prākṛt and Apabhraṃśa poetry) that the syllables l and g should be distinguished quantitatively. This is natural, since it takes twice as much time to utter a long syllable g in comparison with a short syllable l. So, the prosodists assigned the value (mātrā) 1 to l and 2 to g and as usual for a metre or sequence of syllables, defined the value as being equal to the sum of the values of the individual syllables. Problems similar to the combinatorics of Varṇa Vṛttas can then be asked here too and they were. For instance the following tables show the enumeration of metres of values 1, 2, 3 and 4.

l

	g
l	l

	l	g
	g	l
l	l	l

		g	g
	l	l	g
	l	g	l
	g	l	l
l	l	l	l

The number of rows in these enumerations is given by $s_1 = 1, s_2 = 2, s_3 = 3, s_4 = 5$. More generally, the enumeration of all metres of value n is obtained by first writing the enumeration of all metres of value $(n-2)$, appending an additional column with entries g to the right. Below this is placed the enumeration of metres of value $(n-1)$ appended by a column of l to the right. Clearly, this rule for enumeration also leads to the relation for the *saṃkhyā* s_n or the number of rows in the enumeration of all the metres of value n.

$$s_n = s_{n-1} + s_{n-2}$$

This is the way that Indians hit upon the so called Fibonacci numbers, centuries before Fibonacci!

One could then ask questions similar to those for varṇa vṛttas for mātrā vṛttas too. For instance, given an integer n, can one write down explicitly the i-th row in the prastāra of metres of value n without writing down the entire prastāra and conversely? Also how many rows are there in the prastāra of metres of value n which contain i gurus? These questions have been dealt with in ancient texts of Prosody. We refer to Sridharan (2006) for answers to these questions, based on the following analogue of the binary expansion of a number, which may be called the Fibonacci expansion.

Lemma. *Any number is either a Fibonacci number or can be expressed uniquely as a sum of non-consecutive Fibonacci numbers.*

Example 1. The expansion for 5 is simply 5 itself as it is a Fibonacci number. $6 = 5 + 1, 9 = 5 + 3 + 1$ are other examples of the expansion.

The set of rows of a mātrā vṛtta prastāra of metres of value n is in bijection with the set of Fibonacci expansions of numbers from 0 to $s_n - 1$. The k-th row of the prastāra is a mnemonic expressing in a compact way the Fibonacci expansion of $s_n - k$. In particular, hidden in the first row of a mātrā vṛtta prastāra are the following formulae:

$$s_1 + s_3 + \cdots + s_{2r-1} = s_{2r} - 1$$
$$s_2 + s_4 + \cdots + s_{2r} = s_{2r+1} - 1$$

We refer to (Sridharan, 2006) for details.

The other question (lagakriyā) about deciding how many rows are there in the prastāra of metres of value n which contain i gurus, the answer is $\binom{n-i}{i}$, which makes sense for $0 \le i \le \left[\frac{n}{2}\right]$. This is because, when one counts the number of metres in the prastāra of metres of value n which contain i gurus, the number of laghus present must be then $(n-2i)$, which also shows that $i \le \left[\frac{n}{2}\right]$. Further, the number of syllables in such a metre must be $i + (n-2i) = (n-i)$. Therefore the number of such possible metres with i gurus clearly turns out to be $\binom{n-i}{i}$. Summing this number over all possible values of i, we obtain the formula

$$s_n = \sum_{i=0}^{\left[\frac{n}{2}\right]} \binom{n-i}{i}$$

which was apparent from the first staircase that we started with.

Remark 1. Suppose, we look at the first quadrant of the plane and consider those points of the first quadrant, other than the origin, which have integral coordinates. Suppose we assign to the lattice point (i, j) the integer $\binom{i+j}{i}$. Then, the integers $\binom{n}{j}$ occur on the line $x + y = n$ and the integers $\binom{n-j}{j}$ occur on the line $x + 2y = n$. This yields a geometrical interpretation of the fact that the diagonals of the first staircase are the rows of the second and vice versa.

4 A short excursion to the combinatorics of rhythmic patterns (Tāla Prastāra)

The study of combinatorics which developed as we have seen for its usage in prosody continued to be important in many other disciplines. The study of combinatorial questions in music was undertaken by Śārṅgadeva (c.1225) in his celebrated treatise on music, *Saṅgītaratnākara* (Sastri *et al.*, 1943).

These tools were applied both to study the combinatorics of rhythmic patterns (tālas) in music as well as the combinatorics of tānas or sequences of musical tones. In this very short section, we shall give a glimpse of the combinatorics of rhythmic patterns.

Motivated by the mātrā vṛttas where the basic units are l, g, there has been a classification of the rhythmic patterns that occur in music. Here the basic units *l, g* have been replaced by the four units Druta (D), Laghu (L), Guru (G) and Pluta (P). The musicians usually take the durations of these four units to be in the ratio 1:2:4:6. It is therefore appropriate to assume that these units have values $D = 1, L = 2, G = 4$ and $P = 6$ and proceed with the study of rhythmic patterns in the same way as in the case of mātrā vṛttas. This has been systematically done in *Saṅgītaratnākara*. Here each tāla is defined as a sequence of the four basic units D, L, G and P and the value of the tāla is defined to be the sum of the values of all the units occurring in the sequence (these can also be enumerated as an array according to a rule given by Śārṅgadeva). Let S_n denote the number of tālas of value *n*.

We can see that $S_1 = 1, S_2 = 2, S_3 = 3, S_4 = 6, S_5 = 10, S_6 = 19$; and for $n > 6$, we have the easily verified recurrence formula,

$$S_n = S_{n-1} + S_{n-2} + S_{n-4} + S_{n-6}$$

We have here more complicated combinatorial problems similar to those encountered in the case of mātrā vṛttas. For further details on this fascinating subject, see (Sridharan *et al.*, 2010).

5   Śārṅgadeva's study of the combinatorics of permutations

We end this article with an application of combinatorial methods to the permutation of svaras or musical tones (*tāna prastāra*) which is considered in the first chapter of *Saṅgītaratnākara* of Śārṅgadeva. Here Śārṅgadeva considers permutations or tānas of the seven basic musical notes which we denote as S, R, G, M, P, D, N. We shall see that this leads to some beautiful mathematics.

All of us know that given *n* distinct elements, there are *n*! different permutations of them, i.e., there are *n*! different arrangements of *n* different objects. In *Saṅgītaratnākara*, Śārṅgadeva actually presents a rather intricate rule by which we can systematically enumerate all the *n*! permutations as an array or a prastāra

Śārṅgadeva's rule for the construction of the prastāra can be presented in a general context by considering the enumeration of the permutations of *n* distinct elements with a natural order, which we may assume to be $\{a_1, a_2, a_3, \ldots, a_n\}$. The prastāra of these is an *n*! × *n* array of *n* symbols in

$n!$ rows, which will be denoted by $[a_1, a_2, a_3, \ldots, a_n]$. We will also denote by $[a_i]_r$ the $1 \times r$ column vector with all the entries given by a_i.

We explain the rule of enumeration by an inductive procedure. To start with, the prastāra $[a_1]$ of one element is simply the 1×1 column $[a_1]_1$. The prastÄẠra $[a_1, a_2]$ of two elements is the 2×2 array

$$
\begin{array}{cc}
a_1 & a_2 \\
a_2 & a_1
\end{array}
$$

which can also be written as

$$
[a_1][a_2]_1 \\
[a_2][a_1]_1
$$

The prastāra $[a_1, a_2, a_3]$ of three elements is a $3! \times 3$ array, which (in our notation) has the form

$$
[a_1, a_2][a_3]_2 \\
[a_1, a_3][a_2]_2 \\
[a_2, a_3][a_1]_2
$$

Having obtained the prastāra of three elements $[a_1, a_2, a_3]$, we see that the prastāra of four elements $[a_1, a_2, a_3, a_4]$, is a $4! \times 4$ array (with 24 rows) given by

$$
[a_1, a_2, a_3][a_4]_6 \\
[a_1, a_2, a_4][a_3]_6 \\
[a_1, a_3, a_4][a_2]_6 \\
[a_2, a_3, a_4][a_1]_6
$$

The above procedure can clearly be continued inductively. Here we shall explicitly display this prastāra of four elements by denoting them also as the four svaras S, R, G and M (with their natural order) in the following table.

We may also note that the prastāra enumerates all the permutations of 4 distinct elements in an order which is different from the so-called lexicographic order, which is perhaps more commonly used in modern combinatorics. Śārṅgadeva's rule actually enumerates the permutations in the what is sometimes referred to as the so called colex order, the mirror image of lexicographic order in the reverse.

In the prastāra (Table 3), we have numbered the rows serially from 1 to 24. Note that in the first row, the svaras S, R, G, M (or a_1, a_2, a_3, a_4) are in ascending order and in the last (or 24th) row, they are in descending order and all the 4! = 24 permutations of these four symbols are listed in the prastāra.

We claim that the rows are arranged using the so called factorial representation of any number m, $1 \leq m \leq 4!$. Indeed, there is a bijective correspondence between the rows and the integers $1 \leq m \leq 4!$ via the so called factorial representation and this holds for all n (not necessarily for $n = 4$):

Proposition. *Every integer $1 \leq m \leq n!$ can be uniquely represented in the form*

$$m = d_0 0! + d_1 1! + d_2 2! + \cdots + d_{n-1}(n-1)!,$$

where d_i are integers such that $d_0 = 1$ and $0 \leq d_i \leq i$, for $i = 1, 2, \ldots, n-1$.

The problems of naṣṭa and uddiṣṭa (discussed in Section 2 in the context of Prosody) can be solved for tāna prastāra using the above proposition. We shall illustrate this here by some examples. We shall first give an example of naṣṭa. Suppose we ask ourselves the question - what is the 15th row of the prastāra of permutations of four elements? We are to write it down without looking at the prastāra. For this, we need just to write down the factorial representation of 15. It is unique and is given by (later we shall explain how Śārṅgadeva introduces a table called khaṇḍa-meru precisely to find out such factorial representations):

$$15 = 1 \times 0! + 0 \times 1! + 1 \times 2! + 2 \times 3!$$

The coefficients $\{d_i\}$ of this expansion are given by $d_0 = 1, d_1 = 0, d_2 = 1$ and $d_3 = 2$. Now consider the top row $a_1 a_2 a_3 a_4$ (of course we could write all this in terms of the svaras S, R, G, M also). To write down the last element of the 15th row, subtract $d_3 = 2$ from 4 to get 2. The second element of the above sequence a_2 is the last element of the 15th row. How about the penultimate one? We now consider $a_1 a_3 a_4$ (in the increasing order of indices leaving out a_2 which has already been used). Now we subtract $d_2 = 1$ from 3 and get 2. The second element of $a_1 a_3 a_4$ namely a_3 is the penultimate element of the fifteenth row. Thus the last two elements of the 15th row are $a_3 a_2$. Now consider the remaining elements $a_1 a_4$ and we subtract $d_1 = 0$ from 2 to get 2. Hence the next element of the 15th row

Table 3: *Prastāra of permutations of four symbols*

1	S	R	G	M	a_1	a_2	a_3	a_4
2	R	S	G	M	a_2	a_1	a_3	a_4
3	S	G	R	M	a_1	a_3	a_2	a_4
4	G	S	R	M	a_3	a_1	a_2	a_4
5	R	G	S	M	a_2	a_3	a_1	a_4
6	G	R	S	M	a_3	a_2	a_1	a_4
7	S	R	M	G	a_1	a_2	a_4	a_3
8	R	S	M	G	a_2	a_1	a_4	a_3
9	S	M	R	G	a_1	a_4	a_2	a_3
10	M	S	R	G	a_4	a_1	a_2	a_3
11	R	M	S	G	a_2	a_4	a_1	a_3
12	M	R	S	G	a_4	a_2	a_1	a_3
13	S	G	M	R	a_1	a_3	a_4	a_2
14	G	S	M	R	a_3	a_1	a_4	a_2
15	S	M	G	R	a_1	a_4	a_3	a_2
16	M	S	G	R	a_4	a_1	a_3	a_2
17	G	M	S	R	a_3	a_4	a_1	a_2
18	M	G	S	R	a_4	a_3	a_1	a_2
19	R	G	M	S	a_2	a_3	a_4	a_1
20	G	R	M	S	a_3	a_2	a_4	a_1
21	R	M	G	S	a_2	a_4	a_3	a_1
22	M	R	G	S	a_4	a_2	a_3	a_1
23	G	M	R	S	a_3	a_4	a_2	a_1
24	M	G	R	S	a_4	a_3	a_2	a_1

is a_4. Since only a_1 is left, we can conclude that the required 15th row is given by $a_1a_4a_3a_2$. This clearly checks with the 15th row as shown in the table.

Let us try another example. What is the 5th row of the above prastāra? The factorial expansion of 5 is

$$5 = 1 \times 0! + 0 \times 1! + 2 \times 2! + 0 \times 3!.$$

Here $d_0 = 1, d_1 = 0, d_2 = 2$ and $d_3 = 0$. We have $4 - d_3 = 4, 3 - d_2 = 1, 2 - d_1 = 2$. Thus we see as above that the desired row is $a_2a_3a_1a_4$, which clearly checks with the 5th row as shown in the table.

Thus, the factorial representation gives a complete solution to the naṣṭa problem. We shall now see how to tackle the converse problem uddiṣṭa, by means of an example.

Let us consider for instance the row $a_1a_4a_3a_2$ and ask for the corresponding row number. For this purpose we write the symbols in their natural order $a_1a_2a_3a_4$ and notice that the last symbol occurs in the 2nd place in the given row. Hence, the last coefficient in the factorial expansion of the row-number is given by $d_3 = 4 - 2 = 2$. Now, we are left with the rest of the given row namely $a_1a_4a_3$ and write these symbols in their natural order to obtain $a_1a_3a_4$. Here the last symbol again occurs as the second in the given (part of the) row and therefore, the next co-efficient $d_2 = 3 - 2 = 1$. Now we are left with a_1a_4 and here they are already in the natural order and we have $d_1 = 2 - 2 = 0$. And of course, by the proposition stated above $d_0 = 1$ always. Thus the row number is

$$1 \times 0! + 0 \times 1! + 1 \times 2! + 2 \times 3! = 15$$

Indeed we can check from the table above that $a_1a_4a_3a_2$ occurs in the 15th row.

As another example, let us find in what row $a_4a_1a_2a_3$ occurs in the prastāra. Clearly, since a_3, the last element of the row is the third element of a_1 a_2 a_3 a_4, when these symbols are placed in natural order, $d_3 = 4-3 = 1$. Now, the rest of the row is $a_4a_1a_2$ and again since a_2 is the second element of $a_1a_2a_4$, we have $d_2 = 3-2 = 1$. Next we consider the remaining part of the row a_4a_1 and since a_1 is the first element of a_1 a_4 we have $d_1 = 2 - 1 = 1$. Finally $d_0 = 1$ as always and we have the row number given by the expansion

$$1 \times 0! + 1 \times 1! + 1 \times 2! + 1 \times 3! = 10$$

Indeed we can check from the table above that $a_4a_1a_2a_3$ occurs in the 10th row.

The last row, which is $a_4a_3a_2a_1$ corresponds to the factorial representation

$$24 = 4! = 1 \times 0! + 1 \times 1! + 2 \times 2! + 3 \times 3!$$

This is indeed a particular case of the following beautiful formula, that for any $n \geq 1$,

$$n! = 1 \times 0! + 1 \times 1! + 2 \times 2! + \cdots + (n-1) \times (n-1)!$$

Śārṅgadeva bases his discussion of the uddiṣṭa and naṣṭa processes through a tabular figure referred to as the khaṇḍa meru, which is a device to read off the co-efficients $\{d_i\}$ which arise in the factorial representation of any given number and vice versa. Since tānas (musical phrases) involve the seven notes (saptasvara), the tabular figure khaṇḍa meru consists of seven columns with the number of entries increasing from one to seven, from left to right. In the first row, place 1 in the first column (from the left) followed by 0's in the other columns. In the second row, starting from the second column, place the total number (or saṃkhyā) of tānas of 1, 2, 3, etc., svaras. Thus the entries in the second row are the factorials, 1, 2, 6, 24, 120 and 720. In the succeeding rows, place twice, thrice etc. (that is successive multiples) of the factorials, starting from a later column at each stage. Thus, the first column of the Meru consists of just $0! = 1$, and the nth column, for $n > 1$, is made up of the multiples $i.(n-1)!$, with $0 \leq i \leq (n-1)$, as is shown in following table.

Thus, in the tabular figure of khaṇḍa meru, Śārṅgadeva has a table of multiples of all the factorials. He uses this effectively to find out the number of the row in a prastāra that corresponds to a given tāna (the uddiṣṭa process) and conversely (the naṣṭa process) by precisely making use of the factorial representation that we discussed earlier.

In order to illustrate the use of khaṇḍa meru to arrive at the factorial representation of a number, let us consider the case of just four svaras as before. Then the khaṇḍa meru will be the simpler table given below.

Suppose we want to find out the factorial representation of 15. Then we have to locate the largest number in the last column of the khaṇḍa-meru which is strictly less than 15. That is the number 12, which is the third entry in that column and we have to take $d_3 = 3 - 1 = 2$. Now subtract 12 from 15 to get 3. Locate the largest number strictly below 3

Table 4: *khaṇḍa meru*

S	R	G	M	P	D	N
1	0	0	0	0	0	0
	1	2	6	24	120	720
		4	12	48	240	1440
			18	72	360	2160
				96	480	2880
					600	3600
						4320

Table 5: *khaṇḍa meru*

S	R	G	M
1	0	0	0
	1	2	6
		4	12
			18

in the penultimate column. This is 2, which is the second entry in that column. Hence, $d_2 = 2 - 1 = 1$. Now subtract 2 from 3 to get 1. Locate the largest number strictly below 1 in the next column to the left. This is 0 which is the first entry in that column. Hence $d_1 = 1 - 1 = 0$. Finally $d_0 = 1$ as always and we have the factorial representation of 15 as given by

$$1 \times 0! + 0 \times 1! + 1 \times 2! + 2 \times 3! = 15.$$

Hence, the khaṇḍa meru is a "device" which can be used straight away to read off the coefficients $d_1, d_2, \ldots$ which occur in the factorial representation of any given number. Of course, once these co-efficients are determined, the naṣṭa and uddiṣṭa questions can be answered by following the procedure discussed earlier.

To sum up, we trace in this article the origins of Combinatorics in India to Sanskrit Prosody (via the analysis of Vedic verses) and show how this gradually evolved to the mathematical study of rhythmic patterns and permutations of svaras (tones) in music. These can be interpreted in the present day language of combinatorics and lead to some very interesting and beautiful formulae.

We note that these formulae were as a rule encoded by the Indian mathematicians through the listing of patterns and their enumeration through tables. We believe that the idea of using prastāras and merus is a unique Indian approach to the study of combinatorics.

Acknowledgements

The authors thank Prof. Sujatha for kindly inviting them to contribute an article to this volume. They are very grateful to Nivedita for her enthusiastic, timely and invaluable help in the preparation of this article. Sridharan, the very first time that he met Nivedita, who brought to his mind Sister Nivedita, the great disciple of Vivekananda, was convinced that there was something extraordinarily gentle about her and prays that God bless her.

References

Pansikara, Kedaranatha Vasudeva Lakshmana. 1908. *Chandaḥ-Sūtra* of Piṅgalanāga (c.300 BCE): Ed. with the commentary *Mṛtasañjīvanī* of Halāyudha (c.950 CE), Kavyamala Series No. 91, Bombay, Rep. Chowkhamba 2001.

Sastri, Madhusudanna. 1994. *Vṛtta-ratnākara* of Kedārabhaṭṭa (c.1000): Edited with *Nārāyaṇī* of Nārāyaṇabhaṭṭa (c.1555) and *Setu* of Bhāskarabhaṭṭa (c.1675) Krishnadasa Academy, Varanasi.

Sastri, S. Subramanya, Krishnamacharya, V. and Sarada, S. 1943. *Saṅgīta-ratnākara* of Śārṅgadeva (c.1225): Edited with *Kalānidhi* of Kallinātha (c.1420) and *Saṅgīta-sudhākara* of Siṃhabhūpāla (c.1350), 4 Vols. Adyar Library, 1943–1986.

Sridharan, R. 2005. Sanskrit prosody, Piṅgala Sūtras and Binary Arithmetic, in G. Emch, M.D. Srinivas and R. Sridharan Eds., *Contributions to the History of Mathematics in India*, Hindustan Publishing Agency, Delhi. pp. 33–62.

Sridharan, R. 2006. Pratyayas for Mātrāvṛttas and Fibonacci Numbers, *Mathematics Teacher*, **42**:120–137.

Sridharan, Raja, Sridharan, R. and Srinivas, M.D. 2010. Combinatorial Methods in Indian Music: Pratyayas in *Saṅgīta-ratnākara* of Śārṅgadeva, in C.S. Seshadri Ed., Studies in the History of Indian Mathematics, Hindustan Publishing Agency, Delhi. pp. 55–112.

Velankar, H.D. 1962. *Vṛttajāti-samuccaya* or *Kaisiṭṭha-chanda* (in Prākṛta) of Virahāṅka (c. 650): Ed., with commentary by Gopāla (c. 900), Rajasthan Prachya Vidya Pratisthana, Jodhpur 1962.

Weyl, Hermann. 1953. "Universities and Sciences in Germany" (Text of two lectures given in 1945), reprinted in "*The Math. Student*" 21, Vols 1 and 2, 1–26 (March-June).

CHAPTER 27

In Square Circle: Geometric Knowledge of the Indus Civilization

Sitabhra Sinha*, Nisha Yadav** and Mayank Vahia**

*The Institute of Mathematical Sciences,
CIT Campus, Taramani, Chennai 600 113, India.
e-mail: sitabhra@imsc.res.in

**Tata Institute of Fundamental Research,
Homi Bhabha Road,
Mumbai 400 005, India.

1 Introduction

The geometric principles expounded in the *Sulbasutras* (800-500 BCE) have often been considered to mark the beginning of mathematics in the Indian subcontinent (Seidenberg, 1975; Staal, 2008). This collection of *sutras* codify directions for constructing sacrificial fires, including rules for the complex configuration of ritual altars. The bird-shaped *Agnicayan* altar, consisting of five layers of two hundred bricks each, with the bricks being of square, rectangular or triangular shapes of various sizes, is considered by F. Staal to signal the beginning of geometry proper in South Asia (Staal, 2008). It has been dated by him to about 1000 BCE as some of the mantras that are concerned with the consecration of bricks occur in the earliest Yajurveda Samhita, the Maitrayani. The absence of any recorded tradition of geometric knowledge predating these sutras have led some scholars to suggest a West Asian origin for the onset of mathematical thinking in India. However, the discovery of the archaeological remnants of the Indus Valley civilization in parts of Pakistan and northwestern India over the course of last century has revealed a culture having a sophisticated understanding of geometry which predated the *Sulbasutras* by more than a thousand years. It is difficult to ascertain whether there was any continuity between the geometry practised by the Indus civilization and that used by the later Vedic culture; however, it is not impossible that some of the earlier knowledge persisted among the local population and influenced the *sulbakaras* (authors of the *Sulbasutras*) of the first millennium BCE.

2 Indus geometry: the archaeological evidence

The Indus Valley civilization, also referred to as the Mature Harappan civilization (2500-1900 BCE), covered approximately a million square kilometres – geographically spread over what are now Pakistan and northwestern India (Possehl, 2002). It was approximately contemporaneous with Old Kingdom Egypt, as well as, the Early Dynastic Period of the Sumerian Civilization and the Akkadian Empire in Mesopotamia. The Indus culture was characterized by extensive urbanization with large planned cities, as seen from the ruins of Harappa and Mohenjodaro (among others). There is evidence for craft specialization and long-distance trade with Mesopotamia and Central Asia.

The well-laid out street plans of the Indus cities and their accurate orientation along the cardinal directions have been long been taken as evidence that the Indus people had at least a working knowledge of geometry (Sarasvati Amma, 1979; Parpola, 1994). Earlier studies have suggested that not only did these people have a practical grasp of mensuration, but that they also had an understanding of the basic principles of geometry (Kulkarni, 1978). The discovery of scales and instruments for measuring length in different Indus sites indicate that the culture knew how to make accurate spatial measurements (Vij, 1984; Balasubramaniam and Joshi, 2008). For example, an ivory scale discovered at Lothal (in the western coast of India) has 27 uniformly spaced lines over 46 mm, indicating an unit of length corresponding to 1.70 mm (Rao, 1985). The sophistication of the metrology practised by the Indus people is attested by the sets of regularly shaped artifacts of various sizes that have been identified as constituting a system of standardized weights. Indeed, there is a surprising degree of uniformity in the measurement units used at the widely dispersed centers of the civilization, indicating an attention towards achieving a standard system of units for measurement.

However, most of the literature available up till now on the subject of Indus geometry have been primarily concerned with patterns occurring at the macro-scale (such as building plans, street alignments, etc.). The smaller-scale geometric patterns that are often observed on seals or on the surface of pottery vessels have not been analysed in great detail. We believe that such designs provide evidence for a much more advanced understanding of geometry on the part of the Indus people than have hitherto been suggested in the literature.

As an example, we direct the reader's attention to the occurrence of a space-filling pattern on a seal, where a basic motif shaped like a Japanese fan has been used in different orientations to tile an approximately square domain (Figure 1). The apparent simplicity of the design belies the geometric complexity underlying it. The actual repeating pattern is a fairly complicated shape that is composed of four of the basic fan-shaped motifs, each being rotated by 90 degrees (see the shaded object in Figure 2). The transformations that are required to construct this pattern from a rectangle whose length is twice its breadth is shown in Figure 3. Note that, despite the complex shape of the repeating pattern, its area is fairly simple to calculate. By construction, this area is equal to that of the rectangle. The construction also enables us to calculate the area of the fan-shaped basic motif as being $2R^2$ where R is the length from its sharply pointed tip to the midpoint of its convex edge.

The set of operations that are required to generate the shapes described above suggest that the people of the Indus civilization were reasonably acquainted with the geometry of circular shapes and techniques of approximating their areas, especially as we know of several Indus artifacts which exhibit other designs that follow from these principles. In fact, we note that the Indus civilization paid special attention to the circle and its variants in the geometric designs that they made on various artifacts. It reminds us of the special place that the circular shape had in Greek geometry that is aptly summarised by the statement of the 5th century neo-Platonist Greek philosopher Proclus in his overview of Greek geometry: "The first and simplest and most perfect of the figures is the circle" (Proclus, 1992).

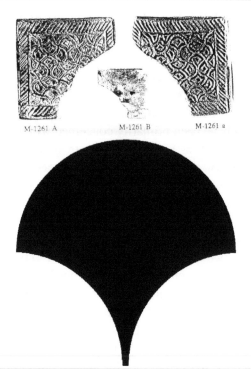

M-1261 A   M-1261 B   M-1261 a

Figure 1: *Space-filling pattern in a seal excavated from Mohenjodaro, DK Area (No. M-1261 in p. 158 of Ref. (Shah and Parpola, 1991)) showing the seal (A), its impression (a) and the boss side (B). On the right, we show the basic motif having the shape of a Japanese fan. [Figure of the seal is reproduced with permission from A. Parpola.]*

The reason for the primacy of the circle in Indus geometry is probably not hard to understand if we focus on the technical means necessary to generate ideal geometric shapes. Unlike rectilinear shapes which are very difficult to draw exactly without the help of instruments that aid in drawing lines at right angles to each other, an almost perfect circle can be drawn by using a rope and a stick on soft clay. Indeed there is evidence for the use of compass-like devices for drawing circles in the Indus Valley civilization. E.J.H. Mackay, who excavated Mohenjodaro between 1927–1932 expressed surprise on finding that "an instrument was actually used for this purpose [drawing circles] in the Indus Valley as early as 2500 BC" (Mackay, 1938). In this context, it may be noted that later excavations

Figure 2: *Close-packing of circles in a four-fold symmetric arrangement having two layers. The circles in the different layers are distinguished by broken and continuous curves, respectively. The two layers are displaced with respect to each other by half a lattice spacing in both the horizontal and vertical directions. The marked out rectangle shows the area which is transformed into the shape indicate by shaded areas.*

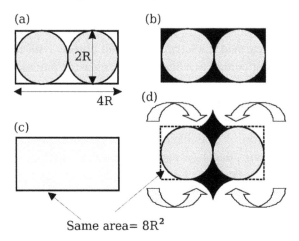

Figure 3: *The set of transformations that produce the repeating motif for the space-filling tiling of Figure 1. Two non-overlapping circles, each of radius R are inscribed within a rectangle of length 4R and width 2R (a). Of the sections of the rectangle not belonging to the circles (shaded black in (b)), four are cut and re-pasted so that they fall outside the original rectangle (c), creating the repeating motif. Note that, its area (= $8R^2$) is the same as that of the original rectangle (d).*

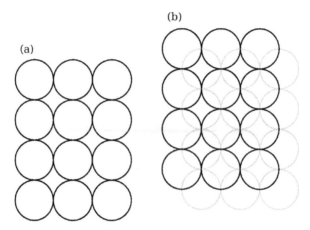

Figure 4: *(a) A periodic lattice consisting of close-packed circles in a four-fold symmetric arrangement. (b) The intersecting circle lattice generated by superposing two lattices of circles displaced with respect to each other by half a lattice spacing in the vertical and horizontal directions.*

at Lothal have unearthed thick, ring-like shell objects with four slits each in two margins that could have been used to measure angles on plane surfaces (Rao, 1985).

Let us return to the issue of how the sophisticated understanding of tiling the plane with shapes having circular edges may have emerged in the Indus civilization. A possible clue can be found from the construction given in Figure 2, viz., the close-packing of circles on a plane in a four-fold symmetric arrangement. Such lattice patterns with the spatial period equal to the diameter of a circle, are easy to generate, for instance, by pressing an object with a circular outline on a wet clay surface repeatedly along rows and columns (Figure 4 (a)). Indeed, certain Indus artifacts bear the impression of circular objects pressed on them to create a design. By exploring the variations of such a basic pattern, a variety of increasingly complicated design motifs can be generated which culminate in the complicated space-filling tiles shown in Figure 1.

One of the simplest variations of the basic circular pattern is to have two layers of such close-packed lattices of circles, with one layer displaced by a length equal to the radius of a circle (i.e., half the lattice spacing) along both the horizontal and vertical directions (Figure 4 (b)). The resulting

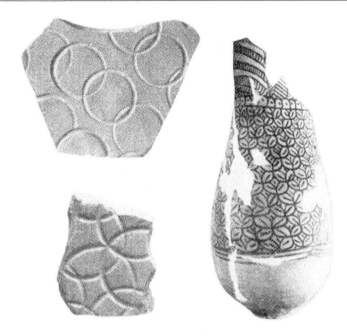

Figure 5: *(Left) The motif of intersecting circles on incised pottery from lower levels of Mohenjodaro, DK Area, G section (Nos. 24 [top] and 21 [bottom] in Plate LXVII of Ref. (Mackay, 1938)).(Right) A decorated intersecting circle lattice used as a space filling pattern on the surface of jar from Mohenjodaro, DK Area, G section (no. 5 in Plate LIV of Ref. (Mackay, 1938)).*

intersecting circle lattice is a frequently occurring design motif in Indus artifacts. See Figure 5 for examples of this pattern, which occasionally occurs in a "decorated" variation (with points or lines used to fill the spaces within the circular shapes). In this context, one may mention that one of the signs that occur in frequently in the Indus seal inscriptions is in the form of two overlapping ellipses. Parpola has suggested as association of this sign with the Pleiades star system (Parpola, 1994). It is a matter of conjecture whether the design of intersecting circles found in so many Indus artifacts has any astral significance.

The next variation we discuss is the imbricate pattern having regularly arranged overlapping edges resembling fish scales. Figure 6 shows how the

Figure 6: *(Left) Imbricate pattern of overlapping circles used for tiling a plane. The basic fan-shaped motif is identical to the one show in Figure 1. (Right) The same pattern as it appears in a present-day urban setting (Photo: Nivedita Chatterjee).*

Color image of this figure appears in the color plate section at the end of the book.

design is used for tiling the plane by having several rows of overlapping circles, each partially obscuring the layer beneath it. It is fairly easy to see that this is a variation of the intersecting circle lattice. One can envisage the design as being made up of multiple rows of circular tiles stacked one on top of the other with successive rows displaced by half a lattice spacing vertically and horizontally. The pattern can be seen frequently on the surface of Indus painted pottery, both in its original and "decorated" forms (Figure 7) (Starr, 1941).

Yet another possible variation obtained from the intersecting circle lattice is shown in Figure 8. If a square, the length of whose sides equal the diameter of each circle, is placed on the lattice so as to completely contain a circle inside it, one obtains a star-shaped motif by focusing exclusively within the region enclosed by the square (Figure 8 (b)). A variant of this four-pointed star is obtained by relaxing the close-packed nature of the lattice so that the circles no longer touch each other. The resulting 'cross'-shaped motif with a small circle added at the center (Figure 8 (d)), sometimes referred to as the quincross sign, is found to occur very frequently in pre-Columbian iconography, especially in Mayan ('Kan-cross')

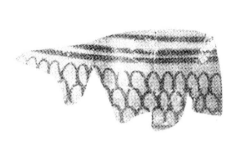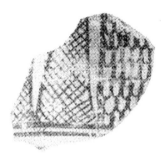

Figure 7: *Imbricate pattern of overlapping circles seen on fragments of painted pottery excavated from Mohenjodaro, DK Area, G section (Nos. 33-34 in Plate LXX of Ref. (Mackay, 1938)).*

and Zapotec ('Glyph E') designs (Sugiyama, 2005). The meaning attached to this motif has been interpreted differently by various archaeologists, and in the context of the Maya, it has been claimed to be a representation of Venus. A possible astral significance of this sign in the context of Indus civilization is probably not far-fetched given its star-like shape. The motif, in both the exact four pointed star-like shape and in its cross-like variation, is found in many artifacts at both Harappa and Mohenjodaro (Figure 9).

The intersecting circle lattice is thus a fairly general mechanism for producing a variety of geometric patterns, ranging from quite simple ones that have been seen to occur across many different cultures to complicated tiling shapes that suggest a deeper understanding of the underlying geometry. This is especially the case for the complicated space-filling tile made from the four fan-shaped motifs shown in Figure 3. It is clear from the construction procedure that the geometrical knowledge underlying the origin of this design is non-trivial and implies a sophisticated understanding of the principles of circular geometry.

3 Conclusion

To summarize, in this article we have argued that the origin of mathematics, and geometry in particular, in the Indian subcontinent may actually date back to the third millennium BCE and the Indus Valley civilization, rather than beginning with the *Sulbasutras* of the first millennium BCE as is conventionally thought. Although the well-planned cities, standardized

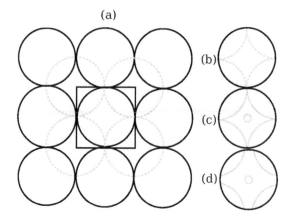

Figure 8: *The intersecting circle lattice (a) can be used to generate a star-like motif (b) with decorated variants (c) and (d).*

Figure 9: *The star-like geometrical motif in ornament-like objects excavated from Harappa (left) and in a seal impression (center) and ivory objects (right) found in Mohenjodaro, DK Area, G section. (No. 8 in Plate CXXXIX of Ref. (Vats, 1940), No. 12 in Plate C [center] and Nos. 9-10 in Plate CXLI [right] of Ref. (Mackay, 1938)).*

system of measurements and evidence of long-range trade contacts with Mesopotamia and Central Asia attest to the high technological sophistication of this civilization, the unavailability of written records has up to now prevented us from acquiring a detailed understanding of the level of mathematical knowledge attained by the Indus people. By focusing on the geometry of design motifs observed commonly in the artifacts excavated from various sites belonging to this culture, we have shown that they suggest a deep understanding of the properties of circular shapes. In particular, designs which exhibit space-filling tiling with complicated shapes imply that the Indus culture may have been adept at accurately estimating the area of shapes enclosed by circular arcs. This speaks of a fairly sophisticated knowledge of geometry that may have persisted in the local population long after the decline of the urban centers associated with the civilization and it may well have influenced in some part the mathematical thinking of the later Vedic culture of the first millennium BCE.

Acknowledgements:

We would like to thank P. P. Divakaran and F. Staal for several insightful comments and suggestions on an early version of the manuscript. S.S. would like to gratefully acknowledge Dr P. Pandian and the staff of IMSc Library for their help in obtaining several rare volumes describing the Indus civilization artifacts.

References

Balasubramaniam, R. and Joshi, J.P. 2008. Analysis of terracotta scale of Harappan civilization from Kalibangan, *Current Science*. **95:**588–589.

Kulkarni, R.P. 1978. Geometry as known to the people of Indus civilization, *Indian Journal of History of Science*. **13:**117–124.

Mackay, E.J.H. 1938. reprint 2000. *Further excavations at Mohenjodaro*, Archaeological Survey of India, New Delhi.

Parpola, A. 1994. *Deciphering the Indus Script*, Cambridge University Press, Cambridge.

Possehl, G. 2002. *The Indus Civilization: A Contemporary Perspective*, AltaMira Press, Lanham, MD.

Proclus. 1992. *A Commentary on the First Book of Euclid's Elements*, (Trans. G.R. Morrow), Princeton Univ. Press, Princeton.

Rao, S.R. 1985. *Lothal*, Archaeological Survey of India, New Delhi.

Sarasvati Amma, T.A. 1979. revised edn. 1999. *Geometry in Ancient and Medieval India*, Motilal Banarasidass, Delhi.

Seidenberg, A. 1975. The ritual origin of geometry, *Archive for the History of Exact Sciences*. 1:488–527.

Shah, S.G.M. and Parpola, A. (eds.). 1991. *Corpus of Indus Seals and Inscriptions*, Vol. 2, Suomalainen Tiedeakatemia, Helsinki.

Staal, F. 2008. *Discovering the Vedas*, Penguin, New Delhi.

Starr, R.F.S. 1941. *Indus Valley Painted Pottery: A comparative study of the designs on the painted wares of the Harappa culture*, Princeton University Press, Princeton, NJ.

Sugiyama, S. 2005. *Human Sacrifice, Militarism and Rulership: Materialization of state ideology at the feathered serpent pyramid, Teotihuacan*, Cambridge University Press, Cambridge.

Vats, M.S. 1940. reprint 1999. *Excavations at Harappa*, Archaeological Survey of India, New Delhi.

Vij, B. 1984. Linear standard in the Indus civilization, in *Frontiers of the Indus Civilization* (Eds. B. B. Lal and S. P. Gupta), Indian Archaeology Society and Indian History & Culture Society, New Delhi, pp. 153–156.

Index

Color Plate Section

Chapter 3

Fermat (1601–1655)

Euler (1707–1783)

Chapter 7

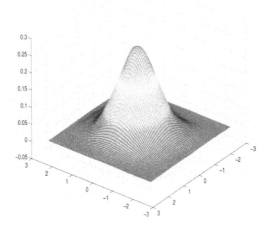

Figure 1

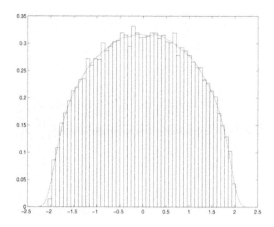

Figure 2

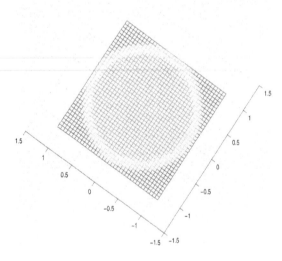

Figure 3

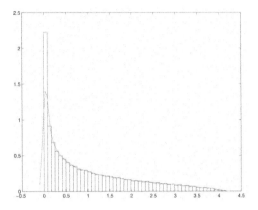

Figure 4

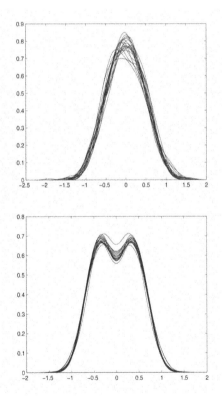

Figure 5

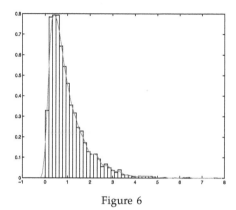

Figure 6

Chapter 13

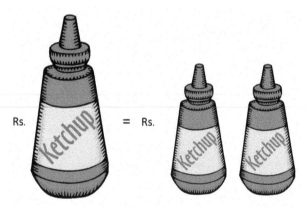

Figure 1

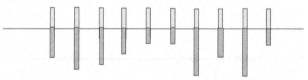

Figure 2

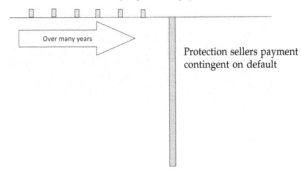

Protection buyer premium payment

Over many years

Protection sellers payment
contingent on default

Figure 3

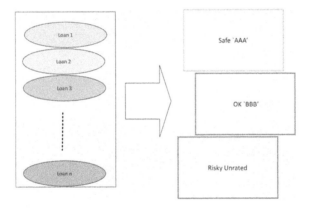

Loan 1

Loan 2

Loan 3

Loan n

Safe `AAA'

OK `BBB'

Risky Unrated

Figure 4

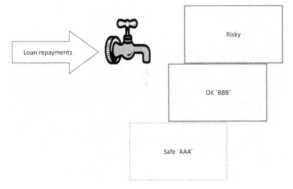

Loan repayments

Risky

OK `BBB'

Safe `AAA'

Figure 5

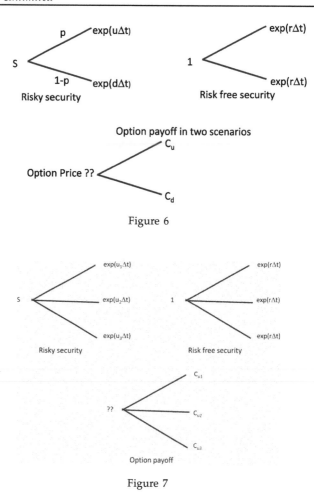

Figure 6

Figure 7

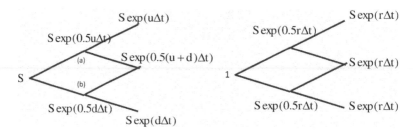

Figure 8

Chapter 20

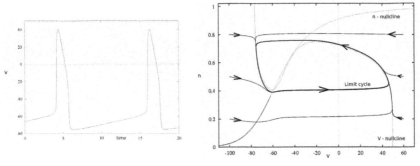

Figure 5

Chapter 21

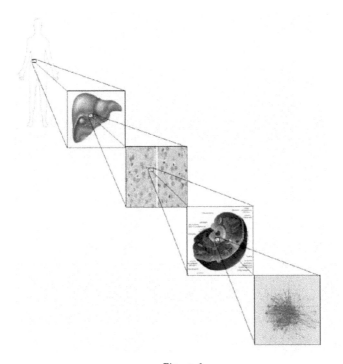

Figure 1

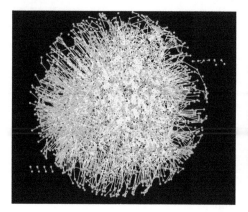
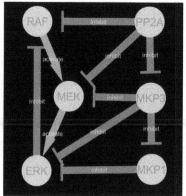

Figure 2

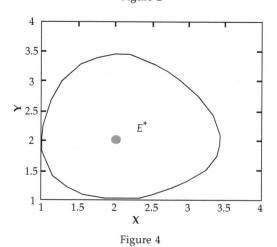

Figure 4

Chapter 22

Real system   **Model**

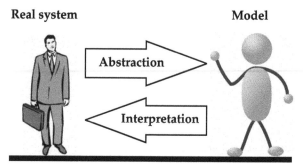

Figure 1

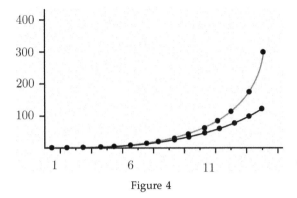

Figure 4

Chapter 23

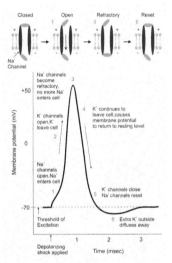

Figure 2

ICA decomposition

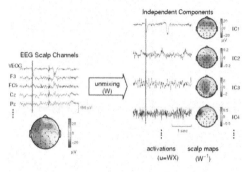

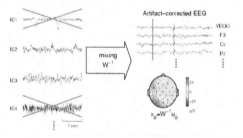

Figure 4

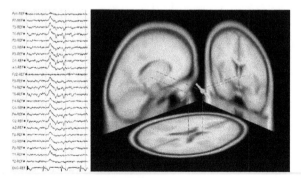

Figure 5

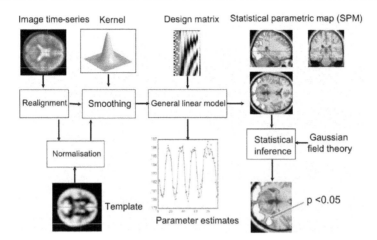

Figure 7

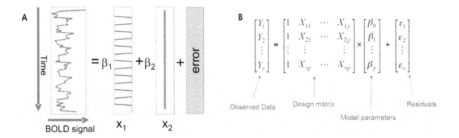

Figure 8

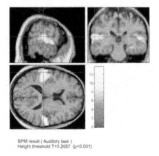

Figure 10

Chapter 27

Figure 6

Milton Keynes UK
Ingram Content Group UK Ltd.
UKHW030900141024
449569UK00025B/1301

9 781578 087044